Artificial Intelligence and Other Innovative Computer Applications in the Nuclear Industry

Artificial Intelligence and Other Innovative Computer Applications in the Nuclear Industry

Edited by

M. Catherine Majumdar
Westinghouse Idaho Nuclear Corporation
Idaho Falls, Idaho

Debu Majumdar
U.S. Department of Energy
Idaho Operations Office
Idaho Falls, Idaho

and

John I. Sackett
Argonne National Laboratory West
Idaho Falls, Idaho

Springer Science+Business Media, LLC

Library of Congress Cataloging in Publication Data

American Nuclear Society Topical Meeting on Artificial Intelligence and Other Innovative Computer Applications (1987: Snowbird, Utah)
  Artificial intelligence and other innovative computer applications in the nuclear industry / edited by M. Catherine Majumdar, Debu Majumdar, and John I. Sackett.
      p.   cm.
  "Proceedings of the American Nuclear Society Topical Meeting on Artificial Intelligence and Other Innovative Computer Applications, held August 31–September 2, 1987, in Snowbird, Utah" — T.p. verso.
  Includes bibliographies and indexes.
  ISBN 978-1-4612-8290-7   ISBN 978-1-4613-1009-9 (eBook)
  DOI 10.1007/ 978-1-4613-1009-9
  1. Nuclear industry—Data processing—Congresses. 2. Artificial intelligence—Industrial applications—Congresses. I. Majumdar, M. Catherine. II. Majumdar, D. (Debu) III. Sackett, John I. IV. Title.
TK9006.A48   1987   88-9396
621.48'3'028563—dc19   CIP

Proceedings of the American Nuclear Society Topical Meeting on Artificial Intelligence and Other Innovative Computer Applications, held August 31–September 2, 1987, in Snowbird, Utah

© 1988 Springer Science+Business Media New York
    Originally published by Plenum Press, New York in 1988
Softcover reprint of the hardcover 1st edition 1988

PREFACE

This conference brought together experts from 15 countries to
discuss application of Artificial Intelligence (AI) techniques
to the nuclear industry. It was apparent from the meeting that
even those active in the field were surprised at the extent of
work and the progress made. There was a strong impression that
application of this technology to nuclear power plants is
inevitable. The benefits to improved operation, design, and
safety are simply too significant to be ignored.

This is a much different conclusion than might have been
reached a few years ago when the technology was new and people
were struggling to understand its significance. We believe
that this meeting reflects a major turning point for the
technology. It has moved from being a topic understood only by
specialists to a situation where users are the most active
people in the field.

A broad array of innovative work is described from all of the
participating countries. The activity in the U.S. is large and
diverse. Although there is no nationally focussed policy for
AI research in the U.S., many of these activities are reported
here. Japan and France have a strong drive to integrate AI
technology into their nuclear plants, and this is reflected in
these proceedings.

There is now a sense of excitement about what can be achieved,
a refreshing departure from the often defensive posture taken
in nuclear power development these days. The excitement stems
from recognition that the technology has the potential to
greatly expand our capabilities as designers and operators. To
take but one example, great strides have been made in
developing systems that can diagnose the condition of the plant
and provide options for response. Artificial Intelligence work
has not yet led to extensive automation, but it will.

There was much discussion at the meeting about the fact that
Artificial Intelligence is a poor title for this work.
Knowledge engineering comes closer to what is being achieved.
We hope that you will find these proceedings both interesting
and useful. This is an emerging field that will affect us all.

J. Curtis Haire,
General Chairman

Debu Majumdar and John I. Sackett,
Technical Program Co-Chairmen

MEETING OFFICIALS

General Chairman   J. Curtis Haire
                          EG&G Idaho, Inc.

Program Co-Chairman   Debu Majumdar
                          U.S. Department of Energy
                          Idaho Operations Office

Program Co-Chairman   John I. Sackett
                          Argonne National Laboratory-West

Arrangements   Richard W. Lindsay
                          Argonne National Laboratory-West

Finance   Orville R. Meyer
                          EG&G Idaho, Inc.

Registration   Carl H. Cooper
                          EG&G Idaho, Inc.

Exhibits   Glen A. Mortensen
                          EG&G Idaho, Inc.

Publications   M. Catherine Majumdar
                          Westinghouse Idaho Nuclear Co.

Public Information   George A. Freund
                          Science Applications, Inc.

Guest Program   Karen Sackett

Registration   Betty Haire and Marge Lindsay
Desk

CO-SPONSORS

Idaho Section of the American Nuclear Society
European Nuclear Society
ANS Human Factors Division
ANS Remote Systems Technology Division

PROGRAM COMMITTEE

Technical Program Co-Chairmen - Debu Majumdar, U.S. Department
of Energy - Idaho Operations Office and John I. Sackett,
Argonne National Laboratory-West

A.D. Alley............. General Electric Company
Hakan Andersson........ Studsvik Energiteknik
Bill Bertch............ General Dynamics, Inc.
Paul Blanch............ Northeast Utilities Service Co.
William V. Botts....... EI International, Inc.
Michael Bray........... EG&G Idaho, Inc.
John S. Brtis.......... Sargent and Lundy
David Cain............. EPRI
Allen Christie......... Westinghouse Electric Corp.
Robert Colley.......... EPRI
Gail A. Cordes......... Intermountain Technologies, Inc.
Brent Dixon............ EG&G Idaho, Inc.
James Dukelow, Jr...... Boeing Computer Service
Robert Engelmore....... Stanford University
Jerry D. Griffith...... U.S. Department of Energy
Dennis Harrison........ U.S. Department of Energy
James P. Jenkins....... U.S. Nuclear Regulatory Commission
C. Edward Johnson...... Westinghouse Idaho Nuclear Co.
Harry Julian........... Volian Enterprises, Inc.
Jan G. Kretzschmar..... SCK/CEN Informatics
Bob Lang............... Middle South Services Company
Henry Makowitz......... EG&G Idaho, Inc.
William Nelson......... EG&G Idaho, Inc.
Pedro J. Otaduy........ Oak Ridge National Laboratory
Howard Rohm............ U.S. Department of Energy
Gary Sandquist......... University of Utah
Donald Schurman........ EG&G Idaho, Inc.
Devin Smith............ S.U.N.Y., Stony Brook
Bob Stiger............. Delian Corporation
Al Sudduth............. Duke Power Company
Gilles Zwingelstein.... Electricite de France

PAPER REVIEWERS

John Sackett........... Argonne National Laboratory-West
Debu Majumdar.......... U.S. Department of Energy
Jan Kretzschmar........ SCK/CEN Informatics
Hakan Andersson........ Studsvik Energiteknik
Paul Blanch............ Northeast Utility Service Co.
John Brtis............. Sargent and Lundy
Henry Makowitz......... EG&G Idaho, Inc.
Mike Bray.............. EG&G Idaho, Inc.
Al Sudduth............. Duke Power Company
Gary Sandquist......... University of Utah
Jim Jenkins............ U.S. Nuclear Regulatory Commission
Brent Dixon............ EG&G Idaho, Inc.
Rich Mark.............. Volian Enterprises
Jim Dukelow............ Boeing Computer Service
Gail A. Cordes......... Intermountain Technologies, Inc.
Pedro J. Otaduy........ Oak Ridge National Laboratory
C. Edward Johnson...... Westinghouse Idaho Nuclear Co.
Bill Nelson............ EG&G Idaho, Inc.
Howard Rohm............ U.S. Department of Energy
Bob Stiger............. Delian Corporation

SESSION CHAIRMEN

A.D. Alley
General Electric Co.

H. Andersson
Studsvik Energiteknik

T. Bjorlo
Institutt for Energiteknikk

P. Blanch
Northeast Utilities Service Co.

W.V. Botts
EI International, Inc.

M.A. Bray
EG&G Idaho, Inc.

J. Brtis
Sargent and Lundy

P.C. Cacciabue
Joint Research Centre Ispra

D. Cain
Electric Power Research
  Institute

A. Christie
Westinghouse Electric Corp.

G.A. Cordes
Intermountain Technologies, Inc.

B. Dixon
EG&G Idaho, Inc.

J. Dukelow, Jr.
Boeing Computer Service

L. Ho
Institute of Nuclear Energy
  Research

E. Hollnagel
Computer Resources International

A. Jaeschke
Nuclear Research Center

J. Jenkins
U.S. Nuclear Regulatory
  Commission

C.W. Johnson
Westinghouse Idaho Nuclear
  Co.

H. Julian
Volian Enterprises, Inc.

T. Kiguchi
Hitachi Energy Research
  Laboratory

J. Kvetzchmar
SCK/CEN Informatics

R.W. Lindsay
Argonne National Laboratory
  West

H. Makowitz
EG&G Idaho, Inc.

P. Malvache
CEA, France

K. Monta
Nippon Atomic Industry Group

J. Mott
EI International, Inc.

W. Nelson
EG&G Idaho, Inc.

J. Panoussian
Framatome

Y. Shinohara
Japan Atomic Energy Research
  Institute

D. Smith
SUNY at Stony Brook

Y. Souchet
CEA, France

A. Sudduth
Duke Power Company

J.B. Thomas
CEA, France

G. Zwingelstein
Electricite de France

CONTENTS

CHAPTER 1
INTERNATIONAL OVERVIEW

CHAPTER 2
ROBOTICS AND MAINTENANCE

CHAPTER 3
ALARM AND SIGNAL VALIDATION

CHAPTER 4
EMERGENCY RESPONSE

CHAPTER 5
PROCESS DIAGNOSTICS AND TRANSIENT ADVISOR

CHAPTER 10
OPERATION ANALYSIS AIDS

CHAPTER 11
PLANT OPERATIONS AND SUPPORT

CHAPTER 12
PROBABILISTIC RISK ASSESSMENT

CHAPTER 13
ADVANCED CONCEPTS AND NON-NUCLEAR APPLICATIONS

TRIBES - A CPC/CEAC <u>TRI</u>P <u>B</u>UFFER <u>E</u>XPERT <u>S</u>YSTEM WRITTEN IN PROLOG

R. B. Lang

Middle South Utilities System Services Inc.
Box 61000
New Orleans, Louisiana 70161

INTRODUCTION

A small personal computer based expert system has been developed
to aid in diagnosing reactor trips initiated by the Core Protection Calculator, CPC, and the Control Element Assembly Calculator,
CEAC. CPCs and CEACs are part of the plant protection system in
some U.S. PWR reactors. The expert system interprets information
from trip buffers created by the CPC and CEAC to determine the reason
for the reactor trip. Harmon and King[1] suggest six logical steps in
developing a small expert system. Each step will be discussed as it
relates to the <u>TRI</u>p <u>B</u>uffer <u>E</u>xpert <u>S</u>ystem, TRIBES. The six steps are:

1. Identify a problem and then analyze the knowledge to be
    included in the system.

2. Select a tool and implicitly commit to a particular consultation
    paradigm.

3. Design the system. Initially this involves describing the
    system on paper. It typically involves making flow diagrams
    and matrices and drafting a few rules.

4. Develop a prototype system using the tool. This involves
    creating the knowledge base and testing it by running a
    number of consultations.

5. Expand, test, and revise the system until it performs
    satisfactorily.

6. Maintain and update the system as needed.

IDENTIFY A PROBLEM

The CPC/CEAC is a system of six digital computers which
monitors nuclear reactor parameters and CEA positions and will
initiate a reactor trip to prevent the violation of fuel design limits.
Online calculations are done for KW/ft and DNBR. In the event of a

trip, current information in the computers is saved in special "TRIP BUFFERS". As part of the normal routine after a trip, the information in these buffers is printed and analyzed to determine the cause of the trip.

There are some problems with the interpretation of the trip buffers. The first problem is the form of the trip buffer. The trip buffer consists merely of a number of numeric point IDs (PIDs) and corresponding values. One would suggest that more descriptive information be dumped in the trip buffers, but this option is limited due to the memory size of the CPC/CEAC computers (64K words), the stringent and expensive QA requirements imposed on changes to nuclear safety software like CPC/CEACS, and the painstakingly slow data dump rate of ten characters/second. A second problem is that not all the data in the CPC/CEAC computers is dumped in the trip buffers and it is sometimes necessary to draw conclusions from incomplete data or data from other plant sources.

The problem is then how to interpret the limited information from CPC trip buffers and other sources to determine the cause of the trip. An expert system seems applicable to the problem because CPC/CEAC trip diagnosing is a narrow and well defined specialty. The problem does not require general background knowledge or common sense. The reasons for CPC trips are carefully formulated and documented in the CPC/CEAC literature. The performance of the expert system can be easily evaluated. There is one reason for the trip and either the system is successful in determining it or it is not. Finally an expert system is a good choice because trip buffer interpretation does require a CPC expert to determine the cause of a trip. The expert's knowledge seems amenable to formulation in rules and the justification subsystem of an expert system allows the novice user to follow the expert system's logic to the trip conclusion.

SELECT A TOOL OR LANGUAGE

Once the problem was selected, the next step was to choose a language or tool in which to build the expert system. There were four main considerations in the choice of a tool or language. They were simplicity, flexibility, cost, and delivery system.

The language chosen was Borland TURBO-PROLOG. TURBO-PROLOG is a fifth generation language with a very short and simple syntax. TURBO-PROLOG requires considerably fewer program lines than other languages such as FORTRAN. TURBO-PROLOG is a compiled language yet it has a very user friendly program development environment.

Since TURBO-PROLOG is a language rather than a tool, TURBO-PROLOG gives the user some added flexibility over an expert system tool to easily customize the knowledge structure or add advanced features. Cost was also an important consideration on this first project. TURBO-PROLOG's cost is under $100.

Because of their wide availability in the nuclear power industry, the delivery system chosen was an IBM PC. A compiled PROLOG ".EXE" file can be produced using TURBO-PROLOG's own built-in fast linker and executed on any IBM XT or AT personal computer in a stand-alone mode.

<u>PROLOG Language</u>

PROLOG (<u>Pro</u>gramming in <u>Log</u>ic) is a symbolic language which was developed in France in the early 1970's. PROLOG has been used for such applications as deductive reasoning, natural language translation, symbolic mathematical equation solving, pattern matching, theorem proving and data base implementation. PROLOG is different from other languages such as FORTRAN in that PROLOG is not procedural. PROLOG does not have the DO loops and GOTOs that are so common in FORTRAN. PROLOG does not specify the sequence of steps to be executed. PROLOG is more "declarative" in that goals are specified and PROLOG uses the relationships and known facts to reach them. To program in PROLOG one must:

1. Specify some facts about objects and relationships.

2. Specify some rules (predicates) about objects and relationships.

3. Ask questions (specify goals) about objects and relationships.

PROLOG's simplicity stems from its simple grammar and most people's intuitive understanding of the simple "and" and "or" logic it is built on. There are some additional features which make PROLOG a very powerful language.

<u>Matching</u>. PROLOG works by attempting to match a goal with the facts in the database. It starts at the top of the database and considers each fact and rule in the order they occur in the program. A match succeeds if the name matches, the number of arguments match, and all the arguments match. Otherwise the match fails.

<u>Instantiation</u>. Instantiation is the process of assigning a value to a variable. A variable without a value is uninstantiated or free. If an uninstantiated variable is matched with a constant or an instantiated variable, it takes on the value of the matching constant or variable and is said to be instantiated or bound.

<u>Recursion</u>. PROLOG lacks the program flow constructs of FORTRAN such as DO loops or GOTO statements. The only program construct available is recursion. Recursion is when a function is defined in terms of itself. For example:

```
/* CONVERT A STRING TO A LIST */
string_namelist(STRING,[H|T]) if fronttoken(STRING,H,RESTSTRING)
and ! and string_namelist(RESTSTRING,T).

string_namelist(_,[]).
```

The predicate string_namelist in this example recursively calls itself each time with a shorter string and list until finally the list is empty and the second clause succeeds. The predicate then backs out of the recursion and returns to calling predicate with the list containing the string broken down into individual words.

<u>Backtracking</u>. Backtracking is a mechanism built into PROLOG whereby, when the evaluation of a given sub-goal is complete, PROLOG returns to the previous sub-goal and tries to satisfy it in a different way. Consider the predicate why_tripped:

```
/* PRINT OUT JUSTIFICATION FOR CPC CHANNEL TRIP */
1.  why_tripped if
    makewindow(3,13,93,"JUSTIFICATION SUBSYSTEM",8,2,15,72) and
    write("\nPrinter? ") and ask(ANSWER) and outflag(ANSWER).

2.  why_tripped if channelno(CHANNEL) and
    write("\nJUSTIFICATION FOR CPC CHANNEL ",CHANNEL," TRIP").

3.  why_tripped if true(Z,STRING)  and
    write("\nRULE ",Z," says ",STRING).
```

In clause #3 of the why_tripped predicate, backtracking is used
to print out all the rules which succeeded. The database is searched
for the first "true" fact and it is printed. Backtracking returns
to find the next "true" and prints it. This continues until all
"true" facts are printed.

<u>TURBO-PROLOG Pluses and Minuses</u>

During the development of the system some pluses and minuses
of TURBO-PROLOG were found which were not foreseen. On the negative
side, the creation of a DO loop to print CPC point IDs in increasing
numerical order was awkward in PROLOG. A solution was found by
retracting, incrementing and asserting a pointer in the database. The
first attempt in reading a formatted file recursively failed due to
stack overflow. This was corrected by reading only a line
recursively rather than the whole file. The organization of a
PROLOG program is hard to follow for someone used to looking at inline
procedural languages because there are many predicates which may be
called by any of the other predicates. The predicates in the program
were organized in alphabetical order so one could locate a needed
predicate easily when tracing through the code. Finally, the TURBO-
PROLOG code segment is limited to 64K bytes. Although 64K bytes was
adequate for this application, it might be a problem for a larger
system.

On the plus side, the TURBO-PROLOG syntax is very easy to learn
and understand. The program development environment is also easy to
use. Interaction with the system is done through EDITING, TRACE,
MESSAGE and DIALOG windows. The compiler and linker are very fast.
The pattern matching and backtracking features of PROLOG simplified
the programming task. The graceful "fail" mechanism of PROLOG made
the resultant expert system very resistant to user input errors. For
example, if the code expects an integer and it receives a real
number the clause simply fails and the program continues. Finally,
TURBO-PROLOG has a very convenient source level trace option for
debugging. TRACE does however increase the code segment requirements
by as much as 100%.

DESIGN THE SYSTEM

The expert system's purpose is to deduce the cause of a reactor
trip from the CPC/CEAC trip buffers. Two options are available for
CPC/CEAC trip buffer input. Either the information can be input
manually or it can be read directly from the CPC and CEAC computers.
Since the CPC and CEACs interface with a teletype, it was decided that
the trip buffers would be dumped directly to the IBM PC as the
primary mode of input. The TREAD program, written in PROLOG, handles
the communications with the CPC/CEAC computers and reformats the data
in the six trip buffer files into one master trip buffer file

containing PROLOG database facts. The TRIBES program handles the determination of the cause of the trip by processing the knowledge base.

The knowledge base consists of two parts. The two components are database facts from the trip buffer and rules (PROLOG predicates) for determining the reason for a trip. One of the first challenges encountered in developing the TRIBES knowledge base was the variability of the trip buffers. Although PID definitions are consistent among CPC plants, the contents and size of the trip buffers are not. The approach decided upon was to use the inference engine to search for a needed fact (PID value) in the knowledge base. Not finding the PID, TRIBES would ask for it to be input manually. If the user was unable to supply the needed value, the hypothesis being evaluated would fail. In addition to the database facts from the trip buffer, additional facts are added or deleted from the database during the program execution by use of the PROLOG "assertz" and "retract" predicates built into the rules.

The rules presented below are examples. The values and logic presented should not be considered specific to any plant or CPC/CEAC design. The rules for determining the cause of the trip are in the form of PROLOG predicates:

```
trip(3,IC)
if true(1,_) and data(9,IC,VALUE,DES) and
PPB1=1870.0 and VALUE<PPB1  and
concat(DES," (PID 9) < 1870 PSIA INDICATES LOW PRESSURE TRIP.",
MESSAGE) and assertz(true(3,MESSAGE)) and
write("\nRANGE TRIP ON LOW PRESSURE") and !.
```

Rule #3 first checks to see if there is a fact that says rule #1 is true. If rule #1 is true, the database is searched for the value and description of PID 9. The value is checked against the trip limit of 1870 psia and if the pressure for the selected CPC channel is less than 1870 psia, then a message is created by combining two strings. A fact stating that rule #3 is true along with the message is inserted into the knowledge base for later use. Finally, "RANGE TRIP ON LOW PRESSURE" is displayed on the screen. The "!" in the rule tells PROLOG to stop processing trip rules if this rule is successful. Otherwise all rules are examined when the "CONSULTATION" option is selected to determine the cause of the trip. A simplified schematic of the trip logic for the CPC is shown in Figure 1.

The rules for determining the cause of the trip had to be flexible. Sometimes the most straight-forward rule might not work because the needed parameter was not in the trip buffer. In those cases it was necessary to invent a secondary rule to diagnose the trip. For example, a trip on less than two reactor coolant pumps would normally be diagnosed by the following rule:

```
trip(31,IC)
if  true(1,_) and data(118,IC,PUMPS_RUNNING,DES) and
PUMPS_RUNNING<2 and
concat(DES," (PID 118) < 2\nINDICATES A TRIP ON LESS THAN TWO
REACTOR COOLANT PUMPS RUNNING.",MESSAGE) and
assertz(true(31,MESSAGE)) and
write("\nTRIP ON LESS THAN TWO REACTOR COOLANT PUMPS RUNNING") and !.
```

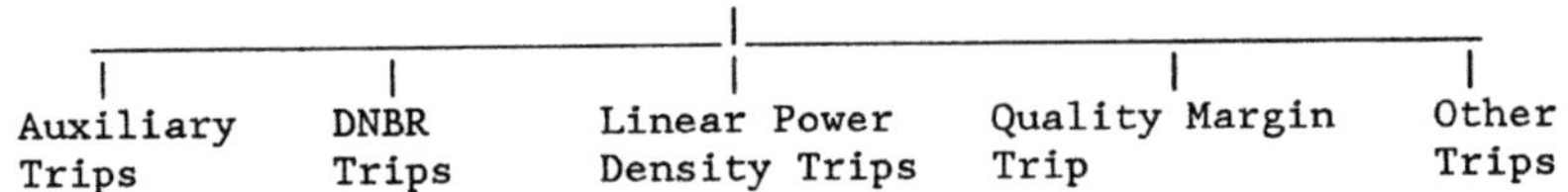

Figure 1. CPC Trip Structure

However, if PID 118 which is the number of pumps running is not in the trip buffer for a particular plant, the pump trip can also be deduced by checking the speed of the four reactor coolant pumps with the following rule:

```
trip(32,IC)
if true(1,_) and check_pump(IC,PUMPS_RUNNING) and
PUMPS_RUNNING<2 and
assertz(true(32,"INVERSE PUMP SPEEDS (PIDS 1-4)
INDICATE TRIP ON LESS THAN\nTWO RC PUMPS RUNNING.")) and
write("\nTRIP ON LESS THAN TWO REACTOR COOLANT PUMPS RUNNING") and !.
```

In this rule each pump is checked to see if it is running. The predicate check_pump reads the four inverse reactor coolant pump speeds (PIDs 1-4) from the database and determines if each pump is running based on the same criteria the CPC uses to calculate PID 118. The predicate check_pump returns the number of pumps running to the rule. If PUMPS_RUNNING is less than 2 then the rule succeeds and the pump trip fact is asserted into the database and a message is displayed on the screen.

Some rules are designed to only be used in manual input mode or in automatic input modes. For example to determine if all control rods are inserted in automatic mode one rule will simply check all rod positions. In manual input mode, rather than asking the user to input each rod position, another rule will simply ask the user, "ARE ALL CONTROL RODS INSERTED? (Y OR N)."

DEVELOP A PROTOTYPE

The prototype of the CPC/CEAC trip buffer expert system was developed on an IBM PC XT with 640K memory and a color monitor. Approximately twenty rules were implemented in the prototype. Figure 2 shows the display of the TRIBES system after startup. The user first selects a trip buffer to read in from disk. Next the "CONSULTATION" option is selected and the user selects the particular CPC channel he wishes to analyze: A,B,C or D. The reason for the trip is then displayed on the screen.

Once the reason for the trip is determined, the "JUSTI-FICATION" option on the main menu can be used to examine the knowledge base to display or print the justification information stored with the "true" facts as shown in Figure 3. From Figure 3 it is clear that there is an auxiliary trip due to low pressure. In addition to the "CONSULTATION" and "JUSTIFICATION" options, there are options to display individual point ID information or print the combined TRIBES trip buffer. Figure 4 shows the results of a search through the TRIBES database for the word "PRESSURE". Information entered manually by responding to TRIBES generated questions can also be saved on disk for future use.

500

```
┌─────────────────────TRIBES - TRIp Buffer Expert System─────────────────┐
│Version 1.4 MSU System Services Inc. 1987                                │
│                                                                         │
│                                                                         │
│                                            ┌──AVAILABLE OPTIONS──────┐  │
│                                            │Read trip buffer into memory │
│                                            │Consultation             │  │
│                                            │Justification            │  │
│                                            │Data for single CPC PID  │  │
│                                            │Data for single CEAC PID │  │
│                                            │Display formatted trip buffer│
│                                            │Display addressable constants│
│                                            │Help Information         │  │
│                                            │Edit or save trip buffer │  │
│                                            │DOS Shell                │  │
│                                            │Exit TRIBES              │  │
│                                            └─────────────────────────┘  │
│                                                                         │
│                                                                         │
└─────────────────────────────────────────────────────────────────────────┘
Select option with arrow key.          10:54   3/14/1987        477 Kb Heap.
```

Figure 2. TRIBES Main Menu Screen

```
┌─────────────────────TRIBES - TRIp Buffer Expert System─────────────────┐
│CPC CHANNEL (A,B,C,D)?a                                                  │
│                                                                         │
│RANGE TRIP ON LOW PRESSURE                                               │
│                                                                         │
│        ┌────────────────JUSTIFICATION SUBSYSTEM──────────────┐         │
│        │Printer? (Y OR N)? n                                  │         │
│        │                                                      │         │
│        │JUSTIFICATION FOR CPC CHANNEL A TRIP                  │         │
│        │RULE 1 says REGION DEPENDENT TRIP FLAG (PID 189) NOT EQUAL 0 INDICATES│
│        │ AUXILIARY TRIP.                                      │         │
│        │RULE 3 says RAW PRESSURIZER PRESSURE (PID 9) < 1870   │         │
│        │PSIA INDICATES LOW PRESSURE TRIP.                     │         │
│        │                                                      │         │
│        └──────────────────────────────────────────────────────┘         │
└─────────────────────────────────────────────────────────────────────────┘
Hit any key to continue.                               133 Kb  Heap.
```

Figure 3. TRIBES Consultation and Justification Screen

```
┌─────────────────────TRIBES - TRIp Buffer Expert System─────────────────┐
│CPC CHANNEL (A,B,C,D)?a                                                  │
│                                                                         │
│RANGE TRIP ON LOW PRESSURE                                               │
│                                                                         │
│             ┌──────────────SINGLE POINT SUBSYSTEM──────────┐           │
│             │Enter keyword for search: pressure            │           │
│             │PID9 RAW PRESSURIZER PRESSURE FOR CPC CHANNELS │           │
│             │    A       B       C       D                 │           │
│             │ 1869.900, 2250.200, 2253.200, 2266.200 PSIA  │           │
│             │FORTRAN Label: PRRAW                           │           │
│             │Minimum of four channels is 1869.9            │           │
│             │Average of four channels is 2159.875          │           │
│             │Maximum of four channels is 2266.2            │           │
│             └──────────────────────────────────────────────┘           │
└─────────────────────────────────────────────────────────────────────────┘
Hit any key to continue.                               132 Kb  Heap.
```

Figure 4. TRIBES Display Single CPC Point ID Screen

EXPAND, TEST AND REVISE

The knowledge base of rules is currently being revised and expanded to handle less common types of trips. Experts on the CPC/CEAC system were contacted for their help in further developing and reviewing the knowledge base. Approximately fifty rules are now in place. Testing on the prototype was done with canned CPC trip buffers entered by hand.

This phase of the development also included testing of the advanced system connected to the CPC single channel test facility through the standard teletype interface. The CPC/CEAC single channel test facility (CSCTF) is an exact hardware copy of the Core Protection Calculator and CEA Calculator which is used for testing and trouble-shooting. The TREAD and TRIBES programs were tested at the reactor site using an IBM compatible TOSHIBA 1100+ portable computer connected to the CSCTF through a 20ma/RS232 interface. An input simulator provided the plant inputs, such as temperatures and pressures, to the CSCTF for testing.

MAINTAIN AND UPDATE THE SYSTEM

There are three areas in which updates will be made to the system. New rules will be added to give additional insight into the cause of CPC trips. The second area of updates will be to revise the rules to reflect changes in the CPC functional specifications or trip limits. Finally, feedback from the system users will be used to improve the system user interface.

CONCLUSION

In summary, a workable CPC/CEAC trip buffer expert system was written in TURBO PROLOG for the IBM personal computer. The system's main advantages over a manual approach are speed, automatic documentation, justification, consistency and thoroughness. In short, the TRIBES system is a tool that can be used by an engineer to diagnose a wide range of CPC trips without referring to detailed CPC/CEAC functional specifications or source listings.

REFERENCES

1. P. Harmon and D. King, "Expert Systems: Artificial Intelligence in Business," John Wiley and Sons, New York, (1985), p. 178.

CRAW - AN EXPERT SYSTEM FOR NUCLEAR REACTOR COVER GAS ALARM ANALYSIS

Bruce D. Zimmerman and John A. Rawlins

Westinghouse Hanford Company
P.O. Box 1970
Richland, Washington 99352

BACKGROUND OF THE APPLICATION

During operation of the Fast Flux Test Facility research reactor, it is important to know when the cladding of a fuel pin or a control rod pin experiences a rupture. Though such ruptures do not pose an immediate safety problem, fission products and fuel or control rod material can leach out of the pin at the rupture point and deposit in other areas of the reactor system. Furthermore, it is necessary to distinguish between a fuel pin or a control rod pin at the time of the rupture because the actions taken by the reactor operators are somewhat different for each.

In order for this determination to be made, new fuel and control rod pins are loaded with "tag gas," that is, unique combinations of certain inert gas isotopes. When a pin ruptures, these isotopes and other isotopes created during irradiation are released into the reactor cover gas. The presence of isotopes in the cover gas triggers an alarm in the control room. The isotopes are sampled by single or multichannel analyzers that are able to ascertain which of a selected set of isotopes are present and the relative concentrations of each. Analysis of data from these detectors allows determination of whether a fuel pin or a control pin is the cause of the cover gas alarm.

In the Fast Flux Test Facility reactor, this fuel or control pin rupture determination is rendered more complicated by the Materials Open Test Assembly (MOTA) experiment. The MOTA experiment consists of several hundred small pressurized capsules made of materials that are subjected to irradiation for test purposes. As part of the experiment, many of these capsules will develop leaks while being irradiated in the reactor. Since much of the data of the experiment consists of recording when these leaks occur, these capsules are also loaded with various tag gases that cause cover gas alarms in the control room. Actions to be taken when a MOTA capsule leak is detected are somewhat different than those for either a fuel or a control rod pin rupture.

The CRAW expert system was developed to provide "on-the-spot" determination of whether a reactor cover gas alarm is caused by rupture of

fuel pin cladding, a control rod pin cladding, or a MOTA experiment capsule.
Specifically, the CRAW expert system was developed for four reasons:

(1) At the time the CRAW expert system was developed, one of three
cover gas alarm experts had to be called for consultation by the Operations staff every time a cover gas monitor alarm was received. Typically,
twenty to thirty alarms are being received during a four-month 24-hour
per day, 7-day per week plant operating cycle. In one instance over 100
alarms were received. The CRAW system is intended to relieve most of the
need for these experts to conduct the routine alarm analysis during
off-hours.

(2) The CRAW system is intended to provide documentation and continuing retrievability of much of the human expertise involved in cover gas
alarm analysis.

(3) The CRAW system is intended to provide for standardization of
decision making. With more than one human expert involved, standardization
can be a problem.

(4) The CRAW expert system was developed and implemented to demonstrate and evaluate the usefulness of such a system in the nuclear plant
operation domain.

Initial development of the CRAW expert system was accomplished using
the Expert-Ease[1] commercial expert system development package. The result
was then translated into ADVANCED BASIC for execution on an IBM-PC®.

The remainder of this report briefly reviews the methodology employed
by the Expert-Ease package and describes the development, structure, operation and testing of the CRAW expert system.

BACKGROUND OF THE METHODOLOGY - EXPERT-EASE

The CRAW cover gas monitor expert system was developed using the
learning system methodology available in the Expert-Ease commercial expert
system development package. The Expert-Ease software functions by being
given a series of cases (various sets of conditions and the associated
conclusions) that describe particular examples from the area to be investigated. Expert-Ease then uses these cases to develop one or more decision
trees which describe the decision logic that is implicit in the given
cases. For the cover gas monitor system problem, Expert-Ease was given a
variety of cases that specify the countrates of various gas tag and fission
product isotopes as measured by the cover gas monitor system equipment,
the day of the week the cover gas alarm is received and whether or not it
is a holiday, and the conclusion that the domain expert says is applicable
for this set of conditions. Expert-Ease is then capable of sorting out
all the various combinations of conditions and associated conclusions
presented in the cases (the "training set") and determining the decision
tree that represents the implicit decision rules the expert is using.

The algorithm that Expert-Ease uses to perform this rule induction
is known as ACLS (Analog Concept Learning System), which is a proprietary
system of Intelligent Terminals Limited, the creators of Expert-Ease.

®IBM-PC is a registered trademark of International Business Machines
Corporation, Boca Raton, Florida.

ACLS is based on the CLS (Concept Learning System) approach developed in
the mid-1960s, and a follow-on system known as ID3 (Iterative Dichotomizer
3), developed in the late 1970s at Stanford University.

The CLS algorithm scans the training set of cases that has been
provided and determines which of the decision variables (called the "attributes" in Expert-Ease, i.e., the various isotope countrates, etc. for the
cover gas monitor problem) seems to have the most explanatory power in
terms of sorting through and organizing the cases. The training set is
then partitioned so that each partitioned set corresponds to one value of
the selected decision variable, and the process is repeated. The first
selected variable becomes the "root node" of the decision tree. As the
process is repeated, the remaining decision variables are organized to
define the remainder of the decision tree. Eventually, all of the decision
variables are assigned a place in the decision tree. As implemented in
Expert-Ease, the decision variable with the most "explanatory power" at
each step of the tree building process is the one that, if selected as the
next partitioning variable, results in the greatest drop in information
entropy. A brief discussion of information entropy and how it is employed
to build decision trees can be found in Reference 2.

The important point is that, through the use of a partitioning strategy
based on some criterion (such as minimizing information entropy), it is
possible for a computer program to induce a decision tree logic structure
from a set of examples and associated conclusions used as input.

DEVELOPMENT OF THE CRAW EXPERT SYSTEM

The CRAW expert system was initially developed using the Expert-Ease
expert system development shell, and was then recoded into ADVANCED BASIC
for execution on an IBM PC. This recoding step was considered desirable
in order to make use of the graphics capability of the IBM PC and to avoid
the expense of purchasing multiple copies of Expert-Ease (no runtime module
was available).

Development of the CRAW expert system began with the lead cover gas
alarm expert providing a list of eight relevant decision variables for
the cover gas alarm (isotope countrates, countrate ratios, and other
parameters) and specifying what actions the expert would recommend for a
variety of cases. These decision variables were then used as the attributes
to allow entry of the cases and associated recommendations into the Expert-
Ease shell.

The problem was divided into six modules as shown in Figure 1, each
of which corresponds to one decision tree. This segmentation, which is
standard procedure for all but the simplest expert systems, is necessary
to avoid the combinatorial explosion of possible decision paths that results
if all attributes are unnecessarily treated as applicable for all cases.
A total of 44 input cases were given to Expert-Ease, which used them to
create six decision trees with a total of 22 decision nodes. These trees
make explicit the logic that is implicit in the cases. The result of the
Expert-Ease analysis was then explained to the expert, who recommended
some additions and changes. These additions and changes were processed
with Expert-Ease to produce modified decision trees.

The modified decision trees were again reviewed by the expert who
requested a few more minor changes. One of the requested changes resulted
in the ACLS (Expert-Ease) algorithm making a major change in the form of
one of the decision trees. Since the domain expert had by this time become
comfortable seeing the decision logic in its previous form, this last

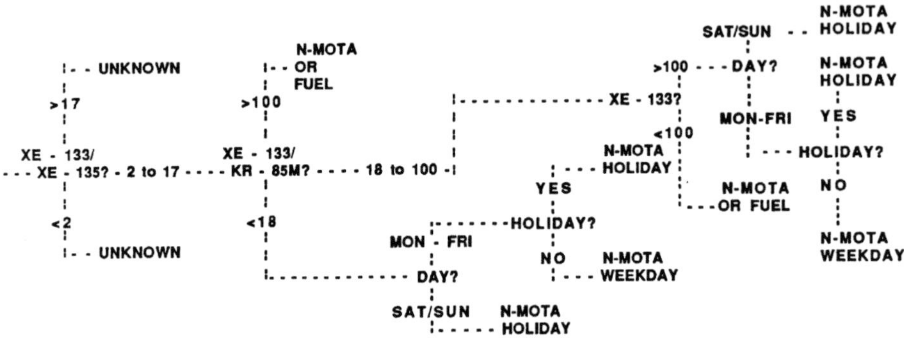

Figure 1. The Decision Modules

round of changes was made manually, directly to the decision trees, in order to avoid major changes.

The final decision trees were recoded into ADVANCED BASIC (along with the necessary code to ask questions, produce graphics, and provide for other program functions), for execution on an IBM-PC. This recoding was performed primarily to allow the inclusion of color graphics interfacing with the decision logic. The color graphics programming was considered important for man-machine interface reasons because the CRAW system is designed to be used by nuclear reactor operations personnel who are not familiar with expert systems or, in some cases, with computer systems in general. Therefore, maximum user-friendliness was needed. The contents of the graphics displays are described further in "Operation." The ADVANCED BASIC version of the CRAW program encodes the decision logic as cascades of "if" statements. The code currently uses 968 ADVANCED BASIC statements.

The entire process of developing the training set of cases, using Expert-Ease to build the decision trees, recoding into ADVANCED BASIC, and the various stages of modification and checking, took approximately four calendar months. However, the actual working time for the knowledge engineer was closer to one-third of this time. Difficulty was encountered in obtaining blocks of the expert's time to work on the project because of other work commitments. This situation slowed work on the CRAW system. However, it is not unusual, since individuals who have achieved "expert" status in some significant domain are usually in considerable demand.

OPERATION

The CRAW gas tag expert system was intended for use by the Fast Flux Test Facility reactor operations staff whenever a cover gas monitor system alarm is received. It functions by asking questions concerning xenon-125, xenon-133, xenon-135, and krypton-85m countrates (tag gas and fission product isotopes), as well as questions about the day of the week and whether it is a holiday. These questions are asked in varying sequences, according to what previous answers have been given. If the answer to a particular question is not required in some circumstances, that question will not be asked. Results of questions asked are used by the program to step through one or more of six decision trees, as needed, until a diagnosis

is made and recommended actions are formulated. The recommended actions
are then printed on the screen.

The six decision trees correspond to six separate decision modules.
The program calls modules in a forward chaining manner, as needed, in
response to conclusions drawn from the answers supplied by the operator
to the questions the program has previously asked. As each module is
called, the decision tree corresponding to the possible decision paths of
that module is drawn on the screen and, as questions are asked and answered,
the decision path being taken by the program is highlighted on the screen
in blinking red.

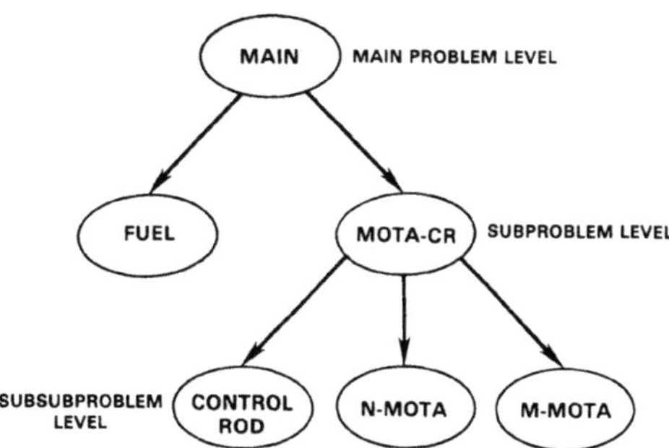

Figure 2. A CRAW Decision Tree Display

[NOTE: "XE-133" refers to the nuclear isotope xenon-133, "XE-135" refers
to xenon-135, and "KR-85m" refers to krypton-85m. Decision nodes marked
with these labels concern the countrates of these isotopes.]

Display of the decision paths available and of the one path being
followed is intended to make the decision process more visible to the
operator (and therefore more believable) and more easily verifiable by
the expert. Figure 2 shows one of the six decision trees that may appear
on the screen when the CRAW system is run. These decision trees are light
green on a blue background. If inappropriate answers are given to questions
asked by the program (such as alphabetic characters where a number is
expected), error trapping recognizes the error and displays a bright red
message on the screen informing the operator that he has made an error
and requesting him to re-enter a response to the question. When a diagnosis
is reached by the CRAW expert system, it displays recommended actions
with red lettering on a white background.

CONCLUSIONS

(1) The CRAW expert system project has demonstrated that the expert
system methodology can successfully capture enough of the complexity of a
decision problem, such as nuclear reactor cover gas alarm analysis, to allow
meaningful operational use. Furthermore, such an expert system can, in
some cases, be created with only several man-weeks of effort.

(2) The Expert-Ease software package was found useful during the initial
stage of CRAW system design when the training cases were being organized
and the initial forms of the decision trees were being worked out. However,
Expert-Ease has no capability to perform mathematical computations (such
as taking the ratio of two countrates), or to access the color graphics
routines available on an IBM PC. Consequently, its usefulness for a problem
domain such as CRAW, with intended implementation on an IBM PC, is limited.
In many instances, recoding of derived decision logic into BASIC, BASICA,
or some other higher-level language is desirable. Another, perhaps better,
alternative is to use a more versatile commercial expert system design
shell, such as the EX-TRAN® package which is a product of the same company
that produced Expert-Ease.

(3) Despite the disadvantages of Expert-Ease (and other similar
products), the important advantage is that the logic is easily set up and
can be easily changed at any time by revising or adding additional cases.
The recoding step performed for CRAW was quite tedious and error-prone
and, once the decision logic is recoded into some language such as BASIC,
changing it is difficult. It is definitely desirable to use commercial
expert system building shells to the extent that these shells are compatible
with other system requirements and to the extent that the cost involved
with purchasing multiple copies of the software is not excessive.

REFERENCES

[1] Expert-Ease, Human Edge Software Corporation, 2445 Faber Place, Palo
Alto, CA 94303, distributed by Jeffrey Perrone & Associates, 3685 17th
Street, San Francisco, California 94114, 1983. Expert-Ease is a registered
trademark of Intelligent Terminals Limited, Glasgow, Scotland.

[2] Feigenbaum and Cohen, editors, The Handbook of Artificial Intelligence,
Vol. 3, article DB3, "Data-driven Rule-space Operators," William Kaufmann,
Inc., Los Altos, California, pp.408.

®EX-TRAN is a registered trademark of Intelligent Terminals Limited,
Glasgow, Scotland.

THE USE OF DEEP KNOWLEDGE QUALITATIVE MODELS AS A BASIS FOR REAL-TIME

DIAGNOSTIC SUPPORT SYSTEMS FOR POWER PLANT CONTROL ROOM OPERATORS

Gareth H. Williams   Allan J. Hudson

Central Electricity   Queen Mary College
Research Laboratories   London
Surrey   England
England

INTRODUCTION

There is a need for diagnostic support systems in power stations to reduce the amount of information presented to the operator in order to prevent 'information overload' and 'mindset' during events in which a large number of plant measurements are changing rapidly. Existing alarm analysis systems have proved to be non-robust, inflexible and incomplete due principally to their event-based nature. Other approaches based on quantitative models are vulnerable to inaccurate measurements and uncertainties in system parameters, and would be both difficult and costly to apply to a whole plant.

The recent commercial development of expert systems and artificial intelligence technology has opened up the possibility of new operator support systems with greater flexibility, robustness, improved man-machine interaction including dialogue and explanation of reasoning, and better transferability between similar applications.

Most diagnostic expert systems already developed have been based on simple relationships between specific faults and specific symptoms. This type of knowledge is termed 'shallow' to distinguish it from 'deep' knowledge describing the behaviour of the plant. The use of deep knowledge will hopefully mean that new diagnostic systems can be developed which have better coverage of faults and require less development effort than can be achieved with existing approaches.

The representation of the plant behaviour in a deep knowledge-based expert system can be either through a quantitative (numerical) or a qualitative (symbolic) model. Qualitative models are by their very nature not as precise as quantitative models but this means that superfluous detail can be omitted. Consequently, they are potentially easier to construct, maintain, explain and execute. However, the use of qualitative models for fault diagnosis can result in a lack of discrimination between some faults due to their limited precision.

This project is concerned with the development of a diagnostic expert system based on a deep knowledge qualitative model. The system described is not intended to be an operational support system, merely to establish appropriate techniques to be used within such a system.

INCREMENTAL QUALITATIVE ANALYSIS

The qualitative modelling technique adopted is based on de Kleer's Incremental Qualitative Analysis (IQA) which is a formalisation of the kind of causal reasoning used by humans to relate trends in physical quantities e.g. the rise in temperature was caused by the increased heat input (de Kleer and Brown, 1984). IQA manipulates qualitative differential equations, called confluences, using a sign algebra (see Fig. 1). These relate the sign-values (+ve, -ve, and 0) identified with changes in quantities over time (i.e. trends) to the sign-values of changes in quantities from steady state values (i.e. deviations). Variables within a confluence must take on the value of one of the three sign-values. However, an additional symbol, ?, is used to represent the case in which the value of a variable is unconstrained. IQA is described in more detail in Herbert and Williams (1986).

The confluences for a system are formally derived from the quantitative (differential) equations for the system to ensure that the resulting qualitative model represents all possible behaviour (completeness) and that each representation is physically possible (realisability) relative to the quantitative model. However, unlike quantitative equations, confluences merely constrain the system state rather than determine it exactly. Consequently, analysis of actual or hypothetical behaviour of the system is equivalent to solving a constraint satisfaction problem.

DIAGNOSTIC BASIS

The diagnostic expert system contains only a description of how the plant behaves during correct operation. This description is, in fact, implicit as it is synthesised from knowledge of the plant structure in terms of components and their interconnections and knowledge of the correct behaviour of these components. This form of representation allows fault diagnosis in which faulty components are identified as components which are not behaving correctly. This implicit specification of faults ensures a better coverage of faults than that associated with shallow knowledge-based expert systems and requires no elicitation of specific faults and symptoms.

Fault detection is performed by trying to achieve consistency between the observed qualitative values and the set of confluences describing the correct operation of the plant. If consistency is achieved then it is assumed that no faults are present, otherwise there must be at least one fault present. Fault diagnosis identifies components whose failure would achieve this consistency. In general, because IQA has some indeterminacy, several alternative diagnoses are produced. Multiple faults can also be diagnosed by identifying combinations of components whose failure would achieve consistency.

PRESSURISER SYSTEM APPLICATION

The first application of the diagnostic expert system was to a PWR pressuriser system (see Fig. 2). This application is described in detail in Herbert and Williams (1987). The main function of this system is to control the pressure of the reactor coolant system. Charging and letdown sub-systems are used to maintain two phase conditions within the pressuriser vessel in order to maintain the efficiency of the pressuriser sprays and relief valves. The knowledge represented using IQA consists of descriptions of the thermodynamic behaviour of the fluid in the pressuriser vessel, the correct behaviour of the plant (including sensors, control systems, relief valves, and the absence of leaks in the vessel), the qualitative measurements and some fixed assumptions about the physical

+	−	0	+	?
−	−	−	?	?
0	−	0	+	?
+	?	+	+	?
?	?	?	?	?

−	−	0	+	?
−	?	−	−	?
0	+	0	−	?
+	+	+	?	?
?	?	?	?	?

=	−	0	+	?
−	T	F	F	?
0	F	T	F	?
+	F	F	T	?
?	?	?	?	?

*	−	0	+	?
−	+	0	−	?
0	0	0	0	0
+	−	0	+	?
?	?	0	?	?

/	−	0	+	?
−	+	?	−	?
0	0	?	0	0
+	−	?	+	?
?	?	?	?	?

Fig. 1. Operation tables for sign algebra

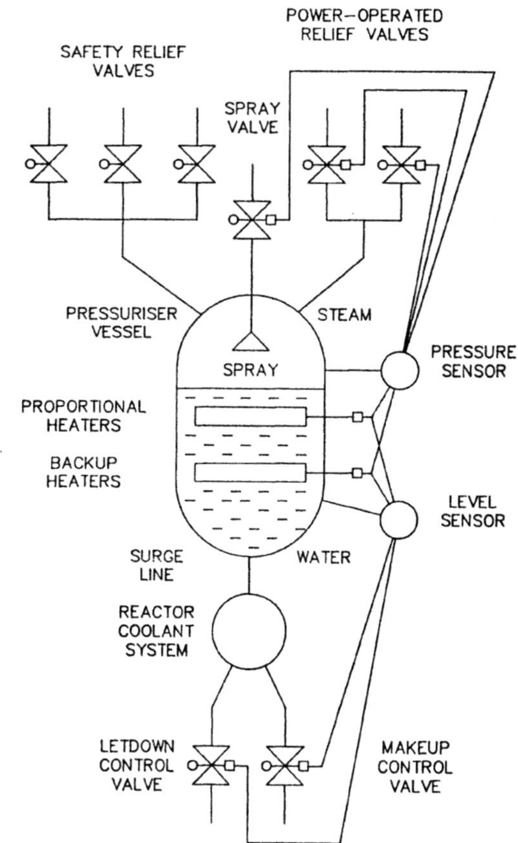

Fig. 2. PWR pressuriser system

quantities involved. The IQA model consisted of about 160 confluences divided amongst 25 components.

The algorithm used for fault detection attempts to satify all of the confluences simultaneously. If this is possible then it is assumed there are no faults, otherwise there is at least one fault present. The order in which the confluences are solved was chosen to minimise the execution time. The algorithm used for fault diagnosis consists of an exhaustive search of the components for the faulty component(s) and is essentially Davis's troubleshooting by synthesis (Davis, 1984). The confluences for each suspect component are suspended and global consistency of the remaining confluences is attempted. The process of constraint suspension examines single failures and then multiple failures until global consistency is achieved. Consequently any number of simultaneous faults may be diagnosed.

This application was implemented in PROLOG on a VAX 11-750 and was validated using transients representing normal operation and selected faults provided by a simulation model of the pressuriser system. The resulting diagnoses were correct and provided adequate discrimination but the diagnostic time was excessive at about five minutes. It was recognised that the exhaustive search diagnostic algorithm used was unsuitable for application to a whole plant. Consequently more sophisticated search techniques were investigated resulting in the development of the improved diagnostic algorithm described in the next section.

IMPROVED DIAGNOSTIC ALGORITHM

The essential feature of the improved diagnostic algorithm is that it uses knowledge of system structure to limit the number of fault candidates which need to be considered. The algorithm is similar to one described by Davis (1984), but differs in that it does not use backward propagation. The reason that backward propagation is not used is that global consistency is more significant in the qualitative models being used than in the highly overdetermined quantitative models used by Davis.

The algorithm consists of four stages:

1) Constraint Propagation

Given a precomputed schedule of components produced by a scheduler which is described below, constraint propagation attempts to bind variables to symbolic values such that confluence and then component consistency are achieved. The effects of these bindings are propagated by attempting to achieve consistency in the next scheduled component.

2) Discrepancy Detection

This occurs when component consistency cannot be achieved for a particular component.

3) Candidate Generation

This produces a list of candidate faulty components on which the discrepancy is dependent. The dependencies between components are determined by the plant structure and the schedule order.

4) Constraint Suspension

Confluences for each candidate faulty component are suspended in turn and constraint propagation resumed until another discrepancy is detected or global consistency is achieved. If global consistency is achieved then the component is included in the diagnosis, otherwise the fault candidates still to be analysed are pruned on the basis of their influence on the original discrepancy.

The order in which the components are tackled during constraint propagation is determined by a scheduler. The aim of scheduling is to minimise the diagnosis time. This is achieved if the confluences are made as determinate as possible during propagation, thus minimising the amount of backtracking required. Consequently, the schedule is produced on the basis of using the maximum amount of sensed data as early as possible in the detection and diagnosis process. An example of a schedule is shown in Fig. 3 in terms of ascending schedule numbers for an example network representing a plant system with five interconnected components. In general, the sensors associated with a component occur earlier in the schedule than the component itself, and heavily constrained components are earlier than lightly constrained components. The scheduling scheme adopted would also enable a meaningful 'causal' explanation to be provided, but as yet an explanation facility has not been implemented.

A dependency record is generated for each component, as the schedule is produced, indicating those components on which it directly depends. Causal pathways for the system are also generated as lists of components in strict order of dependency. In addition, a pathway list is produced for each component as the list of causal pathways on which that component lies. These are used to generate a list of components which are fault candidates for each confluence when that confluence is discrepant. For the above example, component-1 is a fault candidate for a discrepancy at component-4 because, with the schedule shown, component-4 is indirectly dependent on component-1. However, component-5 is not a fault candidate because component-4 is not dependent on component-5. The complete list of single fault candidates for a discrepancy at component-4 is:

(component-4 component-3 component-2 component-1).

The fault candidates considered can also include multiple faults which are generated from combinations of single faults and other multiple faults. For the above example, the joint failure of component-1 and component-3 is also a fault candidate for a discrepancy at component-4, because component-4 is dependent on both component-1 and component-3. However, the diagnostic analysis required to verify this double fault is essentially identical to that required to verify the single failure of component-3. This is because the suspension of the correct behaviour of component-3 removes all dependencies of the discrepancy on component-1. Consequently, this double fault is removed from the list of explicit fault candidates and recorded as a 'redundant' fault associated with the single failure of component-3. In general, multiple fault candidates which require the same diagnostic analysis as single faults or other multiple faults are identified as being 'redundant' to avoid repeated analyses.

For the above example, the complete list of non-redundant fault candidates including arbitrary multiple failures is:

(component-4 component-3 component-2 component-1
component-3 and component-2 component-2 and component-1).

The list of non-redundant fault candidates is pruned during on-line diagnosis using dependency information between confluences and components. Fault candidates which do not influence all of the discrepant confluences are pruned. For the above example, a discrepancy at component-4 in which confluences involving variable Y were discrepant would result in the above non-redundant fault candidate list being pruned to:

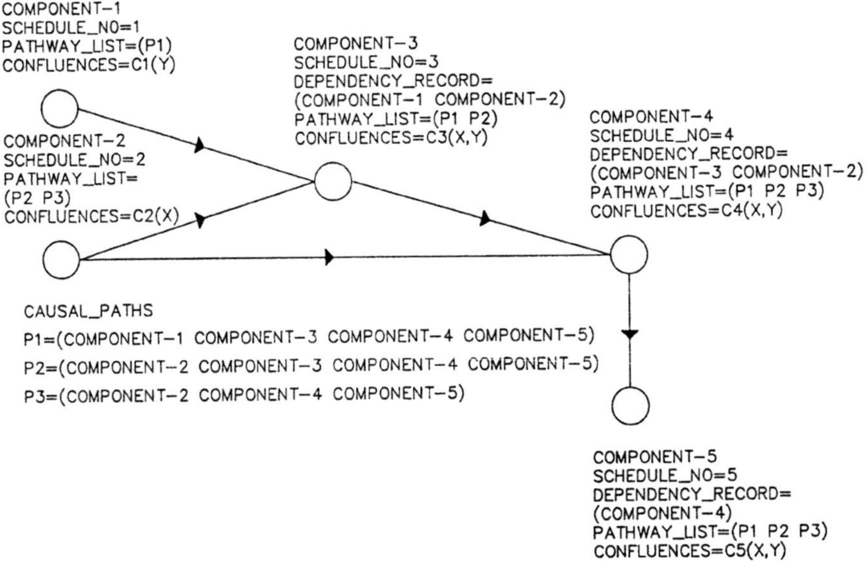

Fig. 3. Example network

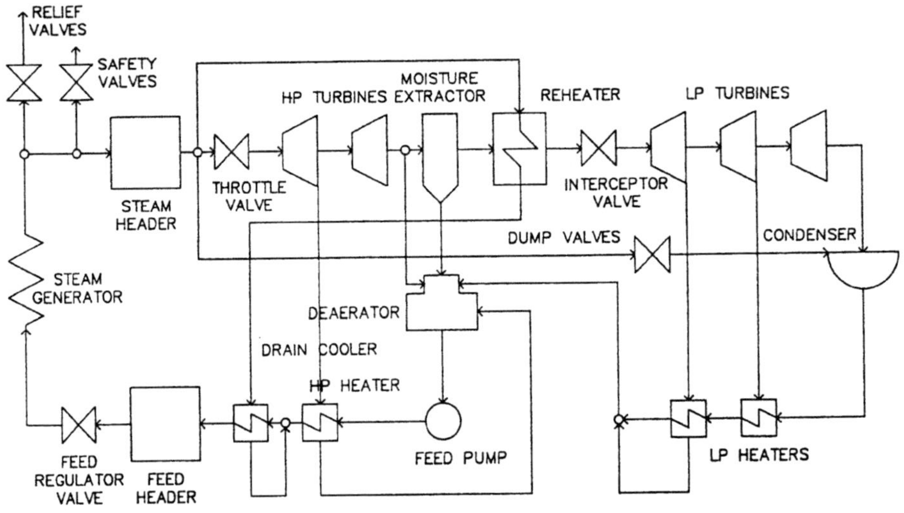

Fig. 4. PWR feed and steam system

(component-4 component-3 component-1).

This is because the confluences for component-2 do not involve variable Y and hence component-2 could not influence the discrepant confluences in component-4.

The non-redundant fault candidate list is also pruned using structural information and information about the success or failure of fault candidates which have already been analysed in removing the discrepancy. Candidates are pruned which only influence the discrepancy via an unsuccessful candidate or which include a successful candidate as part of a multiple fault. Examples of such pruning are given below for the above example with the assumption that no candidates have been removed using dependency information between discrepant confluences and components i.e. the non-redundant fault candidate list is still:

(component-4 component-3 component-2 component-1
component-3 and component-2 component-2 and component-1).

If component-3 was unsuccessful in removing the discrepancy at component-4 then component-1 would be pruned because component-1 only affects the discrepancy via component-3. Conversely, if component-3 was successful in removing the discrepancy at component-4 then the double fault candidate component-3 and component-2 would be pruned.

On-line pruning is performed within the recursive structure of the constraint propagation program. Subsequently, backtracking occurs to each remaining fault candidate, in the reverse order to the schedule, where constraint suspension and constraint propagation are applied.

The improved diagnostic algorithm has been implemented in POPLOG COMMON LISP on a SUN-2 workstation. It uses off-line scheduling of components and generation and grouping of fault candidates in order to speed up on-line fault detection and diagnosis. Currently, the implemented system is being modified to re-use parts of previous diagnoses to enable faster diagnosis.

An early version of the improved algorithm which incorporated some of the features described above was implemented in INTERLISP-D on a XEROX 1108 and applied to the pressuriser application. The resulting diagnosis time of about three seconds was about five times less than that for the exhaustive search algorithm with the same hardware and software. Futhermore additional gains over the exhaustive search algorithm are expected for a complete implementation of the improved algorithm and for applications involving larger plant systems.

FEED AND STEAM SYSTEM APPLICATION

To confirm that the improved diagnostic algorithm is appropriate for larger plant systems a PWR feed and steam system has been chosen as a second application. As well as containing more components than the pressuriser system, the feed and steam system is also behaviourally and structurally more complex (see Fig. 4). This application is currently being implemented on the SUN-2 workstation.

FUTURE WORK

The feed and steam system application has raised several new modelling issues including the absence of a unique normal steady state and the presence of fluid transport time delays. It is intended to tackle the former by introducing additional confluences relating changes in trends together (i.e. second order derivatives). Qualitative integration of historical data will be used to tackle the problem of time delays although the exact form of the algorithm has not yet been decided. It is also intended to investigate the potential for using historical data to provide better discrimination by imposing continuity and state transition constraints on correct plant operation. These constraints would also allow crude predictions to be made which might improve the diagnostic speed in some circumstances.

Other features which it is intended to investigate include the use of shallow knowledge about high consequence faults, common faults and dependent faults, and a hierarchical decomposition of the plant system to allow multi-level diagnosis.

CONCLUSION

The improved diagnostic algorithm described in this paper gives reasonable detection and diagnosis times for a small system, such as the pressuriser system.   Whether it also yields reasonable times for larger plant systems will hopefully be established by current investigations.

ACKNOWLEDGEMENTS

The work described was carried out at the Central Electricity Research Laboratories of the Technology Planning and Research Division of the Central Electricity Generating Board and the Department of Electrical Engineering of Queen Mary College.   The paper is published with the permission of the Central Electricity Generating Board and Queen Mary College.

REFERENCES

Davis, R., 1984, Diagnostic reasoning based on structure and behaviour,
        Artificial Intelligence, Vol. 24, No. 1-3, December, pp. 347-410.
Herbert, M.R., and Williams, G.H., 1986, An examination of qualitative
        plant modelling as a basis for knowledge-based operator aids in
        nuclear power stations, in: Proceedings of the Unicom Seminar on
        Expert Systems and Optimisation in Process Control, Mamdani, A., and
        Efstathiou, J., ed., Gower Technical Press, Aldershot, UK.
Herbert, M.R., and Williams, G.H., 1987, An initial evaluation of the
        detection and diagnosis of power plant faults using a deep knowledge
        representation of physical behaviour, Expert Systems, Vol. 4, No. 2,
        May, pp. 90-99.
de Kleer, J., and Brown, J.S., 1984, A qualitative physics based on
        confluences, Artificial Intelligence, Vol. 24, No. 1-3, December,
        pp. 7-84.

AN EXPERT SYSTEM FOR DIESEL GENERATOR DIAGNOSTICS

Dennis C. Bley, Jack W. Read, Stan Kaplan, James K. Liming,
Neil M. Brosee*, and Donald W. Hanley*

Pickard, Lowe and Garrick, Inc. *Boston Edison Company
2260 University Drive   25 Braintree Hill Office Park
Newport Beach, CA 92600   Braintree, MA 02184

ABSTRACT

The idea of developing artificial intelligence (AI) systems to capture
the knowledge of human experts is receiving much attention these days. The
idea is even more attractive when important expertise resides within a
single individual, especially one who is nearing retirement and who has not
otherwise recorded or passed along his important knowledge and thought
processes. The diesel generators at Pilgrim Nuclear Power Station have
performed exceptionally well, primarily due to the care and attention of one
man. Therefore, we are constructing an expert system for the diagnosis of
diesel generator problems at Pilgrim.

The Pilgrim diesel expert system is based on the knowledge and
judgments of the man who has cared for those diesels since plant startup.
He had almost 30 years experience with the diesel manufacturer before
dedicating his efforts to Pilgrim, where he has followed all diesel work
over the years. To incorporate his knowledge in a way that permits the full
breadth of his knowledge (i.e., thoughts on the likelihood of all possible
sources of trouble), we have developed a Bayesian diagnostic module (BDM) as
the heart of the expert system.

The BDM uses Bayes' theorem to bring the power of probabilistic
thinking to the expert system. Particularly significant benefits of this
approach include:

•   A quantitative statement of our state of knowledge at each step in the
    specification of observed symptoms.

•   The ability to sequentially add symptoms during the diagnostic process,
    narrowing our uncertainty, as required, to increase confidence in the
    diagnosis.

•   The ability to add new symptoms to the model without disturbing the
    integrity of the existing system.

This paper includes a description of the expert system design and operation, examples from the knowledge base, and sample diagnoses, so the reader can observe the process in action.

INTRODUCTION

DG-HELPER is the name we have given to a computer system intended to serve as a trouble-shooting (diagnostic) assistant to power plant maintenance personnel. We have built into DG-HELPER the knowledge and judgments of the man who has cared for those diesels since plant startup. He had almost 30 years experience with the diesel manufacturer before dedicating his efforts to Pilgrim, where he has followed all diesel work over the years. DG-HELPER represents an effort to encapsulate his knowledge of the Pilgrim diesel design, operating history, and possible failure modes, along with their likelihood and observable symptoms. In DG-HELPER, you will find not only the standard trouble-shooting guidance tabulated in technical manuals but also those tricky, unusual problems learned through years of experience and a quantitative statement of the probability of each possible problem--the kind of information that leads the real expert to the source of trouble long before others have culled out the alternatives.

THE BAYESIAN DIAGNOSTIC MODULE - A PROBABILISTIC APPROACH TO KNOWLEDGE BASE AND INFERENCE ENGINE DEVELOPMENT

The idea of a diagnostic expert system is of high interest. What is perhaps different about our approach is the structure of the knowledge base; that is to say, the form in which the plant-specific information is encoded. The traditional AI approach to this problem is to encode the knowledge in the form of a "rule base," which is a set of rules (called production rules) of the form, "If A, then B." This approach works well for those physical systems and situations for which our knowledge has this absolute, black and white, syllogistic character. For systems about which our knowledge is uncertain and about which we must reason in terms of levels of confidence, the absolute approach is not adequate. For systems like this, attempts have been made to graft onto the rule base a treatment of uncertainty. Best known perhaps is the certainty factor method of Shortliffe and Buchanan[1], which uses modified production rules of the form:

"If A, then B with certainty C."

Numerous other approaches for including uncertainty in expert systems have also been suggested, including probability,[2] fuzzy logic,[3] Dempster-Shaeffer Theory,[4] and others.

We have taken a different approach that combines the use of production rules, when these are valid, with the use of Bayes' theorem when significant uncertainty is present.* At the heart of the Bayesian approach is the idea of "updating" or quantifying the impact of new evidence on previous degrees of confidence. The thought here is that since Bayes' theorem is the

---

*Since Bayes' theorem reduces to production rules in the special case in which there is no uncertainty, we may describe the whole approach as simply "apply Bayes' theorem."

fundamental law for logical reasoning under uncertainty, it ought to be
possible to structure a knowledge base around this theorem and that, if this
were done, the resulting expert system would reason about uncertainty in an
exactly correct manner as an ideal human expert would.

We have developed a Bayesian diagnostic module as the heart of the
expert system. Kaplan, Frank, and Bley,[5] provide a complete theoretical
foundation and description of the BDM. That foundation is summarized in the
following paragraphs.

Let us speak of the possible problems of the diesel generator as points
in the state space, X_i, of the machine. Further, we will characterize the
evidence (or symptoms) by a set of "feature variables" or "evidence
variables" v^k, $k = 1 \ldots K$, each of which may take on one of several possible
values. Let v^k represent the jth possible value of the variable v^k. The
information describing the observable evidence is then capatured and encoded
by specifying the values, v^k_j, taken on by the variables, v^k.

In other words, the evidence E now consists of the set of values v^k_j.

$$E = \left\{ \ldots v^k_j \ldots \right\}^* \tag{1}$$

This provides us with a computationally workable way of discretizing
the evidence. That in turn provides the key to the use of Bayes' theorem to
build our diagnostic module.

Imagine a tableau (Table 1) in which we list across the top the various
possible discrete state space points, x_i. Down the left side of the table,
we list the discrete evidence variables, v^k. For each variable v^k, we list
also the possible values, v^k_j that it may take on. In each box of the
table, we enter the number $p(v^k_j | x_i)$; i.e., our probability that the kth
feature variable, v^k, would take on the value v^k_j, given that the plant is
in state x^i. This table, in effect, is a presentation, of the likelihood
function $p(E|x)$, as we shall now show.

---

*For example, let the variable v^1 stand for "air pressure," which we allow
to have the values: $v^1_1 = 250$, $v^1_2 = 210$, $v^1_3 = 175$, and $v^1_4 = 100$.
Let v^2 be the variable "excessive belt wear" for which we allow the values:
$v^2_1 =$ "severe event," $v^2_2 =$ "worn in one way," and $v^2_3 =$ "none." With just
these two variables, the evidence could be

$$E = \left\{ v^1_4, v^2_2 \right\}$$

In general, then, we write the evidence vector as

$$E = \left\{ v^1_{j_1}, v^2_{j_2}, \ldots v^k_{j_k}, \ldots v^K_{j_K} \right\}$$

where $v^k_{j_k}$ stands for the particular value taken on at this time by the
variable v^k. We use the double subscript in the above equation to emphasize
that the j's need not be the same for different k's. Thus, for example, j_1
need not be the same as j_2. With this understood, we shall drop the double
subscript from here on.

Table 1. Bayesian Tableau

Evidence Variable v^k	Possible Values	Plant State				
		x_1	x_2	$\cdots$	x_i	$\cdots$
v^1	v^1_1 $\vdots$ v^1_j $\vdots$ v^1_J					
$\vdots$						
v^k	v^k_1 $\vdots$ v^k_j $\vdots$ v^k_J				$p(v^k_j\|x_i)$	
$\vdots$						
Prior					$p(x_i)$	

As we have said, we may think of the evidence, at any moment, as encoded in the set of values v^k_j; i.e., in the specializations taken on by the variables v^k at that moment.

$$E = \left\{ v^1_j, v^2_j, \ldots v^k_j \ldots v^K_j \right\} \tag{2}$$

Therefore, if we now make the assertion

$$p(E|x_i) = p(v^1_j \ldots v^K_j|x_i)$$

$$= p(v^1_j|x_i)\, p(v^2_j|x_i) \ldots p(v^K_j|x_i) \tag{3}$$

$$= \prod_{k=1}^{K} p(v^k_j|x_i)$$

then, we can implement Bayes' theorem directly from Table 1 in the form,

$$p(x|E)_i = \frac{p(x_i) \prod_{k=1}^{K} p(v_j^k|x_i)}{\sum_i p(x_i) \prod_{k=1}^{K} p(v_j^k|x_i)} \tag{4}$$

Thus, given the values v_j^k of the observables, we can calculate the probability distribution over the x_i in a very simple way using Equation (4) and the numbers $p(v_j^k|x_i)$ stored in Table 1 (and also using the prior $p(x_i)$, which can be thought of as included in the last row of Table 1 as shown). We see then that Table 1 is effectively the diagnostic knowledge base. In this table, we encode and embody our understanding of the behavior of the plant and the relationship of the observables to the plant state. This is done in advance and with much discussion and analysis, gathering the knowledge of the experts. With this knowledge codified in the form of Table 1, it is a simple matter at any instant, j, to infer the state of the machine from the observables v_j^k at that instant.

Table 1, therefore, and Equation (4) can be thought of as together constituting the Bayesian diagnostic module.

The above discussion constitutes just an outline presentation of the basic idea. Needless to say, there are many details, many ifs and buts, and many clever little tricks required to apply this in practice. There will be many objections, especially surrounding assertion (3). All such objections can be handled, as they must, because of the fundamental nature of Bayes' theorem. It is best, however, not to try to discuss the details and the tricks in an abstract framework. It is much more communicative and convincing to deal with problems and objections as they come up in specific examples.

THE PILGRIM DIESEL GENERATOR EXPERT SYSTEM

The diesel generator expert system project for Pilgrim Nuclear Power Station has been structured in two phases. Phase 1 is a pilot study to define the overall structure of the system and demonstrate feasibility, and Phase 2 will complete the knowledge base and develop a user interface to meet the needs and desires of the plant staff. Only the pilot study is complete. It produced an operable, limited-scope version of the diesel generator troubleshooting (diagnostic) expert system called DG-HELPER.

The overall structure of DG-HELPER is sketched in the flow chart of Figure 1. Here, the mechanic begins with a signal to start the diesel and, by observing straightforward aspects of machine performance, zeros in on the area of trouble. In those cases in which a number of problems could be causing degraded performance, the mechanic is referred to an appropriate expert system. The logic of Figure 1 has been coded into a series of

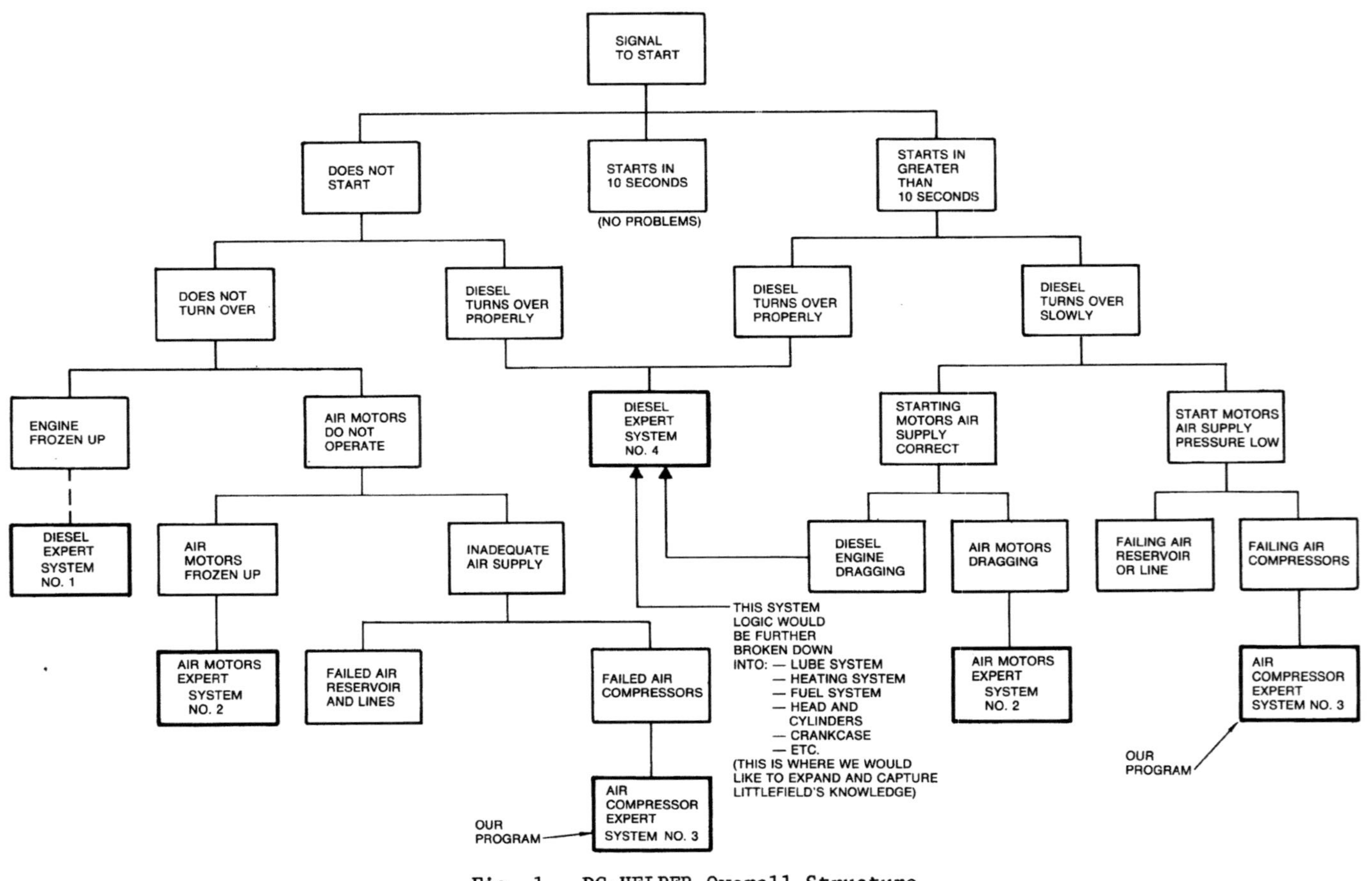

Fig. 1. DG-HELPER Overall Structure

questions in a simple rule system which will operate on an IBM PC and which automatically calls the appropriate expert system. At this time, only expert system number 3 (DG-HELP3) for the air compressor has been developed.

The Bayesian tableau for the BDM of DG-HELP3 is shown in Table 2. Seventeen evidence variables have been defined against twenty-nine machine states. Selection of those states and variables was an arduous iterative task, as was the assignment of priors and likelihood. In fact, the careful reader will observe that additional work in structuring the variables and their possible values still remains.

We began with a short list of machine problems from the diesel generator technical manual supplemented by first thoughts of the Pilgrim expert and our own experience. Long discussions of unusual past problems led to a gradual expansion of the list of machine states. The process involved analyzing anecdotes, suggesting possibilities, reviewing old logs and notebooks, etc. It was slow, but immensely rewarding.

For evidence variables, we again began with the obvious ones; e.g., air discharge pressure, oil pressure, and belt wear. As we assigned likelihood and tested the system, we found cases leading to inconclusive diagnoses. Then, it was necessary to probe the expert's memory; i.e., what unusual characteristic clearly distinguished a particular problem from other possibilities. For example, could he tell if a mechanic had used the wrong viscosity lubricating oil? Yes, there would be excessive visible oil in the discharge air, and the oil pressure would be low. Thus, although many problems can cause low oil pressure, when we add the variable V15, Oil in Discharge Air, a firm diagnosis can be accomplished.

The problem of identifying key evidence variables was especially important for very severe and unlikely problems. In such a case, the prior probability of being in the state is very low and strong evidence is needed to obtain a valid diagnosis. Traditional technical manuals lead the mechanic through a thorough and time-consuming disassembly, inspection, and test process that DG-HELPER can bypass. (An example follows shortly that demonstrates this situation.) Much of the work in developing information for Table 2 involved this test and the refueling process.

DG-HELP3 is currently implemented as a LOTUS 1-2-3® spread sheet, using macros to organize the system and perform the calculations of Equation (4). This system has been suitable for demonstrating feasibility, but has severe drawbacks that must be corrected in the final version. Nevertheless, a description of a DG-HELP3 LOTUS session may clarify operation of the BDM concept.

When DG-HELP3 is loaded, the menu of Figure 2 appears on the screen. Use menu item 1 to begin a new diagnosis.

The other menu items are self-explanatory, and we illustrate their use through an example.

- First, press "1" to begin. DG-HELP3 then presents the screen of Figure 3.

- Now, select a symptom. Suppose the first indication of a problem is that the air pressure is low; say it is about 180 psig. Move the cursor to V1, value 3, "150 to 200"; i.e., line 151. Select this symptom; DG-HELP3 then repositions the screen to permit selection of additional symptoms. Rather than identifying other symptoms, let us see what we learn from just this one.

Table 2.   DG-HELP3 Bayesian Tableau

DIESEL STARTING AIR (COMPRESSED AIR) VARIABLES	VALUES	PRESSURE SWITCH FAILED	PRESSURE SWITCH--BAD CALIBRATION	CONNECTION AIR LEAKS	RESTRICTED AIR INTAKE	DIRTY COOLING SURFACES	BELTS TOO LOSE (SLIPPING)	BELTS TOO TIGHT	PULLEY NOT OK OR BURR	PULLEY OUT OF ALIGNMENT	LOOSE PULLEY
V0 PRIOR	PRIOR	0.0257	0.0257	0.0257	0.0017	0.0001	0.2667	0.0001	0.0017	0.0017	0.0257
V1 AIR DISCHARGE PRESSURE	1 225 TO 250	-1.0000	-1.0000	0.9500	1.0000	1.0000	-1.0000	1.0000	1.0000	1.0000	1.0000
V1 AIR DISCHARGE PRESSURE	2 200 TO 225	-1.0000	0.9000	0.0050	-1.0000	-1.0000	0.0500	-1.0000	-1.0000	-1.0000	-1.0000
V1 AIR DISCHARGE PRESSURE	3 150 TO 200	-1.0000	0.0000	0.0050	-1.0000	-1.0000	0.1000	-1.0000	-1.0000	-1.0000	-1.0000
V1 AIR DISCHARGE PRESSURE	4 LESS THAN 150	1.0000	0.0200	0.0400	-1.0000	-1.0000	0.0500	-1.0000	-1.0000	-1.0000	-1.0000
V2 LOW OIL PRESSURE (25 PSI NORMAL)	1 15 TO 20 PSI	-1.0000	-1.0000	-1.0000	-1.0000	-1.0000	-1.0000	-1.0000	-1.0000	-1.0000	-1.0000
V2 LOW OIL PRESSURE (25 PSI NORMAL)	2 LESS THAN 15 PSI	-1.0000	-1.0000	-1.0000	-1.0000	-1.0000	-1.0000	-1.0000	-1.0000	-1.0000	-1.0000
V3 COMPRESSOR NOISE	1.1 SEVERE KNOCKING (HEAVY)	-1.0000	-1.0000	-1.0000	-1.0000	-1.0000	-1.0000	-1.0000	-1.0000	-1.0000	0.1000
V3 COMPRESSOR NOISE	1.2 LIGHT KNOCKING (TINNY)	-1.0000	-1.0000	-1.0000	-1.0000	-1.0000	-1.0000	-1.0000	-1.0000	-1.0000	-1.0000
V3 COMPRESSOR NOISE	1.3 ROLLING RUMBLE	-1.0000	-1.0000	-1.0000	-1.0000	-1.0000	-1.0000	-1.0000	-1.0000	-1.0000	0.9000
V3 COMPRESSOR NOISE	2 AUDIBLE AIR ESCAPE	-1.0000	-1.0000	1.0000	-1.0000	-1.0000	-1.0000	-1.0000	-1.0000	-1.0000	-1.0000
V3 COMPRESSOR NOISE	3 SQUEALLY BELT	-1.0000	-1.0000	-1.0000	-1.0000	-1.0000	1.0000	-1.0000	-1.0000	-1.0000	-1.0000
V3 COMPRESSOR NOISE	4 COMPRESSOR SOUNDS UNLOADED	-1.0000	-1.0000	-1.0000	-1.0000	-1.0000	-1.0000	-1.0000	-1.0000	-1.0000	-1.0000
V4 EXCESSIVE SHAFT ENDPLAY	1 YES	-1.0000	-1.0000	-1.0000	-1.0000	-1.0000	-1.0000	-1.0000	-1.0000	-1.0000	-1.0000
V5 COMPRESSOR, HOT SMELL	1 YES	-1.0000	1.0000	-1.0000	1.0000	1.0000	-1.0000	-1.0000	-1.0000	-1.0000	-1.0000
V6 MOTOR CURRENT	1 HIGH	-1.0000	-1.0000	-1.0000	-1.0000	-1.0000	-1.0000	0.2000	-1.0000	-1.0000	-1.0000
V6 MOTOR CURRENT	2 NORMAL	1.0000	1.0000	1.0000	1.0000	1.0000	1.0000	0.8000	1.0000	1.0000	1.0000
V6 MOTOR CURRENT	3 LOW	-1.0000	-1.0000	-1.0000	-1.0000	-1.0000	-1.0000	-1.0000	-1.0000	-1.0000	-1.0000
V7 LOW OIL LEVEL INDICATED	1 YES	-1.0000	-1.0000	-1.0000	-1.0000	-1.0000	-1.0000	-1.0000	-1.0000	-1.0000	-1.0000
V7 LOW OIL LEVEL INDICATED	2 NO	1.0000	1.0000	1.0000	1.0000	1.0000	1.0000	1.0000	1.0000	1.0000	1.0000
V8 HIGH OIL LEVEL INDICATED	1 YES	-1.0000	-1.0000	-1.0000	-1.0000	-1.0000	-1.0000	-1.0000	-1.0000	-1.0000	-1.0000
V9 STALL ON START	1 YES	-1.0000	-1.0000	-1.0000	-1.0000	-1.0000	-1.0000	-1.0000	-1.0000	-1.0000	-1.0000
V10 DURATION OF COMPRESSOR OPERATION	1 EXCESSIVE	-1.0000	-1.0000	0.0500	1.0000	-1.0000	0.1000	-1.0000	-1.0000	-1.0000	-1.0000
V10 DURATION OF COMPRESSOR OPERATION	2 NORMAL	1.0000	1.0000	0.9500	-1.0000	1.0000	0.9000	1.0000	1.0000	1.0000	1.0000
V11 NUMBER OF COMPRESSOR STARTS	1 EXCESSIVE	-1.0000	0.5000	0.9500	-1.0000	-1.0000	-1.0000	-1.0000	-1.0000	-1.0000	-1.0000
V11 NUMBER OF COMPRESSOR STARTS	2 NORMAL	-1.0000	-1.0000	-1.0000	1.0000	1.0000	1.0000	1.0000	1.0000	1.0000	1.0000
V11 NUMBER OF COMPRESSOR STARTS	3 TOO FEW	1.0000	0.5000	0.0500	-1.0000	-1.0000	-1.0000	-1.0000	-1.0000	-1.0000	-1.0000
V12 EXCESSIVE BELT MOVEMENT	1 YES	-1.0000	-1.0000	-1.0000	-1.0000	-1.0000	1.0000	-1.0000	-1.0000	-1.0000	-1.0000
V13 EXCESSIVE PULLEY MOVEMENT	1 YES	-1.0000	-1.0000	-1.0000	-1.0000	-1.0000	-1.0000	-1.0000	1.0000	-1.0000	-1.0000
V14 EXCESSIVE BELT WEAR	1 SEVERE WEAR	-1.0000	-1.0000	-1.0000	-1.0000	-1.0000	0.9000	0.9000	0.1000	0.1000	0.5000
V14 EXCESSIVE BELT WEAR	2 WORN IN ONE WAY	-1.0000	-1.0000	-1.0000	-1.0000	-1.0000	0.1000	0.1000	0.9000	0.9000	0.5000
V15 OIL IN DISCHARGE AIR	1 HEAVY AMOUNT & DARK COLOR	-1.0000	-1.0000	-1.0000	0.1000	-1.0000	-1.0000	-1.0000	-1.0000	-1.0000	-1.0000
V15 OIL IN DISCHARGE AIR	2 LIGHT AMOUNT, YELLOW COLOR	-1.0000	-1.0000	-1.0000	0.9000	-1.0000	-1.0000	-1.0000	-1.0000	-1.0000	-1.0000
V15 OIL IN DISCHARGE AIR	3 NORMAL AMOUNT, MILKY LOOK	1.0000	1.0000	1.0000	1.0000	1.0000	1.0000	1.0000	1.0000	1.0000	1.0000
V16 CRANKCASE BREATHER BLOWBY	1 YES	-1.0000	-1.0000	-1.0000	-1.0000	-1.0000	-1.0000	-1.0000	-1.0000	-1.0000	-1.0000
V17 IRREGULAR CYLINDER WEAR	1 YES	-1.0000	-1.0000	-1.0000	-1.0000	-1.0000	-1.0000	-1.0000	-1.0000	-1.0000	-1.0000

HP CYLINDER DISCHARGE VALVE LEAKING	BLOWN GASKETS	WORN RINGS AND CYLINDER WALLS	PISTON HITTING HEAD	HP/LP CYL. INLET VLVS OR UNLOADER MECH. FAILED OPEN	RESTRICTED DISCHARGE	LOW OIL IN CRANKCASE	EXCESSIVE OIL IN CRANKCASE	OIL PUMP NOT FUNCTIONING	LOW OIL VISCOSITY	RESTRICTED BREATHER VALVE	UNLOADER MECH. FAILED CLOSED	LOOSE VALVE ASSEMBLIES	WRONG TYPE OF OIL
0.0257	0.2667	0.0017	0.0001	0.0017	0.0017	0.2667	0.0001	0.0017	0.0017	0.0257	0.0257	0.0017	0.0001
0.0010	0.0400	-1.0000	1.0000	-1.0000	1.0000	1.0000	1.0000	1.0000	1.0000	1.0000	1.0000	0.0400	0.0400
0.8500	0.0050	0.8500	-1.0000	0.0001	-1.0000	-1.0000	-1.0000	-1.0000	-1.0000	-1.0000	-1.0000	0.0050	0.0050
0.1000	0.0050	0.0500	-1.0000	0.0200	-1.0000	-1.0000	-1.0000	-1.0000	-1.0000	-1.0000	-1.0000	0.0050	0.0050
0.0490	0.9500	0.1000	-1.0000	0.9800	-1.0000	-1.0000	-1.0000	-1.0000	-1.0000	-1.0000	-1.0000	0.9500	0.9500
-1.0000	-1.0000	-1.0000	-1.0000	-1.0000	-1.0000	0.7000	-1.0000	0.1000	0.8000	-1.0000	-1.0000	-1.0000	-1.0000
-1.0000	-1.0000	-1.0000	-1.0000	-1.0000	-1.0000	0.3000	-1.0000	0.9000	0.2000	-1.0000	-1.0000	-1.0000	-1.0000
-1.0000	-1.0000	-1.0000	0.0500	-1.0000	-1.0000	0.7000	-1.0000	0.5000	-1.0000	-1.0000	-1.0000	0.3000	-1.0000
-1.0000	-1.0000	-1.0000	0.9500	-1.0000	-1.0000	0.3000	-1.0000	0.5000	-1.0000	-1.0000	-1.0000	0.7000	-1.0000
-1.0000	-1.0000	-1.0000	-1.0000	-1.0000	-1.0000	-1.0000	-1.0000	-1.0000	-1.0000	-1.0000	-1.0000	-1.0000	-1.0000
-1.0000	1.0000	-1.0000	-1.0000	-1.0000	-1.0000	-1.0000	-1.0000	-1.0000	-1.0000	-1.0000	-1.0000	1.0000	-1.0000
-1.0000	-1.0000	-1.0000	-1.0000	-1.0000	-1.0000	-1.0000	-1.0000	-1.0000	-1.0000	-1.0000	-1.0000	-1.0000	-1.0000
-1.0000	-1.0000	-1.0000	-1.0000	-1.0000	-1.0000	-1.0000	-1.0000	1.0000	-1.0000	-1.0000	1.0000	-1.0000	-1.0000
-1.0000	-1.0000	-1.0000	-1.0000	-1.0000	-1.0000	-1.0000	-1.0000	-1.0000	-1.0000	-1.0000	-1.0000	-1.0000	-1.0000
1.0000	1.0000	1.0000	-1.0000	1.0000	1.0000	-1.0000	-1.0000	-1.0000	-1.0000	-1.0000	-1.0000	1.0000	1.0000
-1.0000	-1.0000	-1.0000	-1.0000	-1.0000	0.8000	-1.0000	-1.0000	-1.0000	-1.0000	-1.0000	1.0000	-1.0000	0.8000
1.0000	1.0000	1.0000	1.0000	1.0000	0.2000	1.0000	1.0000	1.0000	1.0000	1.0000	-1.0000	1.0000	0.2000
-1.0000	-1.0000	-1.0000	-1.0000	-1.0000	-1.0000	-1.0000	-1.0000	-1.0000	-1.0000	-1.0000	-1.0000	-1.0000	-1.0000
-1.0000	-1.0000	1.0000	-1.0000	-1.0000	-1.0000	1.0000	-1.0000	1.0000	-1.0000	1.0000	1.0000	-1.0000	-1.0000
1.0000	1.0000	1.0000	1.0000	1.0000	1.0000	-1.0000	1.0000	1.0000	1.0000	1.0000	1.0000	1.0000	1.0000
-1.0000	-1.0000	-1.0000	-1.0000	-1.0000	-1.0000	-1.0000	1.0000	-1.0000	-1.0000	-1.0000	-1.0000	-1.0000	-1.0000
-1.0000	-1.0000	-1.0000	-1.0000	-1.0000	-1.0000	-1.0000	-1.0000	-1.0000	-1.0000	-1.0000	-1.0000	-1.0000	-1.0000
-1.0000	-1.0000	-1.0000	-1.0000	-1.0000	-1.0000	-1.0000	-1.0000	-1.0000	-1.0000	-1.0000	-1.0000	-1.0000	-1.0000
-1.0000	-1.0000	-1.0000	-1.0000	-1.0000	-1.0000	-1.0000	-1.0000	-1.0000	-1.0000	-1.0000	-1.0000	-1.0000	-1.0000
-1.0000	-1.0000	0.3000	-1.0000	-1.0000	-1.0000	-1.0000	0.3000	-1.0000	0.7000	0.3000	-1.0000	-1.0000	-1.0000
-1.0000	-1.0000	0.7000	-1.0000	-1.0000	-1.0000	-1.0000	0.7000	-1.0000	0.3000	0.7000	-1.0000	-1.0000	-1.0000
1.0000	1.0000	1.0000	1.0000	1.0000	1.0000	1.0000	1.0000	1.0000	1.0000	1.0000	1.0000	1.0000	1.0000
-1.0000	-1.0000	1.0000	-1.0000	-1.0000	-1.0000	-1.0000	1.0000	-1.0000	-1.0000	-1.0000	-1.0000	-1.0000	-1.0000

NOTE: -1 is a Code for Likelihood Equal to 0.

```
================================================================
           << DIESEL AIR COMPRESSOR DIAGNOSTICS >>
                      BDS MAIN MENU
================================================================


                  ALT-M RETURNS TO MENU
                  ALT-C SELECTS SYMPTOMS


        1)   RESET TABLES, BEGIN SYMPTOM SELECTION
        2)   CONTINUE SYMPTOM SELECTION
        3)   PERFORM CALCULATIONS
        4)   EXIT THIS MACRO AND RETURN TO WORKSHEET

                      CHOOSE ONE
```

Fig. 2. DG-HELP3 Main Menu


```
B141: (F3) [W29] \-                                          READY


                    A                          B
141 ---------------------------------141 ------------------------------
142 DIESEL STARTING AIR (COMPRESSED AIR) 142
143                                    143
144               VARIABLES           144               VALUES
145 ==============================145 ==============================
146 VO   PRIOR                       146             PRIOR
147 ==============================147 ==============================
148                                    148
149 V1  AIR DISCHARGE PRESSURE       149 1  225 TO 250
150 V1  AIR DISCHARGE PRESSURE       150 2  200 TO 225
151 V1  AIR DISCHARGE PRESSURE       151 3  150 TO 200
152 V1  AIR DISCHARGE PRESSURE       152 4  LESS THAN 150
153                                    153
154 V2  LOW OIL PRESSURE (25 PSI NORMAL) 154 1  15 TO 20 PSI
155 V2  LOW OIL PRESSURE (25 PSI NORMAL) 155 2  LESS THAN 15 PSI
156                                    156
157 V3  COMPRESSOR NOISE             157 1.1  SEVERE KNOCKING (HEAVY)
158 V3  COMPRESSOR NOISE             158 1.2  LIGHT KNOCKING (TINNY)
159 V3  COMPRESSOR NOISE             159 1.3  ROLLING RUMBLE
160                                    160
16-Jun-87  09:26 AM                            CALC
```

Fig. 3. DG-HELP3 Symptom List

- Go back to the main menu and select "3"; i.e., perform calculations.
  You now reach a portion of the spread sheet showing results. The
  bottom line of this display ("FINAL PROBABILITY") provides the
  probability that each root cause (top line) is responsible for the
  problem, given the symptom selected. Here, we find the following
  possibilities:

Root Cause	Probability
Belts Too Loose	0.811
High Pressure Cylinder Discharge Valve Leaking	0.078
Pressure Switch - Bad Calibration	0.063
Blown Gaskets	0.041
Connection Air Leaks	0.004
Worn Rings and Cylinder Walls	0.003
High Pressure and Low Pressure Cylinder Inlet Valve or Unloader Mechanism Failed	0.001
Loose Valve Assemblies	0.001
Wrong Type of Oil	0.001

The most likely cause is loose belts. That would be easy to check, but
suppose we also noticed that the machine was making an unusual noise--a
light knocking noise. Let us refine the diagnosis.

- Go back to the main menu and select item "2" to continue symptom
  selection. Now, select V3 (compressor noise), value 1.2 (light
  knocking), by moving the cursor to line 158.

- Return to the main menu and recalculate ("3") to obtain a new results
  area. Look through the table (using right arrow "→") to find that
  all probabilities are zero except for the cause, "loose valve
  assemblies," which has a probability of 1.0. DG-HELP3 is telling us
  that no other single root cause can produce both symptoms V1.3
  (150 to 200-psig discharge pressure) and V3.1.2 (light knocking).

The above example demonstrates the BDM approach. We find much promise
in the methodology and plan to move ahead with development of the complete
DG-HELPER system. Two major tasks remain. The first is to develop Bayesian
tableaus for expert systems 1, 2, and 4 of Figure 1. Here, we will follow
the same approach to elicit information from the diesel generator expert as
we did in DG-HELP3 for the air compressor.

The second task will be to develop a suitable user interface by working
with the maintenance department people who will be the real users of the
system. The large tabular display of Figure 3 needs to be replaced by more
effective menus and graphical displays. We could add algorithms to help the
user decide what symptoms (variables) would provide the best additional
discrimination to a currently inconclusive diagnosis; i.e., what to look at
next. Algorithms to help the user better understand the diagnosis and
identify potential sources for error could also be helpful. Such
information is currently available within the spread sheet, but

understanding requires deep familiarity with the tableau and the Bayesian
methodology.

The goals of the pilot study have been met. The overall structure of
DG-HELPER is established, and system feasibility has been demonstrated. We
are ready to move into Phase 2.

ACKNOWLEDGMENT

The authors are indebted to Mr. Albert Littlefield, diesel generator
consultant to Boston Edison Company. It is his expertise that is central to
DG-HELPER. We are especially thankful for his patience with us during the
long hours talking through scenarios, and testing and revising the Bayesian
diagnostic module.

REFERENCES

1. E. H. Shortliffe and B. G. Buchanan, "A Model of Inexact Reasoning
    in Medicine," Mathematical Biosciences 23:351-379 (1975).

2. J. Pearl, "Fusion, Propagation, and Structuring in Bayesian
    Networks," presented at the Symposium on Complexity of
    Approximately Solved Problems, Columbia University, New York,
    New York (1985).

3. L. A. Zadeh, "The Role of Fuzzy Logic in the Management of
    Uncertainty in Expert Systems," Fuzzy Sets and Systems
    11:199-227 (1983).

4. G. Shafer, A Mathematical Theory of Evidence, Princeton Univeristy
    Press (1976).

5. S. Kaplan, M. V. Frank, and D. C. Bley, "COPILOT: An Expert System
    to Aid Reactor Operations - Conceptual Framework and Sample
    Applications of the Bayesian Diagnostic Module," PLG-0532,
    Pickard, Lowe and Garrick, Inc., prepared for U.S. Department of
    Energy, Oakland, California (1987).

EARLY DETECTION AND DIAGNOSIS OF DISTURBANCES IN

NUCLEAR POWER PLANTS

T.J. Bjørlo*, Ø. Berg*, R.E. Grini*, and
M. Yokobayashi**

(*) Institutt for energiteknikk   (**)Japan Atomic Energy
    OECD Halden Reactor Project   Research Institute

INTRODUCTION

The surveillance and control of nuclear power plants comprises a number
of tasks and functions which have to be shared between the operators and the
control and instrumentation systems. The trend in control room design
towards a higher degree of computerisation of the control and instrumentation systems and replacement of conventional instrument panels by VDU-based
man-machine communication systems opens possibilities for improving the
support given to the operators in their cognitive tasks. At the OECD Halden
Reactor Project these possibilities are explored through a research and
development programme centred around the NORS/HAMMLAB experimental control
room facility. The full-scale PWR simulator, NORS, coupled with the HAlden
Man-Machine LABoratory (HAMMLAB), which includes the experimental control
room as well as an established research methodology and staff, constitutes a
unique basis for the design, development and validation of operator support
systems, as well as for more basic operator performance experimentation. The
aim of the system development work at the Halden Project is to design, build
and validate computer-based systems which can assist and support the operators in their various tasks and through this improve the total performance
and safety of complex plant operation. One such system, the core surveillance system SCORPIO, has been implemented at the Ringhals NPP in Sweden[1].
Currently, the Halden Project is developing an integrated disturbance
handling system for use at nuclear power plants. This paper describes the
activities on fault detection and diagnosis within this development project.

DISTURBANCE HANDLING

A major task for the operators of nuclear power plants is surveillance
of plant behaviour such that disturbances and faults in the process are detected and corrected at an early stage. The handling of disturbances often
requires the operator to perform a series of complex tasks, as illustrated
in Fig. 1. The first step is to detect the existence of an abnormal process
state. Then the operator must perform a diagnosis to decide the probable
cause of the disturbance. When the cause is identified the operator must
plan the counteractions needed to restore the normal plant state and evaluate the possible consequences of these actions. Finally the correct counteractions must be implemented. The operator is offered little support by
surveillance and control systems currently in use in nuclear power plants in

carrying out these different tasks. It is mainly during the detection phase
that the operator is assisted by the plant alarm systems. The development of
computer-based surveillance and control systems makes it possible to assist
the operator in all the different phases of disturbance handling. The processing power of modern plant computers makes it feasible to include on-line
simulators (mathematical models) of the plant as part of the surveillance
system. This, in combination with symbolic reasoning based on expert knowledge, can provide improved operator assistance during plant disturbances.
As indicated in Fig. 1, expert systems may be useful tools for the operator
both in diagnosis and action planning. By integrating expert systems with
more "intelligent" fault detection systems than todays alarm systems and
with computerised procedure systems[2], the handling of plant disturbances may
become an easier task for the operators.

FAULT DETECTION

<u>General</u>

Early detection of faults and plant disturbances in nuclear power
plants reduces the risk of disturbances developing into severe plant conditions (shutdown or accidents) since the operators have more time for diagnosis and counteractions. Further, early detection of the disturbance usually means better localisation of the problem area in the plant, thereby
facilitating the diagnostic task. The traditional way of informing operators
about possible problems is through alarm systems based on limit checking of
process variables, which should stay within prescribed limits. In many cases
a disturbance in a plant subsystem may propagate into neighbouring subsystems
before the operator is alerted by the alarm systems. Therefore, the operator
is confronted with a large number of alarms within a short period of time
which makes the diagnostic task difficult. Alarm filtering techniques may
reduce this problem to some extent by focussing on essential alarms[3,4].

An alternative method for fault detection is illustrated in Fig. 2. The
method is based on mathematical reference models describing the dynamic behaviour of the process in normal operating conditions (no disturbances or
faults in the process). By comparing measured process variables with corresponding calculated variables from the reference models in real-time, the
time to detect disturbances can be reduced compared to traditional alarm
systems. By splitting the reference models into a number of submodels where
the input variables to each individual submodel are <u>measured</u> output variables from the preceding subprocess (Fig. 2), a good localisation of the
problem area in the plant is obtained. By this technique, propagation of
faults in the detection algorithms outside the particular subsystem containing the fault, is avoided thus reducing the diagnostic task. However, also
this method requires additional rules for detailed diagnosis in order to
discriminate among various possible failures within a subsystem which may
cause an observed deviation between reference models and measurements. For
instance, errors in the control system or instrumentation may turn out to be
the real problem, but this type of failure should also be detected as early
as possible.

<u>A Demonstration System for Early Fault Detection</u>

In order to demonstrate the potential of an early fault detection
system based on comparison of reference models and measurements, a prototype
system has been developed and implemented on the NORS PWR-simulator at the
Halden Project. The feedwater trains of NORS have been chosen as the process
section for this system, including condensate systems, low pressure preheaters, feedwater tank, feedwater pumps and high-pressure preheaters. Thus
the demonstration system comprises a fairly complex sequence of connected

subsystems which makes it suitable for demonstrating fault detection of various kinds. In the following the high-pressure preheaters (HPPs) will be used to illustrate the features of the demonstration system.

The HPP system for one of the feedwater trains is shown in NORS display format, RD10 (Fig. 3). It comprises a three stage heat exchanger system denoted RD11, RD12 and RD13. The warmest water from the superheater is collected in the tank RN13 and transported to RD11. The main heating, however, comes from saturated steam from the turbine transferring its condensation heat to the feedwater. The condensate on the hot side in the heat exchangers is fed in the opposite direction to feedwater thus heating the feedwater before the condensate enters the feedwater tank. This results in a high degree of coupling between the heat exchangers which renders fault detection and diagnosis difficult within this system. In the display format, RD10, the bright white, unbracketed numbers are measurement values, while the less intense numbers in brackets are calculated by reference models described below.

Each heat exchanger tank is equipped with a control valve which regulates the water level in the tank. A reference model describing the dynamic behaviour of the level controller (PI-type) with input from the water level measurement is used to check the measured valve position. In this way faults in any of the level control loops, e.g., stuck valve, erroneous measurements or controller malfunction, can be detected. This is a difficult task for an operator assisted only by conventional alarm systems, since no alarms would be triggered before the control valve is closed or fully open.

Leakage from the feedwater to the tanks is another possible fault in this system. The pressure in the feedwater at full power operation is about 72 bar, while in the tanks it ranges from 10 to 23 bar. If there were a tube rupture in the RD11 tank, feedwater would leak into this tank and cause a higher water level. This would be compensated for by the level controller which would open the outlet valve more, leading in turn to a higher level in tank RD12; consequently the outlet valve of RD12 would open more thus increasing the level of tank RD13. This would eventually lead to an increased flow out of the preheaters to the feedwater tank. The characteristics of this fault are that:

-   No alarms are given from the conventional alarm system as long as the leak is small.

-   The controllers compensate automatically for the increased water levels and a distinction between leakages and normal changes due to process instabilities or different operating conditions is very difficult to make without calculations.

In the modelling of this process part only measurement signals available in display format RD10 are used as input to the reference models. Only the flow out of the last tank is measured, and there are no measurements of flows from the turbine. The models calculating the flows are described in the following:

-   The flow through each of the valves is calculated by assuming that it is proportional to the valve position and the square root of the pressure drop over the valve. If there is a leak these flows will increase, therefore another method is needed to calculate flow when there is no leak. This is done as follows:

-   The reference flow out of a tank is calculated by summing all the flows into that tank, taking account of the level change and finally using the heat transfer balance between the feedwater and the warm turbine

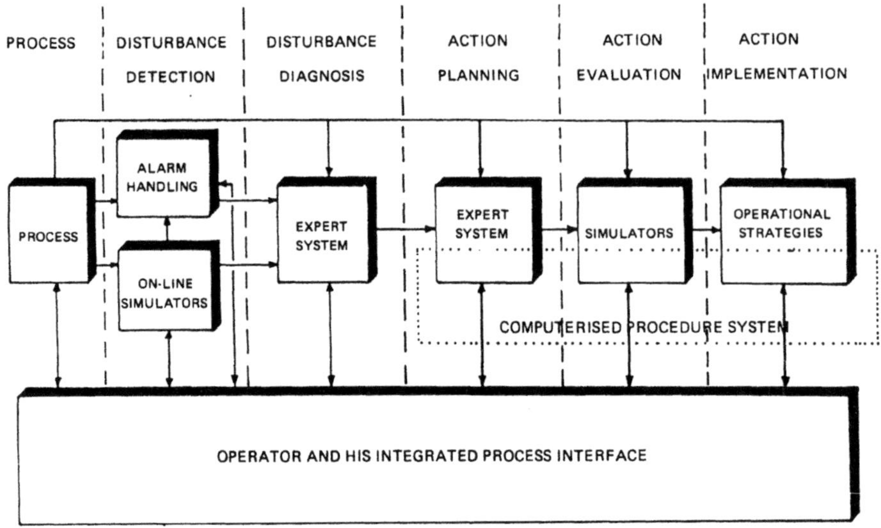

Fig. 1. Different phases in plant disturbance handling

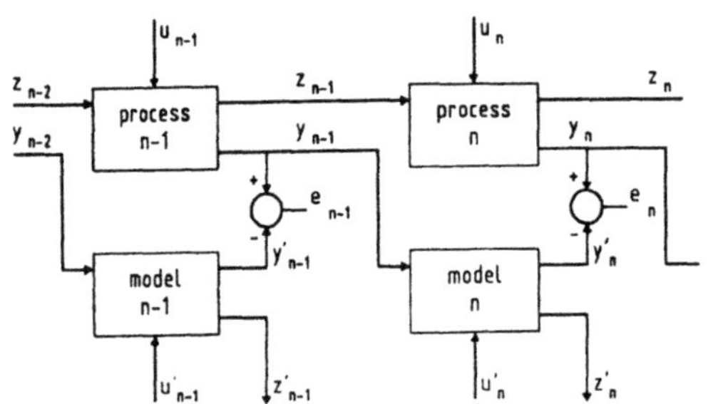

Fig. 2. Process divided into several sub-processes with correspondng sub-models. Faults are detected by error signals e=y-y'. Process measurements y are fed into the models to avoid sub-model errors to spread into other sections.

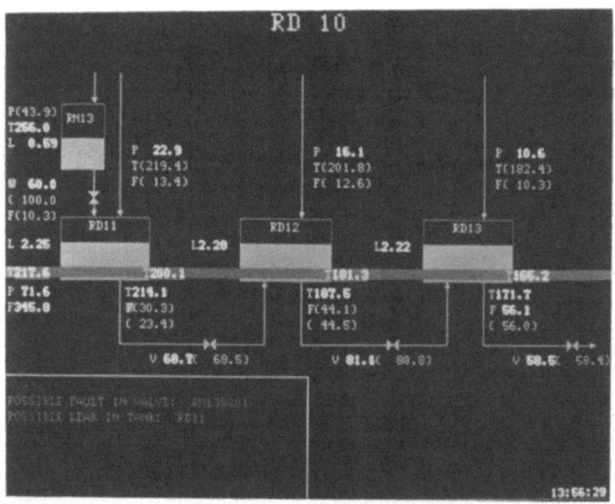

Fig. 3. Detail picture of the high pressure preheaters with measurements. Variables calculated by reference models are shown in parenthesis.

water. This results in an equation which gives us the reference flow
out of the tank without the leakage term and can thus be compared to
the first calculated flow for leakage detection.

<u>Experimental Results</u>

The demonstration system is connected on-line to the NORS simulator,
and the information from the system is presented to the operators on a
colour graphic screen. This man-machine interface has a two-level display
hierarchy with a global overview picture guiding the operator to the lower
level detailed displays containing much more information on the fault conditions discovered by the system. The demonstration system has been evaluated
in experiments where malfunctions have been introduced in the NORS simulator, and the operators' task has been to detect and diagnose the problem.

Fig. 3 is a black and white copy of the detailed colour display for the
HPP system in the case where two independent faults have been introduced in
this part of the process. As can be seen, the faults have been detected by
the early fault detection system. In the display, suggestions concerning
probable causes for the observed discrepancies between measurements and
reference models are presented in a message field (lower left corner). Two
messages are present: <u>Possible fault in valve RN13S001</u>, and <u>Possible leak
in tank RD11.</u>

Although the simultaneous occurrence of the 2 faults represented here
might be rare, this example illustrates how faults can be detected and
diagnosed. The size of the leak in this case, is so small, that no alarms
are received from the traditional alarm system. The two faults both influence other process components and measurements in this display format.
However, due to the high degree of model decomposition, each component
(subsystem) is checked with individual reference models which adapt to the
changed conditions through the measured input variables. In this way the
problem areas in the HPP system are localised. This is a good example of the
principles outlined in Fig. 2.

Experiments in which faults have been introduced in other parts of the
feedwater system have also demonstrated that the reference models follow
plant dynamics and adapt to new operating states. To determine the accuracy
of the models and the potential detection level for this system in real
operation in the presence of process and instrumentation noise, it would be
desireable to test the system against recorded data from a nuclear power
plant, and work to this end has been initiated at the Halden Project.

DIAGNOSIS OF DISTURBANCES

<u>General</u>

Diagnosis is one of the most popular areas for the application of
expert system techniques. However, most of the systems developed such as
MYCIN[5], EXPERT[6], etc. are basically designed to solve static problems. They
are consequently not well suited for diagnosis of dynamic phenomena in industrial process control. Therefore, Japan Atomic Energy Research Institute
(JAERI) has developed the expert system DISKET[7] for diagnosis of disturbances in a dynamic system such as a nuclear power plant. The Project has,
in cooperation with JAERI, implemented this diagnostic tool as part of the
surveillance system of the NORS simulator, and is testing out this expert
system in experiments. Particular emphasis is placed on developing an efficient man-machine interface based on a fully graphic operator station.

<u>Description of DISKET</u>

The diagnosis system DISKET consists of an inference engine and a
knowledge base (Fig. 4). The inference engine identifies the cause and type
of plant disturbance using the stored knowledge base and on-line process
data from the plant (in our case the NORS-simulator). DISKET was originally
written in UTILISP at JAERI. At the Halden Project the system has been
translated into FORTRAN and implemented on a supermini-computer which is
part of "the plant computer system" of the NORS-simulator. The knowledge
base contains the relational data (so-called rules) needed for diagnosis,
and is made up of relevant information about the plant such as actuation
conditions of components and signals etc., and characteristics of transient
responses obtained from transient analysis. The inference execution and the
display of diagnostic results can be controlled interactively by the operator through a CRT-terminal.

The knowledge base of DISKET consists of two parts, a data definition
part and a rule part. The former includes knowledge about hypotheses (accident causes) and consequences (process observations). The latter part includes knowledge about relations between causes and consequences formulated
as rules, and can be divided into several knowledge units for effective data
handling.

The rules in DISKET can be divided into three main categories on the
basis of the logical combinations of observations and hypotheses contained
in the <u>condition part</u> of the rules. Denoting hypotheses and observations by
A,B,C... we get the following categories of logical combinations:

(a)   All observations or hypotheses are satisfied, A & B & C &...
(b)   n out of N observations or hypotheses are satisfied
      n/N: A,B,C,...
(c)   Order of observations or hypotheses are satisfied
      A AFTER B AFTER C AFTER...

The logical operator AFTER is introduced in DISKET to handle sequences
of hypotheses and observations in the rule representation. This is one of
the features of DISKET for treating time-varying characteristics of transients. DISKET also uses an experience-based <u>Certainty Factor (CF)</u> to describe the degree of belief in a hypothesis contributed by given observations as illustrated below:

<condition part>   <action part>
(condition of process data) → (accident hypothesis, CF)

Certainty factors take values between -1.0 and 1.0. Positive values
define the degree of increased certainty of the hypothesis in the action
part being true when the condition part is satisfied, while negative values
of CF indicate increased probability of the hypothesis being false when the
condition part is satisfied. CF=-1.0 means a definitely false hypothesis,
while CF=1.0 means absolute confidence in the hypothesis. The introduction
of certainty factors in DISKET means that more flexible and robust diagnoses
are achieved. If DISKET receives wrong process data, or process data are
missing for some reason, diagnosis can still be performed. However, the
certainty factor of a given hypothesis will in such cases be lower than if
all observations were available.

Fig. 5 shows the structure of the accident hypotheses in DISKET. As can
be seen, the structure is hierarchical with three levels. The degree of
detailed diagnosis increases from top (first grade hypotheses) to bottom
(third grade hypotheses) of the hierarchy. The inference method is forward
chaining, that is, hypotheses are evaluated starting from the top level.

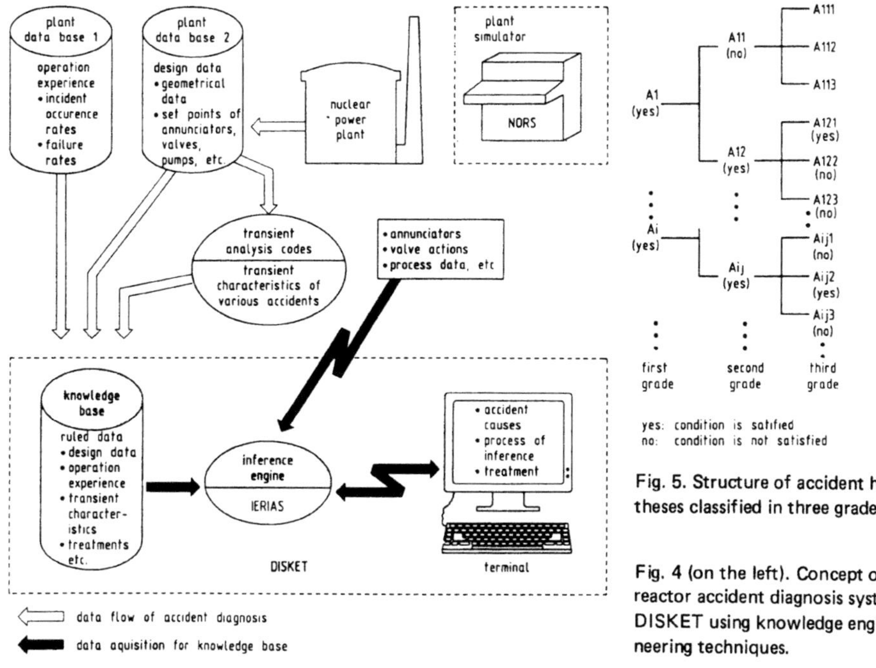

Fig. 5. Structure of accident hypotheses classified in three grades.

Fig. 4 (on the left). Concept of reactor accident diagnosis system DISKET using knowledge engineering techniques.

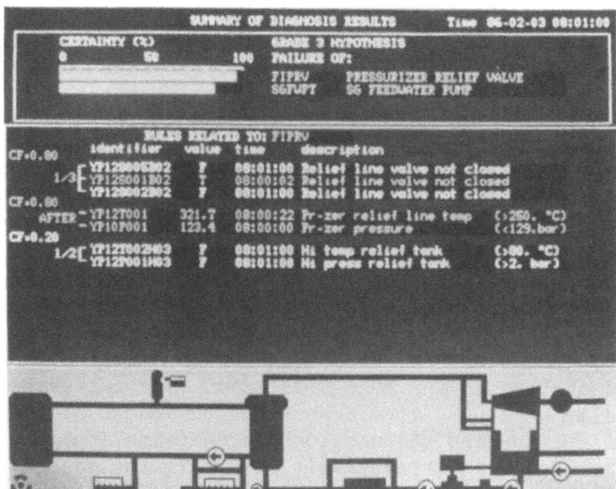

Fig. 6. Results of inference execution one minute into a transient in the case of two independent failures.

First grade hypotheses are always evaluated. However, if the condition of
the hypothesis is not satisfied with a high enough certainty factor at this
level, the associated hypotheses on the lower levels in the hierarchy are
not evaluated.

This inference method has two advantages, firstly, CPU-time is saved
compared to backward chaining methods, and secondly, parallel evaluation of
hypotheses makes it possible to investigate simultaneously occurring and
mutually independent faults. The method is therefore well suited for real-
time diagnosis of process disturbances.

<u>Results</u>

A knowledge base for fault diagnosis on the NORS-simulator has been
prepared. It comprises about 160 hypotheses and 300 rules obtained by analysis of NORS transients. DISKET is coupled on-line to NORS and performs
diagnosis when malfunctions are introduced in the simulator. The man-machine
interface comprises a full-graphic colour screen and an associated touch
panel keyboard. Inference execution is initiated by the operator through
pressing a button on the touch panel.

Fig. 6, which is a black and white copy of the colour display, shows
the results of inference execution one minute into a transient in the case
of two independent failures, namely,<u>failure in pressuriser relief valve</u>, and
<u>all feedwater pumps tripped</u>. At this time in the transient the process ob-
servations are sufficient for DISKET to proceed to third grade (lowest
level) hypotheses with sufficient certainty, as illustrated in Fig. 6. Two
hypotheses are identified with high certainty, namely, FIPRV, <u>failure in
pressuriser relief valve</u>, and SGFWPT, <u>steam generator feedwater pump trip</u>,
as shown in the top frame of the picture. In the bottom frame the failure
parts of the plant are shown by red (bright fields in black-white copy)
marks in the plant schematic. In the middle frame of the picture, the values
of process data and conditions for rules supporting the hypothesis are displayed on request. In Fig. 6, data supporting the hypothesis FIPRV are shown
in this frame.

CONCLUSIONS

A fault detection system based on the use of dynamic reference models
of the plant as part of the surveillance system has been built for the feedwater system of the NORS-simulator. Two main features characterising this
approach have been demonstrated. Firstly, the method makes it possible to
detect faults before a traditional alarm system is triggered even in dynamic
(transient) plant situations. Secondly, by a proper model decomposition
scheme and use of available measurements of process variables, the problem
area can be confined to the particular reference models covering the faulty
process parts thus resulting in a better geographical localisation of the
fault than usually obtained by traditional alarm system.

In addition, a fault diagnosis system, DISKET, based on expert system
techniques has been coupled on-line to the NORS-simulator. This rule-based
expert system has successfully been used to diagnose the causes of plant
disturbances resulting from introducing malfunctions in the simulator. By
structuring the accident hypotheses in three levels of detail the search
space is limited while still maintaining the feature of analysing parallel
problems (simultaneously occurring, independent faults). The process observations used as input to this expert system have mainly consisted of limit
violations provided by the traditional alarm system of the NORS-simulator.
The knowledge base is so far of fairly limited size containing about 300
rules. CPU times for inference execution are short enough for on-line,

real-time diagnosis of faults in this demonstration system.

The fault detection system based on reference models and the diagnosis
system complement each other in several ways. The reference models describing the process behaviour have proved useful for early detection of
process faults and limitation of the problem area for subsequent diagnosis.
Observed deviations between process measurements and reference models could
therefore be a valuable information source for the expert system performing
the diagnosis. By coupling the expert system to the fault detection system
the search space in the inference execution can be considerably reduced,
which is very important in the diagnosis of process disturbances due to the
need for real-time response.

Further plans at the Halden Project therefore aim for the development
of an integrated disturbance handling system which includes both model-based
and rule-based reasoning for efficient detection, diagnosis and prognosis of
disturbances. The detection module will be based on reference models as outlined in this paper. The diagnosis module will normally be triggered by the
detection module when a discrepancy between model calculations and measurements is observed in a plant subsection and diagnostic rules associated with
this subsection will be activated. Thus the number of fault hypotheses will
be limited compared with the plant-wide rules used in DISKET today, and will
be based partly on data (early warnings) other than the alarms and symptoms
in the currently used rules.

Another extension in the planned system will be the use of process
models as fast predictors in various plant situations to identify critical
time points in the development of the fault situation. This will provide
additional guidance for the operator in handling plant disturbances.

REFERENCES

1.   Ö. Berg, S. Hval, U.S. Jörgensen, "The Core Surveillance System SCORPIO
     and its Validation against Measured Pressurised Water Reactor Plant
     Data", Atomkernenergie-Kerntechnik 45 (1984) 271.
2.   F. Öwre, "Computer Aided Procedure Execution", ANS International
     Topical Meeting on AI and other Innovative Computer Applications in
     the Nuclear Industry, Snowbird, Utah, August 31 - September 2, 1987.
3.   P. Visuri, "Multivariable Alarm Handling and Display", A joint ANS, ENS
     and JAENS Meeting on Thermal Reactor Safety, Chicago, Illinois, 1982.
4.   E.C. Marshall, C.S. Reiersen, F. Öwre, "Operator Performance with the
     HALO II Advanced Alarm System for Nuclear Power Plants - A Comparative
     Study", ANS International Topical Meeting on AI and other Innovative
     Computer Applications in the Nuclear Industry, Snowbird, Utah, August
     31 - September 2, 1987.
5.   E.H. Shortliffe, "Computer-Based Medical Consultation, MYCIN", American
     Elsevier, New York 1976.
6.   S.M. Weiss, C.A. Kulikowski, "EXPERT, a System for Development
     Consultation Models", CBM-TR-97, 1979.
7.   K. Yoshida, M. Yokobayashi, T. Aoyagi, Y. Shinohara, A. Kohsaka,
     "Development and Verification of an Accident Diagnostic System for
     Nuclear Power Plant by Using a Simulator", ANS International Topical
     Meeting on Computer Applications for Nuclear Power Plant Operation
     and Control, Pasco, Washington, September 8 -12, 1985.

MACROSCOPIC MASS AND ENERGY BALANCE FOR

NUCLEAR PLANT DIAGNOSTICS USING FUZZY LOGIC

Jeré A. Hassberger and John C. Lee

Department of Nuclear Engineering
University of Michigan
Ann Arbor, Michigan 48109

INTRODUCTION

This paper presents a method for computer-assisted nuclear power plant failure diagnostics. The method is based on the analysis of macroscopic mass and energy balances in thermal hydraulic control volumes using fuzzy logic. The method is intended to provide assistance to operators during unanticipated transient conditions.

Expert Systems are computer programs which attempt to apply human-like reasoning to solve problems using rules. Expert systems were first applied to medical diagnostics[1] and since have been used to solve diagnostic problems in many areas, including mechanical pump failures[2] and chemical processing[3,4]. In nuclear plants, emergency procedures can be coded as "if ... then" rules and used to prompt operator responses during off-normal events. These rules represent the experience of a well trained operator. This approach is referred to by Kramer[3] as the "shallow" expert system approach. He views shallow knowledge as being "shortcuts through or the experiential compilations of the underlying principles of the problem domain." In the nuclear domain, this approach is workable for a majority of plant operating transients, particularly where the transient is relatively simple and the diagnosis straightforward. The weakness of rule-based systems is one of completeness; it is simply not possible to anticipate and formulate rules to cover every conceivable plant situation. Having no general principles, just thousands of heuristics, shallow knowledge expert systems are unable to analyze new situations.

One method of providing expert systems with crucial first principle knowledge is by integrating them with best-estimate simulation programs. The unique feature of simulation-based reasoning is that the associated computer algorithms contain a mathematical model of the process under consideration. Data from the plant can be input into a simulation program at the critical moment and the effect of various faults or actions can be tested. Theoretically, this allows us to take full advantage of the known functional relationships between variables when processing the values of their measurements. This is actually similar to the

procedure a skilled operator follows in comparing functionally
related measurements for consistency with his "mental model" of
how the plant should behave. In diagnosing process faults in
chemical plants, Kramer[3,4] makes use of "governing equations"
which carry with them similar information on the relationships
among process variables.

THE MACROSCOPIC BALANCE METHOD

In the following, we show how the simulation of nuclear
plant dynamics can be used in diagnostics, and then show how the
method can be translated into an expert system. The technique is
based on the simplest, first principle physical relationships one
might think of: the balance of macroscopic mass and energy
inventories in the nuclear plant. The mass and energy balance
equations for each discrete control volume can be solved
providing an analytical estimate of mass and energy inventories
throughout the plant. Experimental estimates of these
inventories can also be obtained directly from process
measurements. Large discrepencies between analysis and
measurement indicate the possibility of a fault. The macroscopic
balance method is used to analyze the effect of each possible
fault on plant-wide mass and energy inventories. This technique
differs from other expert system approaches to nuclear plant
diagnostics[5] in that mass and energy inventories are chosen as
symptoms, as opposed to directly measurable variables, e.g.,
pressure and temperature. Mass and energy inventories offer a
direct causal link with component failure modes as well as a
direct relationship to first principle balance relations.
Interpretation of a pattern of constraint violations is treated
by an expert system using a fuzzy logic inferencing procedure.

<u>Constraints Based on Macroscopic Balance Equations</u>

The balance equations for macroscopic mass and energy
inventories for a control volume are

$$\frac{dM}{dt} = W_{in} - W_{out}, \tag{1}$$

$$\frac{dU}{dt} = (Wh)_{in} - (Wh)_{out} + Q, \tag{2}$$

where M is the total mass inventory of the control volume, U is
the total energy inventory, W_{in} and W_{out} are the inlet and outlet
mass flow rates, h_{in} and h_{out} the inlet and outlet enthalpies and Q
the heat generation rate within the control volume. The total
mass and energy inventories of each control volume in a system
can be calculated by integrating Eqs. (1) and (2). In addition,
measurements of constituent variables such as pressure P,
temperature T and fluid level can be related to the total mass
and energy inventories of a given control volume V through the
equation of state

$$M_y = \rho(P, T) \cdot V, \tag{3}$$

$$U_y = M_y \cdot h(P, T) - PV. \tag{4}$$

Taken together, Eqs. (1) through (4) place a set of constraints

on the measured mass and energy inventories. These constraints
can be expressed by the relations

$$\Delta M = M_y - M = 0, \qquad (5)$$

$$\Delta U = U_y - U = 0. \qquad (6)$$

We define a failure as an unexpected phenomenon, not
accounted for in the macroscopic balance equations, which causes
a violation of these constraints. For example, a break in a
piping system will introduce loss terms in the mass and energy
equations for the control volume containing the pipe break,
causing the predicted mass and energy inventories to be larger
than measured. The basis of the macroscopic balance method of
diagnostics is to associate with each possible failure mode its
effect on the constraints on mass and energy inventory. Table 1
shows a matrix giving the effect of six possible failures in the
control mechanisms of a pressurized water reactor (PWR)
pressurizer on the mass and energy inventories of the pressurizer
control volume. This matrix can also be expressed as rules of
the form

"if f then C_1 and C_2 and C_3 ..."

where the antecedant is a failure mode and the consequences are
constraint violations. Diagnosis can proceed from these rules
using a backward chaining inference algorithm which starts with
the consequences to infer the antecedants. Formally, this
inferencing procedure will be treated using fuzzy logic.

Table 1. Logic Matrix for Control Mechanism Failures in a PWR Pressurizer.

		constraints			
	failures	$\Delta M{<}0$ b_1	$\Delta M{>}0$ b_2	$\Delta U{<}0$ b_3	$\Delta U{>}0$ b_4
a_1	*relief valve fails open*	X		X	
a_2	*relief valve fails closed*		X		X
a_3	*spray fails on*		X		X
a_4	*spray fails off*	X		X	
a_5	*heater fails on*				X
a_6	*heater fails off*			X	

Formulation of the Reasoning Algorithm Using Fuzzy Logic

Let $X = \{x\}$ be a universe of events. A fuzzy set B in X is
characterized by a mapping function $\mu_B(x): X \rightarrow [0,1]$ by which each
x is assigned a number in the closed interval $[0,1]$. These
numbers indicate the extent to which x has the properties or
characteristics that are attributed to B. In general, $b_i \equiv \mu_B(x_i)$
is considered to be the degree of membership of element x_i in set
B. Given a set $X = \{x\}$ and another set $Y = \{y\}$, a fuzzy relation
R is a mapping $\mu_R(x,y): (X,Y) \rightarrow [0,1]$, where $r_{ij} \equiv \mu_R(x_i,y_j)$ is the
degree to which the relation expressed by R is true for the
ordered pair (x_i,y_j).

In the macroscopic balance method, the set X is the set of constraints and the vector b is what we observe as symptoms, i.e., $b_i = 0, 0.5, 1$ imply, respectively, that the ith constraint is not violated, partially violated or fully violated. It is assumed for the moment that the violation of each constraint can be established based on fixed tolerance levels. Further, the vector a is taken to be the extent of the failures giving rise to the observed symptoms, i.e., $a_j \equiv \mu_A(y_j) = 0, 0.5, 1$ imply, respectively, that the jth failure did not occur, partially occurred or fully occurred. If we take R to be the causal relationship between the set of failures Y and the set of symptoms X, then the composition rules of inference state that

$$b_i = \max_j (r_{ij} \wedge a_j), \tag{7}$$

or simply

$$b = a \circ R. \tag{8}$$

Here r_{ij} is understood as the extent to which failure y_j is known to cause symptom x_i. Thus, the *causality matrix* R determines whether or not there is any relationship between the jth fault and the ith symptom.

The derivation of Eq. (8) assumes independence of symptoms for multiple failures. With this assumption, the diagnostics problem is to solve Eq. (8) for the vector a having observed the components of the vector b, given that the relationship matrix R is already known based on causal reasoning. This is known as the inverse problem of fuzzy mathematics. The following theorem, first proven by Sanchez[6], states the necessary and sufficient condition for the existence of a solution of the inverse problem.

Theorem 1:

Given $R=[r_{ij}]$ $(m \times n)$, $a=[a_j]$ $(n \times 1)$ and $b=[b_i]$ $(m \times 1)$,

$$a \circ R = b \quad \Leftrightarrow \quad (R^T \alpha b) \circ R = b \text{ and } a \leq R^T \alpha b. \tag{9}$$

Here, the α-composition is defined as

$$(R^T \alpha b)_j = \min_k (r_{kj} \alpha b_k), \tag{10}$$

where the α-product $a \alpha b$ is 1 if $a \leq b$, b if $a > b$. Applied to Eq. (8), Theorem 1, if satisfied, guarantees that not only is $R^T \alpha b$ a solution, it is also the least upper bound on the set of all solutions to the problem $b = a \circ R$. In general, a number of lower bound solutions may exist. These lower bounds can also be obtained analytically through a similar theorem proven by Pappis and Sugeno[7].

HYPOTHESIS TESTING

This section describes the algorithm for both choosing among several failure hypotheses and establishing the extent or severity of the failure. The first step is to estimate state variables and fault parameters for each of the failure modes hypothesized by the macroscopic balance relations. This is done using a Kalman filter[8].

<u>The Kalman Filter</u>

The Kalman filter is a mathematical technique for obtaining the optimal estimate of a system state given the information contained both in the equations of system dynamics and measurements made on the system. The balance equations for macroscopic mass and energy inventories for a control volume as given by Eqs. (1) and (2) can be written in matrix notation as

$$\frac{d\boldsymbol{x}(t)}{dt} = \boldsymbol{A}(t)\,\boldsymbol{x}(t) + \boldsymbol{w}(t), \qquad (11)$$

where $\boldsymbol{x}$ is the state vector $[M,\ U,\ W_{in},\ W_{out},\ Q]^{T}$ and $\boldsymbol{w}$ is a zero-mean white random process added to model errors or uncertainties in the state equations. The measurement vector $\boldsymbol{y}$ is related to the state vector according to the equation

$$\boldsymbol{y}(t) = \boldsymbol{H}(t)\,\boldsymbol{x}(t) + \boldsymbol{v}(t), \qquad (12)$$

where $\boldsymbol{H}$ is the state-space to measurement-space transformation matrix and $\boldsymbol{v}$ is a zero-mean white random process used to model noise in the measurements. In this case, the measurement vector is taken to be $\boldsymbol{y} = [M_y,\ U_y]^{T}$. The Kalman filter algorithm[9] computes the optimal estimate of the state vector at each time step t_k, based on both a prior analytic estimate $\boldsymbol{x}(t_k)$ obtained by integrating Eq. (11) and a measurement $\boldsymbol{y}(t_k)$ made on the system. The filter is optimal in the sense of minimizing the variance of the error in the final state estimate by properly accounting for both the errors in the measurements and errors in the state equations.

The accuracy of the filtered state estimate depends on the quality of the prior state estimate. In our algorithm, the prior estimate is calculated in simulation mode using the nuclear thermal-hydraulic code TRANMB[10]. This simulation code uses the Extended Implicit Continuous Eulerian (EICE) integration scheme[11] to solve for mass and energy distributions in a nuclear steam supply system based on a microscopic (distributed parameter) formulation of the mass, energy and momentum balance equations. This microscopic treatment used to compute the prior state estimate yields a more accurate solution than would be otherwise possible using the macroscopic equations. The macroscopic equations are better suited for diagnostics and filtering, however, due to their simplicity and approximate linear form.

<u>Decision Algorithm</u>

After state and fault parameters have been estimated for each of the possible failure modes, the problem of choosing the operational mode which best accounts for the actual behavior of the plant (i.e., the system identification problem) must be solved. This is done using pattern recognition techniques to match the plant data to the optimal state estimates obtained for each hypothesized mode of operation.

If G possible failure modes are hypothesized, then at each point in time, $G+1$ hypotheses (including the no failure hypothesis) must be examined to verify the existence of a failure. For each anticipated failure, a distance can be calculated that gives a measure of the probability of hypothesis g being true relative to each competing hypothesis. The normal

probability distribution function for the gth failure mode is
given by

$$F_g(\boldsymbol{y}) = (2\pi)^{-n/2}|\Sigma_g|^{-1/2} \exp[-1/2(\boldsymbol{y}-\boldsymbol{Hx}_g)^T\Sigma_g^{-1}(\boldsymbol{y}-\boldsymbol{Hx}_g)], \quad (13)$$

where Σ_g is the covariance matrix for category g in the
n-dimensional measurement space (obtainable from the Kalman
filter). The observation vector $\boldsymbol{y}$ is said to satisfy hypothesis
g if the "distance" between $\boldsymbol{y}$ and $\boldsymbol{Hx}_g$ is a minimum for all
hypotheses. For Gaussian data, the generalized distance[12],

$$d_g(t) = [\boldsymbol{y}(t)-\boldsymbol{Hx}_g(t)]^T \Sigma_g^{-1}(t)[\boldsymbol{y}(t)-\boldsymbol{Hx}_g(t)], \quad (14)$$

is used because it operates on the mean square error. To provide
a statistical interpretation to the generalized distance, note
that $d_g(t)$ is a chi-square distribution. Thus, the probability
of $\boldsymbol{y}$ belonging to a category with mean $\boldsymbol{x}_g$ and covariance Σ_g is
simply given by the cumulative function of the chi-square
distribution found in standard statistical tables.

MACROSCOPIC BALANCE ANALYSIS OF THE TMI-2 ACCIDENT

To demonstrate the macroscopic balance method, we consider
the pressurizer transient during the initial stages of the Three
Mile Island (TMI) accident of March 28, 1979. In this sequence,
a pressure transient results from the loss of feedwater (LOFW)
flow to the secondary side of the steam generator in a PWR. The
loss of heat sink causes an initial surge of flow into the
pressurizer due to the resulting primary pressure rise.
Pressures and temperatures decrease following reactor shut down
due to the reduction in energy input on the primary side, causing
flow to surge out of the pressurizer. The transient is further
complicated by the failure of a power operated relief valve
(PORV) to close after the pressure has dropped below the closure
set-point at 2220 psia. Failure on the part of the operators to
properly detect and recognize this PORV failure eventually led to
core uncovery and fuel damage at TMI.

For this test problem, the first 100 seconds of the TMI
transient were simulated using TRANMB to generate pressure,
temperature and level data. The measurement process was
simulated by adding random noise to these data. Expected mass
and energy inventories were generated using TRANMB assuming the
PORV does not fail. Figure 1 shows the differences ΔM and ΔU
between the expected inventories and inventories based on the
simulated measurements. At 5 s into the transient, the set point
for opening the PORV is reached due to the pressure increase
caused by the initiating LOFW event. This is not a fault, but
rather represents the expected system response to the LOFW
transient. At this point, the expected inventories track the
measured inventories reasonably well. Following the automatic
scram, pressure drops as a result of the drop in energy input and
the PORV closure set point is reached at 10 seconds, but the
valve remains open. As a result, the expected mass and energy
inventories begin to diverge from the measured inventories. The
constraints on ΔM and ΔU are slightly positive initially due to
an overprediction in outsurge flow during the rapid pressure
drop, but both begin to fail negative around 40 s as the total
amount of mass lost through the open relief valve becomes
significant. At this point, the fuzzy expert system hypothesizes
a failure of the relief valve to close.

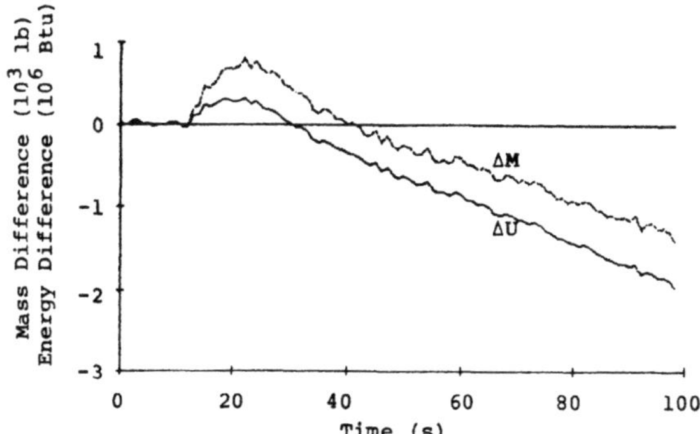

Figure 1.   Mass and Energy Differences vs. Time

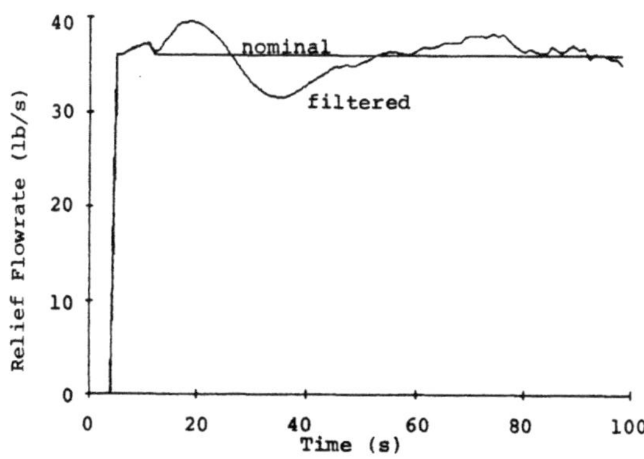

Figure 2.   Relief Flow vs. Time

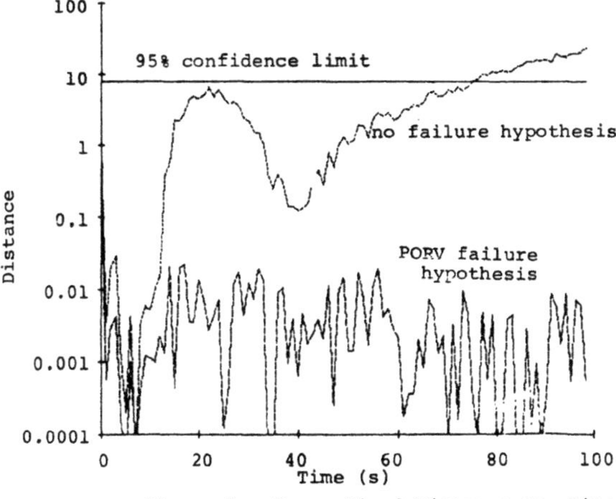

Figure 3.   Generalized Distance vs. Time

The PORV hypothesis is tested through simulation. In this simulation, relief flow rate is assumed to be unknown and is filtered to match the measured data. Figure 2 shows the optimal estimate of relief flow vs. the nominal relief flow used to generate the simulated measurements. The generalized distance for both the PORV failure and no failure hypotheses are shown in Figure 3 along with the 95% confidence threshold for the chi-square distribution. Based on these results, the PORV failure in this sequence can be detected, identified and confirmed within the first 100 s of this transient. Furthermore, the Kalman filter yields a reasonable estimate of the magnitude of this failure.

CONCLUDING REMARKS

Macroscopic mass and energy balance equations appear to provide simple diagnostic rules in a manner suitable for failure hypothesis testing through Kalman filtering. Although these rules are expected to be more general and compact than those based on directly measurable variables, e.g., pressure and temperature, the present method still requires accurate simulation models for power plant components.

ACKNOWLEDGEMENT

This research was supported by the U.S. Department of Energy under contract DE-AC02-85NE37945.

REFERENCES

1. E. H. Shortliffe, *Computer-Based Medical Consultations: MYCIN*, American Elsevier/North Holland, New York (1981).
2. K. Reinschmidt, "Expert Pump Diagnostic System," presented at the EPRI Workshop on AI Applications to Nuclear Power, Palo Alto, CA (1985).
3. M. A. Kramer, "Integration of Heuristic and Model-Based Inference in Chemical Process Fault Diagnostics," *IFAC Kyoto Workshop on Fault Detection and Safety in Chemical Plants*, Kyoto, Japan, September, 1986.
4. M. A. Kramer, *AIChE Journal*, **33**, 130 (1987).
5. J. A. Hassberger and J. C. Lee, "Intelligent Simulations for On-line Transient Analysis," *Proceedings of the International Topical Meeting on Advances in Reactor Physics, Mathematics and Computation*, Paris, April, 1987.
6. E. Sanchez, *Inform. Control*, **30**, 38 (1976).
7. C. P. Pappis and M. Sugeno, *Fuzzy Sets and Systems*, **15**, 79 (1985).
8. R. E. Kalman and K. S. Bucey, *J. Basic Eng.*, **22**, 51 (1965).
9. A. H. Jazwinski, *Stochastic Processes and Filtering Theory*, Academic Press, New York (1970).
10. M. M. Methnani, *Nuclear Power Plant Simulation with a Moveable Boundary Steam Generator Model*, Ph.D. Dissertation, University of Michigan (1983).
11. M. W. Crump and J. C. Lee, *Nucl. Sci. Eng.*, **77**, 192 (1981).
12. P. C. Mahalanobis, *Proc. Natl. Inst. Sci., India*, **12**, 49 (1936).

POWER PLANT EXPERIENCE WITH ARTIFICIAL INTELLIGENCE BASED,

ON-LINE DIAGNOSTIC SYSTEMS

R. L. Osborne and M. Coffman

Westinghouse Electric Corporation
Diagnostics and Monitoring Systems Department
Service Technology Division
Orlando, Florida

INTRODUCTION

The utility industry is entering a period when generation equipment availability will become increasingly critical due to the lack of new power plants being planned and built. The increasing percentage of all electric homes adding to peak demands will require more plant equipment to be used in a cyclic duty mode. Availability is on the increase with forced and planned maintenance hours decreasing. Factors that are contributing of this improvement are new units coming on-line with the latest in technology coupled with the installation of retrofit components containing that same technology such as the Rigi-Flex generators and ruggedized turbine rotors. In conjunction with hardware advances, technology advancements in monitoring and diagnostics are permitting the identification of potential malfunctions so that corrective actions can be taken, thus preventing lengthy outages. It is this last area that this paper will address.

SYSTEM OPERATION

In order to maximize understanding of the remainder of the paper, the following definitions of monitoring and diagnostics will be used. Monitoring is defined as the gathering, displaying, alarming and storing of data (temperatures, flows, load etc.). Diagnostics, on the other hand, is the identification of the condition of instrumentation and equipment (a bearing is failing, an electrical conductor is broken, a thermocouple has failed, etc.). Note that no variables were used in the definition of diagnostics.

Technology advances have led the industry to smart monitors and computerized monitoring systems which trend, display, and store data but lack the ability to diagnose conditions. As more of these systems have been installed, utilities have requested help for operators now awash in a sea of data. Specifically, assistance in reducing this data to actionable information is needed. Work has been concentrated on improving the actionable information available to operators from monitors. Existing monitors at site have been integrated with an Artificial Intelligence (AI) system that contains the knowledge of experts in the areas of equip-

ment diagnosis, performance and service. This capability has been implemented by utilizing expert system technology developed in conjunction
with Carnegie-Mellon University and Westinghouse Research and Development
personnel [1].

To illustrate what an expert system is and how it works, consider the
approach used by a medical doctor who relies on his training and experience to turn the initial data available into a set of diagnoses that vary
in confidence levels. From this set of preliminary diagnoses, the doctor
proceeds thru additional testing techniques and data acquisition to develop a higher confidence diagnosis resulting in a recommendation directed at preventing further complications. This doctor is a medical expert
system but in many ways he is limited:

1)   He can only diagnose one patient at a time.

2)   He is not on duty 7-days-a-week, 24-hours-a-day.

3)   He is restricted for immediate diagnoses to his
     knowledge & skill alone.

Keep that medical expert in mind and let's look at how power plant
equipment information might be diagnosed. Figure 1 shows data coming
from a generator through an information network into a local plant data
center in the control room, and follows that data as it is diagnosed by
an AI expert system. The plant data center contains a color CRT that has
touch screen capability. Figure 2 shows how a CRT display would appear
if the operator touched the screen indicating he wanted to look at the
generator capability curve. This shows the current operating point relative to unit capability and is typical of the data screens available to
the operator.

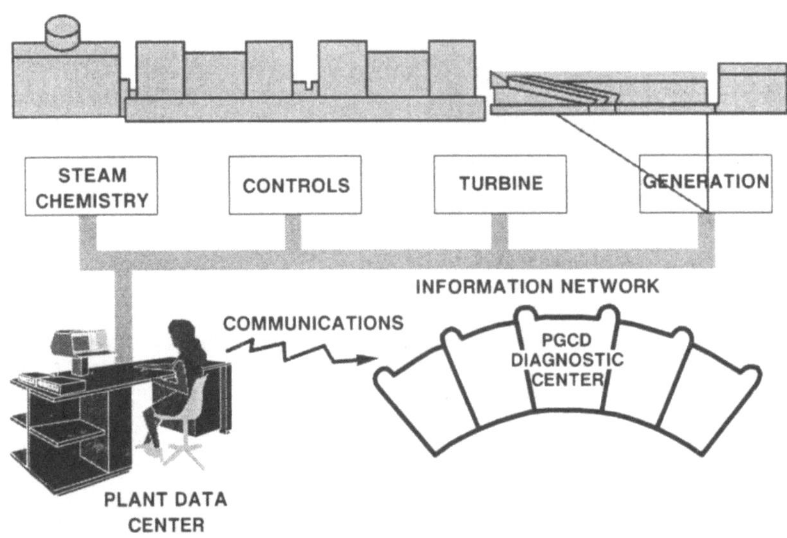

Figure 1. On-Line Diagnostic System

Just like our medical expert used patient's symptoms, the GenAID system uses on-line data such as this from the generator whose operation at this time appears normal.

In this case however the end-turn vibrations - the vertical motion of the coils at the end of the winding basket - from some of the 12 fibre optic sensors were running slightly elevated from their normal levels based on previous history and expected values. In addition, the GenAID system analyzes data from coil temperatures and finds that two of the temperatures have increased above normal value and are continuing to increase although no alarm points have been triggered. The diagnostic rule base also evaluates the effects of cooler balance and past temperature calibration corrections. At this point the diagnostic touch button on the lower left of Figure 2 would be blinking red indicating a new diagnosis has been made by the expert system. This generator expert system has the combined experience and knowledge of many design engineers, service engineers and field engineering personnel loaded into its rules. Like the doctor using a patient's data, the generator's data has been diagnosed with the conclusion that there is high confidence that cracked strands exist in phase "C", and recommends to the utility that the unit be removed from service for planned surgery within the next two weeks.

The same concepts and technology have been applied to another area critical to power plant reliability - the steam chemistry area. A chemistry module [2] can be added to the GenAID system just described or installed alone. Control of secondary side chemistry is a major factor in controlling corrosion in steam generators, turbine-generators, as well as other plant components. The location of monitored points are main steam, condensate pump discharge, feedwater, blowdown, heater drains, and makeup. The system contains software that monitors the chemical parameters on a continuous basis and alerts the operator whenever alarm thresholds have been exceeded or a new diagnosis is present. A major reason why this offers a higher degree of protection is that studies have

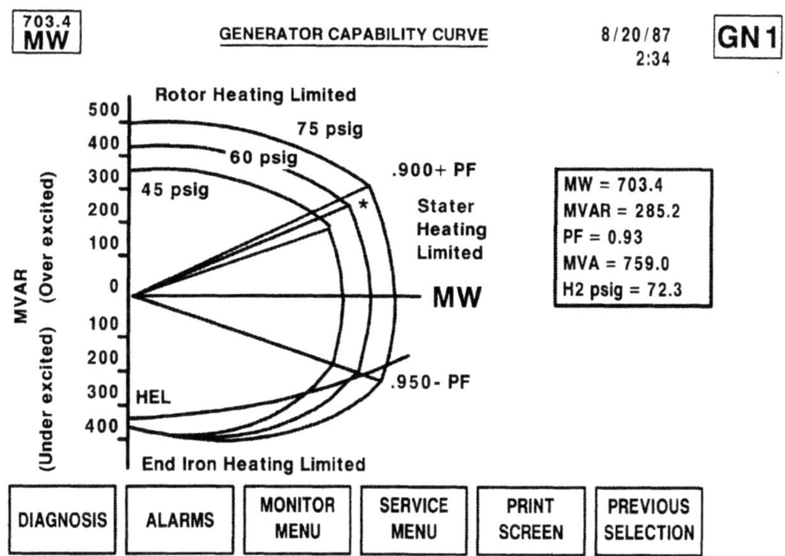

Figure 2. Typical Data Display

shown errors in data interpretation can occur from conventional grab
sampling approaches. Significant signal deviation can be masked, depending on the frequency of grab sample collection. By having continuous
data, upsets of minimal duration can be identified and corrected before
significant damage results.

Leaving chemistry, let's look at a fossil steam turbine cycle screen
shown in Figure 3. This screen displays items controlable by the operator. The operator can establish a target value when he recognizes a
piece of equipment may be limited in performance. Otherwise, the program
sets the target values. The display shows current operating conditions
and the resulting heat rate loss associated with the present operation.
The operator can now identify that the major elements contributing to the
heat rate loss are the high back pressure and the high attemperation
spray flow.

Besides operator-influenced performance information, diagrams such
as shown in Figure 4 are provided on an element by element basis to identify component specific performance. The unit's data is also sent to the
Diagnostic Center where it is diagnosed as shown in Figure 5. In this
case blade path seal damage is indicated with inspection and possible
replacement at the next scheduled outage recommended. The utility can
perform an economic analysis to determine the most beneficial time for
corrective maintenance to be performed. With advanced ordering of specific repair materials, and pre-outage planning, down time can be minimized.

FIELD EXPERIENCE

The components of the monitoring and AI diagnostic systems and the
technology behind them have been covered. Let's now look at some real
life conditions in generators that have been diagnosed using the GenAID
system connected on a 24 hour a day, 7 day a week basis to the diagnostic
center rule base in Orlando, Florida. This AI program has been making
diagnosis in minutes that in the past required days or weeks, thus making
available to all connected units the benefit of any single units experience through an ever increasing rule base, and continuing to move from a
reactive into predictive mode.

The first field example involved the fiber optics end-turn vibration
monitor which takes signals from sensors located at the tip of the winding basket. By utilizing these signals, the GenAID system was able to
diagnose excessive basket vibration that if uncorrected would have resulted in rapid wear of the winding insulation in the area where the coil
is in contact with its support member. Recommendations for operation
parameter changes were implemented, and the vibration reduced to safe
levels.

In the next case, the diagnostics system recognized increasing hy-
drogen gas dew point and diagnosed the gas dryer was malfunctioning. The
utility was contacted and given recommendations to isolate the dryer and
take immediate corrective action to decrease the dew point inside the
generator to avoid the consequence of a retaining ring failure that could
initiate from improper environmental conditions and subsequent crack
formation.

Another example involved a problem with cracked strands in the generator coils at the end of the winding. In this situation, recommendations were given to continue running until the next scheduled outage
unless additional cracking was diagnosed, indicating a deteriorating

situation. This unit was followed very closely during this operating period until it safely reached its scheduled outage window.

In this last example, sporadic, high frequency vibration was detected on the exciter end of a winding which appeared as fiber optics signals of ever increasing amplitude and decreasing period, all of which followed an outage caused by a severe system fault exterior to the generator. The expert system rule base did not contain rules at that time that dealt with the unusual frequency pattern observed, but the diagnosis of abnormal vibration was made and both service and development engineering personnel participated in an active session analysing the units online data. From this, it was concluded that component impacting was occurring, and contingency plans were developed. During a scheduled 43

		CURRENT	TARGET	HEAT RATE DEVIATION (BTU/KWH)			
PARAMETER	UNITS	VALUE	VALUE	-100	0	100	
THROTTLE PRESSURE	PSIG	2410	2400				-4.
THROTTLE TEMPERATURE	DEG F	970	985				+29.
REHEAT TEMPERATURE	DEG F	1012	1000				-10.
BACK PRESSURE	INHGA	3.63	2.87				+71.
SUPERHEAT SPRAY	KLB/H	0	0				0.
REHEAT SPRAY	KLB/H	67	0				+52.
					TOTAL		+138.

Figure 3. Fossil Steam Turbine Cycle Screen

hour outage over the following weekend, looseness in the clamping bolts holding the parallel ring assembly was located and corrected, verifying the manual diagnosis and resulting in new rules for the expert system. The unit returned to service and all diagnostic messages regarding this event cleared. With the contingency planning and subsequent repairs that occurred as a result of the rapid diagnosis of this problem, this unit was able to be at full load during the highest peak load ever experienced on the utility system the following week.

In each of the examples just reviewed, a potentially serious equipment failure did not occur. The AI based systems provided rapid, around the clock expert assistance to aid the utility in increasing equipment availability.

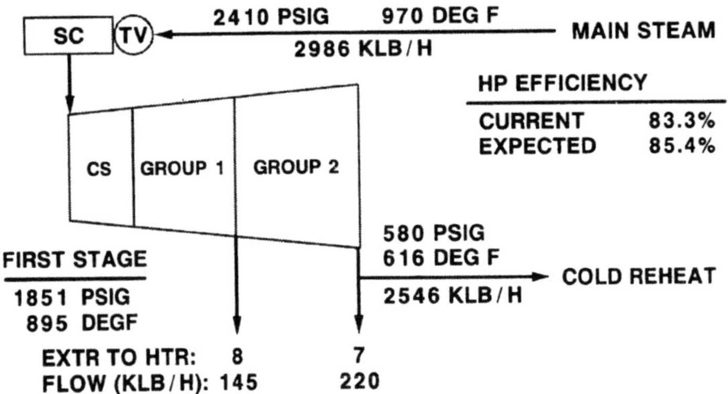

Figure 4. HP Turbine Diagram Screen

FINANCIAL JUSTIFICATION

Presently there are a total of 17 individual on-line, AI based diagnostic systems in commercial operation or on order applied to old and new units, both fossil and nuclear. In order to quantify the benefits which justify such purchases, the availability and forced outage rate before installation and after two years of operation of the GenAID systems at T. U. Electric will be reviewed. In actual practice many things can influence availability and forced outage rates such as the purchase of retrofit hardware, outage planning and diagnostic capabilities. The on-line diagnostic system aided T. U. Electric in all of these areas. It helped identify which units retrofit hardware should go on first. While waiting for hardware installation, on-line diagnostics assisted the operator in protecting the units against forced outages and high unavailability. When outages did occur they were held to a minimum by knowledge of where the problem was, thus minimizing long costly outages.

- **CONDITION**
  BROKEN BLADEPATH SEALS
- **RECOMMENDATION**
  SCHEDULED OUTAGE
- **BENEFIT**
  ADVANCE ORDER MATERIALS
  REGAIN 3.5% HP EFFICIENCY

Figure 5. Diagnosis and Recommendation

The bottom line: prior to 1984 when on-line diagnostics were being installed, the average availability on the units was 95.2% and the forced outage rate was 1.4%. In the period 1985 to 1986 the average availability had increased to 96.1% and the forced outage rate fallen to 0.2%. Thus an increase of 0.9% in availability and a decrease of 1.2% in the forced outage rate occurred. These units represent 4450 MW of generation. This means an average of 14,600 MW-Days of additional generation was available to T. U. Electric each year. If the average cost of replacement power was $1000 per MW-Days this translates to a saving of over 14.6 million dollars per year. By the continued use of on-line diagnostics, similar savings should be realized year after year.

REFERENCES

1.   R. L. Osborne. "Online, Artificial Intelligence - Based Turbine Generator Diagnostics", *AI Magazine,* Vol. 7, No. 4, Fall 1986 pp. 97-103.

2.   J. C. Bellows, "An Artificial Intelligence Chemistry Diagnostic System", Proceedings of 45th International Water Conference, Pittsburgh, PA., October 22-24, 1984.

AN ON-LINE DIAGNOSTIC EXPERT SYSTEM

L. Felkel

Gesellschaft für Reaktorsicherheit (GRS) mbH
Forschungsgelände
8046 Garching, F.R.G.

INTRODUCTION

Expert System technology has become the dominant force behind Artificial Intelligence. Expert Systems have mushroomed in many application areas in an attempt to solve information problems that have thus far resisted computer-based solutions for various reasons. Almost every Expert System article begins by stating the differences between "modern" Expert Systems and conventional programs: The major difference being that Expert Systems provide for the separation of data (Knowledge) from procedure (Inference) [1].

An ever increasing number of Expert Systems are being based upon state-of-the-Art AI technology. They exhibit an extended utility and improved processing power when compared with traditionally organized systems because Expert Systems employ "Symbolic Reasoning" methods rather than numerical or algorithmic methods. However, in [2] it is shown that the true value of AI technology lies in an enormous increase in software development productivity for systems relying upon domain knowledge.

The (semantic) content of an Expert System Knowledge Base is formalized in terms of facts and "condition/action" rules. First, this assumes that most "experiential" knowledge is capable of such a formal expression. A small Knowledge Base, consisting of about 10 rules, can probably demonstrate the feasibility of the Expert System approach, but such a demonstration prototype can often not be extended to meet the practical requirements of a real-world problem situation.

Most of the early impressive Expert System applications were developed for the medical industry where experiential knowledge had been accurred over a very long period of time. The application of Expert System experience to other problem domains can be difficult, especially when the Expert System must handle continually changing data coming from a technical control or instrumentation process.

A typical nuclear power plant process control application must deal with a signal processing capability ranging between 2,000 and 5,000 analog (physical) signals and between 10,000 and 15,000 digital signals. During a power plant emergency the signal dynamics. i.e. the rate of data flow change, can reach 500 signals during the first second and the digital signal rate over the next 10 seconds can be 2,000 signals. Under normal plant operating conditions the signal rate averages approximately 500 digital signals per hour. Analog signal sources are periodically scanned between 1 and 60 seconds where a signal rate change is expected (on the average) once per scan.

ON-LINE EXPERT SYSTEMS

On-line Expert Systems, as opposed to Off-line Expert Systems, maintain a continuous connection to some dynamic process. Such Expert Systems require

-    a complete analysis of the diagnostic procedures involved in fault detection and

-    the diagnostic and predictive process must be synchronized with the physical process.

This is particularly important when considering a continuously changing process (time dependency problem).

We must assume that a dynamic process should never be disturbed so that causes and consequences need not be searched for. Emergency situations are (hopefully) rare. The problem is what should be done with the Expert System in the idle time. People need to be well trained when the emergency occurs in order to use the Expert System effectively. This is not the case with Medical Expert Systems where this experience is acquired by its continual use on different subjects i.e. patients.

A Nuclear Power Plant is an intricate system of physical processes. Not only the qualitative nature of their behaviour must be considered, but in many situations, their quantitative nature must also be assessed (Naive physics provides only partial truths in practical situations).

The large amount of dynamic physical and status data, as well as the enormous number of interrelationships, poses a serious complexity problem.

Both the dynamic process and the Expert System consultation continuously progress with time. The dynamic data flow to the Expert System and the resulting diagnosis and prediction tasks of the Expert System must be kept consistent with one another with respect to logical, quantitative and temporal reasoning. This must never impair the required real-time notification and assessment functions of the Expert System (real-time problem).

As opposed to Off-line Expert Systems, On-line Expert Systems must be viewed as a component of a control loop within the overall operation of a dynamic process. In addition, application of an On-line Expert System imposes more rigorous requirements on the overall operation, such as, Completeness, Reliability, Operability, Robustness and User friendliness. Some of these requirements may contradict the general specifications of Off-line Expert Systems where, for example, processing with incomplete data (incompleteness) or the absence of algorithms (unreliability) may still lead to acceptable reliability.

ON-LINE REAL-TIME SYSTEM HARDWARE

By considering the past activity of the computer market, it is obvious that there are no severe limitations for the application of AI technology to industrial settings. A careful requirement assessment, however, is necessary to insure that the system attain the expected computational power for worst-case situations. Several hardware solutions currently exist on the market (see [3]).

556

The work presented here attempts to combine research results on GENERIS, an on-line disturbance analysis systems [4] and, for example, Expert System explanation facilities. This facility could not be satisfactorily realized in the system due mainly to lack of manpower. A project to incorporate this facility into the GENERIS system is being sponsored by the Federal Ministry for Research and Technology (BMFT) under contract number RS 714.

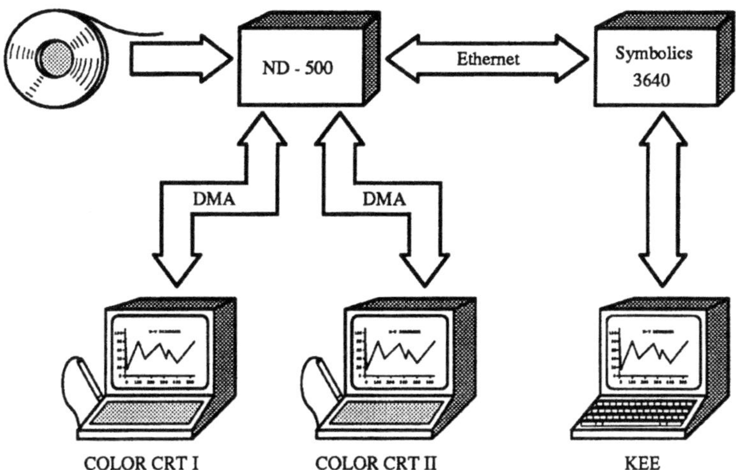

Fig. 1:   Hardware configuration

The hardware involved in the project is outlined in fig. 1. The configuration consists of 2 machines: a 7-mips NORSK DATA ND-500 and a Symbolics 3640. Both machines communicate over an Ethernet-link using the TCP/IP protocol. The ND-500 runs the well known GENERIS system and the Symbolics runs the Knowledge Engineering Environemnt (KEE) which will support the GENERIS explanation facility.

To test the system original signal data streams recorded during operation of the Biblis Unit B 1300 MWe PWR are delivered from a tape station.

Although integrated hardware solutions such as interfacing of a LISP-processor and a general-purpose processor sharing a single bus, are generally preferable, it does not seem to be a necessary requirement for the project because the data exchange between the computers is not extraordinairly large.

THE TIME DOMAIN

There is no doubt that time plays an important role in the dynamics of physical processes. When considering temporal reasoning processes, however, there are certain degrees of complexity that one might want to apply to a specific situation. Several concepts exist such that some are more practical [5,6,7], and others [8] offer a fairly exhaustive treatment of temporal logic. The type of temporal reasoning varies with respect to the type of the physical process under consideration. For example, absolute time must be taken into account when considering manufacturing systems because a manufactured product is usually constrained by its date of delivery.

The operation of a nuclear power plant is essentially cyclic in the sense that

-       some states will recur and
-       other states will not recur identically,
        but in an equivalent sense.

This is essentially due to the nature of control loops.

Each control loop has a cycle time that determines the possible reasoning contraints including some maximum time interval.

The maximum time interval i.e. the set of all time points or Time Domain must be of a specific granularity. In [8] the real numbers are used to provide a continuous open-ended time axis for temporal reasoning. In practical applications, however, this ideal time axis becomes infeasible.

The granularity of the Time Domain then is influenced by two factors:

-       Resolution, or the minimal relative distance between the occurrence of two events and

-       Length, or the duration reflected by the distance between the smallest and the largest time points of two events.

Under these assumptions the Time Domain must be an ordered set of time points. For practical purposes this set should be a finite collection of discrete time points. This set can be translated into absolute time points by anchoring it to a Base-Time. Since it is possible to redefine the Base-Time, the complete set can be gradually shifted along the time axis in appropriate steps (depending on the application requirements).

The application described in this report uses a Time Domain resolution of two milliseconds (ms). Considering a nuclear power plant operation over a 48 day period is more than sufficient for disturbance analysis and prediction. This requirement can be realized, indexing a relative time point in the range of a 32-bit computer word:

-       using 2 ms real-time units one can cover 48 operational days.

By recalibrating the resolution down to 2 microseconds each, the Time Domain would cover an operational interval between one and two hours. This enables us to handle smaller time differences at the expense of a smaller operational interval.

If the occurrence of an event falls outside of the current Time Domain, the time of the event's occurrence can no longer be used in a consultation involving temporal reasoning. However, the value of the event is retained thus still allowing logical reasoning.

REAL-TIME CONSULTATION

When we consider classical Off-line Expert Systems and look at their computational behaviour, then we discover that they stand idle (no CPU processing) most of the time waiting for work.

Upon activating a consultation, input data are acquired by a read-cycle. After sufficient data has been stored, a reasoning process is invoked (either interleaved with input gathering or not) to draw conclusions from the data.

Upon completion of the reasoning process, the output data are generated by a write-cycle. The Expert System again enters the idle state until another consultation is activated. The end user usually controls the activation of such consultations.

In real-time applications there is a continuous data stream transfer at speeds far beyond manual control or manual intervention. The environment is continuously changing at a rate that would prohibit accurate consultations because the current consultation data may become obsolete (replaced) before they can be processed. Therefore, real-time applications must abandon the concept of a single consultation cycle (or integrate them into a perpetually active "super consultation").

The time required to generate the results of a consultation must be much smaller than the time over which the results of the consultation are available. This means that there are fast physical processes that allow only simple reasoning in order to meet real-time constraints. One approach to optimizing such contraints is to use idle CPU time for performing computations which may be required in the next consultation. This strategy is pursued in the system described below. Consultation processing takes place in advance instead of the CPU being idle. This relieves the processing constraints when a consultant is requested.

ON-LINE CONTINUOUS FORWARD-CHAINING

Forward-Chaining is defined as an event-driven procedure. In on-line applications Forward-Chaining becomes especially significant, since the events are generated by a time-dependent source. As mentioned above these events are mapped onto a discrete, finite Time Domain. The Forward-Chaining rule processing is actually an event propagation model (PM) which remains active at one designated instance of time until the event ceases to cause further changes in the model.

For example, if event A occurs causing a value change from "0" to "1", but event B contains a value of "1" and an event C has been defined as the disjunction of event A and B (i.e. C ← OR(A,B)), then propagation terminates, since no change in event C can be effected.

Fig. 2 shows an example of an event propagation model for a buffer tank (feedwater tank of a 1300 MWe PWR). It consists of a collection of input signals (either original process events or events synthesized in other models) combined with logical or arithmetical operations (AND-gate: & , OR-gate: ≥1 , limit check: ∫ , sum: Σ , absolute difference: absdiff, on/off switch: ⅄) and operations on the Time Domain (time delay: ⊢—⊣).

This example computes the absolute difference between all incoming water quantities and all outgoing water quantities. It tests whether or not the water flow is in a state of constant equilibrium for at least a 30 second time period.

The Propagation Model can be represented as a directed graph (mathematical) or as a set of Knowledge Base rules (conceptual). Rules are capable of expressing arithmetical and logical operations. Time can be considered as a quantitative ordering of events whenever appropriate. It can be determined that event A occurred before event B, and event A occurred n time units before B. Two sample rules that characterize the temporal realtionships are stated below:

$$(\text{RULE 1}) \qquad LT(DTIME(A,B),\emptyset) \rightarrow X$$

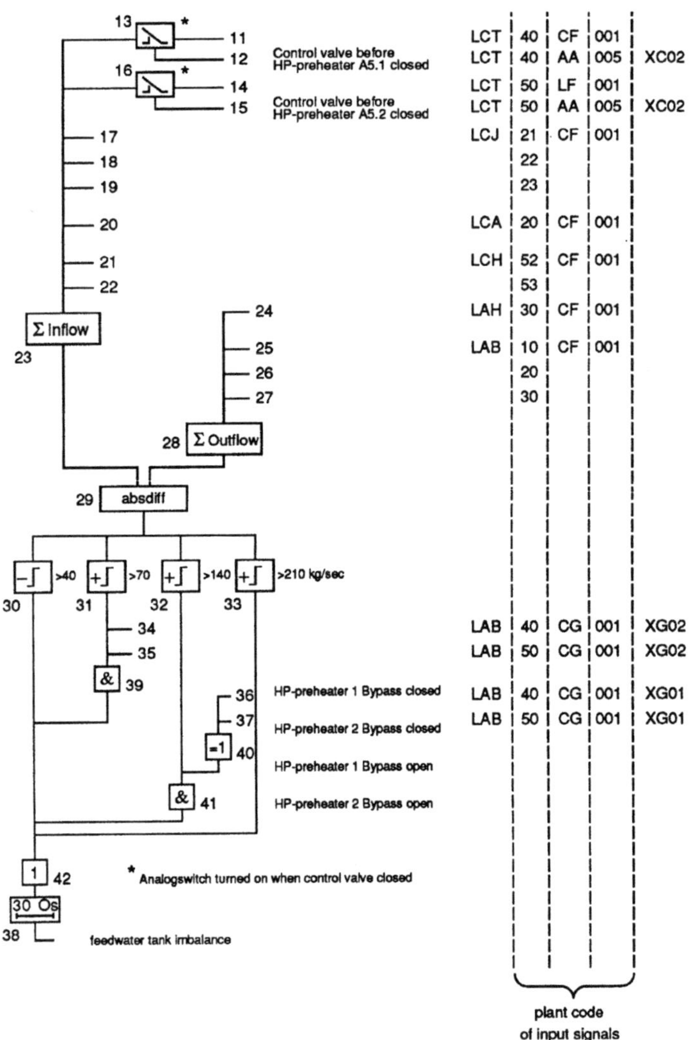

Fig. 2:   Propagation model

Rule 1 characterizes the time difference between events A and B
such that the diffenrential time (DTIME) for the occurrences of
event A and event B must be less than (LT) zero. This simply means
event A occurred before event B.

$$(\text{RULE 2}) \qquad \text{AND } (\text{LT}(\text{XTIME}(C),\text{XTIME}(B)),$$
$$\text{GT}(\text{XTIME}(C),\text{XTIME}(A))) \rightarrow Y$$

Rule 2 characterizes the time difference between events A and B
with respect to some intermediate event C such that

-    XTIME(C) represents the expected time for the occurrence of
     event C,

-    GT(X,Y) represents the arithmetical "greater" relation.

These fundamental relationships apply the temporal logic concepts proposed in [5,6]. A less tedious notation for these relationships would be highly desirable.

When control loops of a propagation model are modelled, as is often the case in disturbance analysis applications, the resulting models and their corresponding graphical representations contain "cycles". This is a highly undesirable condition because it results in extremely complex propagation algorithms. Fortunately, there are no control "cycles" or feedback without a time delay. Thus the algorithm can be suspended until this lag has elapsed. The effect of the feedback can be handled in a way similar to other incoming events at the appropriate point in the Time Domain.

The data structure used to manage this approach is the Internal Event Queue (IEQ). Fig. 3 illustrates the handling of internal events using this approach. A Process event arrives in the Process Event Queue activating the associated entry node (a process event exists for every non-constant entry node).

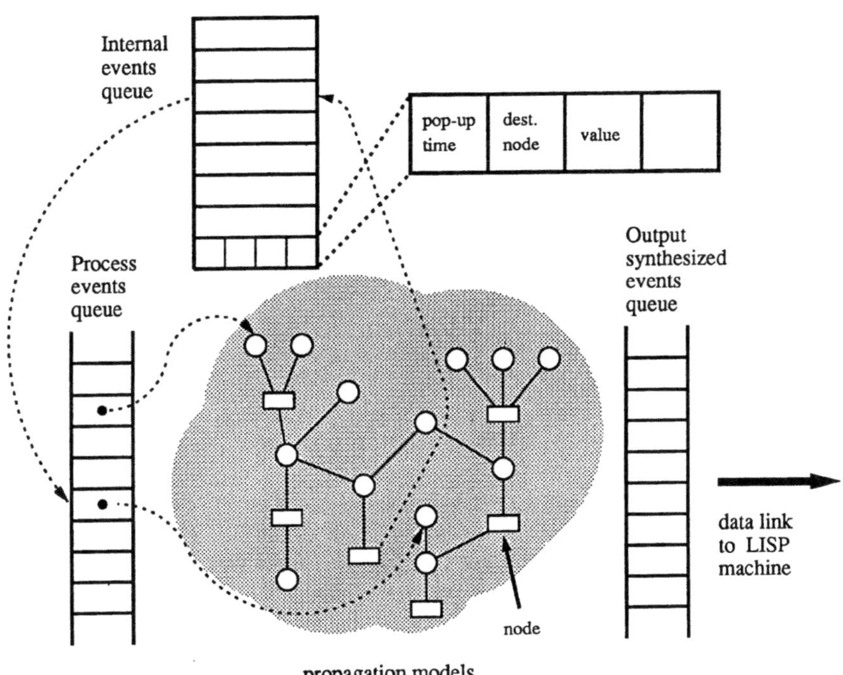

Fig. 3:   Handling of events

The process event propagates through the PM network possibly triggering a delay node (usually an exit node). The attibutes of a delay node specify a Pop-up Time, i.e. the point in time when control must be returned to the Propagation model and a Destination node, the node in the Propagation model where control will be returned to.

Every triggered Delay node is entered in the IEQ such that the entry with the most recent pop-up time is placed in front of the IEQ. Since all

Process events are ordered chronologically it can be easily determined where the front element of the IEQ is to activate the propagation model next.

The Delay node may be reactivated before the associated IEQ entry moves to the front of the queue. In this case several reactions, according to the attributes of the associated node, are possible, e.g.:

-    the IEQ entry may have to be cancelled,
-    a duplicate entry may have to be inserted into the IEQ.

If the granularity of the Time Domain is relatively small, 10**6 time points, say, then a data structure can be designed to eliminate the sorting overhead whenever a new entry must be inserted.

HYPOTHETICAL REASONING

Every node in the propagation model consists of its occurrence time and its physical value. A hypothetical propagation may be simulated by the appropriate tagging and insertion of generations of hypothetical events into the process event queue. The predecessor/successor ordering of the data structure remains fixed. Each node that has been visited receives additional information which consists of

-    its time of occurrence,
-    its value and
-    hypothesis generation.

The synthesized events in the output queue receive corresponding information.

The problem with hypothetical reasoning is not so much finding a concept and implementation. Rather, it is a human engineering problem. To set up a hypothesis and test its consequences requires, first of all, the provision of an appropriate and consistent initial environment. Since the implications of the events are fairly complex it may be a tedious, time-consuming and error-prone task creating and firing a necessarily great number of artificial events and after some time even keeping track of where they come from and what they were intended to. One possible solution might be aggregating pre-defined "hypothesis event sets" and set them in action returning only whether the hypothesis would hold in the current environment. Many experiments will have to be carried out to near a solution.

BACKWARD CHAINING AND EXPLANATION

Usually we require an expert system not only to tell us what has happened or what is going to happen, but also why something has happened the way it did. Also in this case the propagation model lends itself to useful application. When we backtrace from a synthesized event to its initiators and, even more important, to its enablers, we can chain a lot of information as to how the synthesized event has been created.

The information, however, needs to be given in textual rather than in a "value"-format (i.e.,true, false etc.) and the textual synthesis, i.e. putting together the eventual explanation is a nontrivial task. Here we are presently carrying out experiments using Intellicorps knowledge engineering environment (KEE) as the software basis.

CONCLUSIONS

As our experience with on-line information systems, Experts Systems and Artificial Intelligence tools grows, we retreat from the first euphoria that AI could help us solve the problem we were unable to solve with conventional programming. The major effort of the development time goes into building the knowledge-base. There is no such thing as a generic knowlege-base for nuclear power plants as there is, for example, for the diagnosis of a Boeing 747 aircraft.

AI-methods, tools and hardware are still in a state which does not optimally lend itself to real-time application. The ability of developing prototype systems to investigate variants otherwise too costly to justify is one advantage that we gladly accept. Last, but not least the tools provide you with a flexible and adaptable user interface (desktop window systems) etc. The development of such tools in a project would be prohibitive and room for experimentation would be limited.

ACKNOWLEDGEMENT

The author is indebted to Paul P. Orgas who helped him with numerous comments and language problems.

REFERENCES

1. F. Hayes-Roth, D.A. Waterman, D.B. Lenat, Building Expert Systems, Addison-Wesley, 1983.

2. J. Doyle, Expert Systems and the "Myth" of Symbolic Reasoning, IEEE Trans on Software Engineering, Vol. SE-11, No. 11, Nov. 1985.

3. R.L. Moore et al., A Real Time Expert System for Process Control, ANS-Meeting Pasco, Washington, Sept. 1985.

4. W. Bastl, Integrated Disturbance Analysis - A Basis for Expert Systems in Nuclear Power Plants, this conference.

5. J.F. Allen, Maintaining Knowledge about Temporal Intervals, CACM, Vol 26, No. 11, pp832-843, 1983.

6. M.B. Vilain, A System for Reasoning about Time, Procs. AAAI, Pittsburgh, 1982, pp 197-201.

7. A. Fusoaka, H. Seki, K. Takahashi, A Description and Reasoning of Plant Controllers in Temporal Logic, IJCAI 1983, Karlsruhe, West Germany.

8. D. Mc Dermott, A Temporal Logic for Reasoning about Process and Plans, Cognitive Science 6, pp 101-155, 1982.

DEVELOPING AN EXPERT SYSTEM FOR FAULT DIAGNOSIS OF A TURBO GENERATOR

GROUP USING AN OPS5 PRODUCTION RULES PROGRAMMING ENVIRONMENT

Claudio Balducelli   Massimo Gallanti

ENEA-Term/Mep,   CISE S.p.A.
P.O. box 2400   V. Reggio Emilia 29
00100 Rome (Italy)   Milano (Italy)

1. INTRODUCTION

The Turbogenerator is one of the most complex and critical subsystems
of a thermal production plant; it is composed of 3-5 coaxial stages (low,
medium, and high pressure turbines, and generator), each one supported by
a couple of oil film bearings. Almost all known turbogenerator
malfunctions result in an alteration of the vibrational state of the
machine during operation.   Thus turbogenerators are factory equipped by
vibration sensors, triggering alarms when vibration amplitude goes beyond
a prefixed threshold.

Since mid-seventies   techniques   for   an   early   diagnosis   of
turbogenerator malfunctions have been developed at CISE [1]. These
techniques require an advanced sensory and monitoring equipment, based on
proximity sensors installed in the bearings of the machine; they grant
however early detection of malfunctions, as well as an accurate diagnosis
of the malfunction cause.

Diagnosis techniques based on vibration analysis require to monitor a
large number (over fifty) of variables. Some of them concern the overall
state of the machine (rotational speed, power produced, bearing
alignment); other are relevant to the single bearing and provide detailed
information about the shaft vibration (amplitude of the overall vibration;
amplitude and phase of the first and second harmonics), as well as about
the oil film state (temperature and pressure); therefore it is necessary
to provide the generator with additional sensory equipment, other than the
one provided by the factory. Once the sensors are installed, the machine
is periodically inspected during its life by means of the analysis of the
acquired data; a processing system performs, on-line, a spectral analysis
of the vibration; sampled and computed data remain available on a mass
memory for the diagnostic analysis, usually performed by comparison of
plots (versus time or rotational speed) of many different variables.

Diagnostic   skill   mainly   relies   on   'empirical'   knowledge
(relationships, often expressed in form of condition-action pairs, that
have   been   acquired   through   previous   experience   about   machine
malfunctions). These relationships can hardly be derived by a deep
understanding of the problem or referred to a model of the dynamic
behavior of the system. Scarse availability and long training required by
human experts also suggests the adoption of a knowledge-based approach to
the design of turbogenerator diagnostic equipment.

2. SYSTEM FUNCTIONS DESCRIPTION

The above motivations led CISE, under the commitment of ENEA, to the development of TVM, an expert system for diagnosing early malfunctions of a turbogenerator. The first implementation of TVM was based upon SAGE, an expert system shell developed by System Designer (UK). ENEA researchers then recoded TVM using OPS5 [2], a production rule system offering a more flexible and powerful programming environment, in which was also improved a graphic interface displaying a trace of the diagnostic reasoning.

TVM has been designed as an intelligent aid for the apprentice diagnostic technician; it drives the apprentice through the process of analysing the data base acquired during a measurement campaign, selecting the diagrams to be consulted, and the symptoms to be evaluated on each plot, in order to check a given malfunction hypothesis, like mass unbalance of the rotor, shaft bending, cracks, rubbing due to bearing displacement or misalignment. More specifically, expert diagnosticians are in charge of three tasks:

1) drafting a reference chart called <u>baseline</u>, when the machine enters its operational life, or reenters into operation after a major repair. The baseline is a set of typical data, e.g. plots of 1st and 2nd harmonics amplitude and phase versus time and rotating speed, characterizing the machine behavior in a good operating condition, and is consequently taken as a reference to judge deviations from correct behaviour;
2) when a major malfunction is detected by the operating personnel, on the basis of evident symptoms, interpreting previous and current data, in order to single out the malfunction cause and give advise to the conduction and maintenance personnel;
3) after periodic measurement campaigns, ordinarily held during prescheduled, plant start-ups and shut-downs, judging the health state of the machine, and possibly diagnose early malfunctions.

While task 1) basically involves a well-defined, purely procedural knowledge, tasks 2) and 3) rely on experimental knowledge, that is typically declarative, and, as discussed above, mainly judgemental and informal. It can be partitioned into <u>different areas</u>, relevant to:

a) data interpretation, i.e. extracting significant qualitative features from numerical data;
b) diagnosis, i.e. relating the above qualitative symptoms to a definite fault cause;
c) machine operation and maintenance. Here however other significant knowledge source are involved, apart from vibration diagnosticans, e.g. machine designer, plant managers, maintenance and repair experts.

TVM focuses on diagnosis. It is a pure consultation system, thus data interpretation, in the above meaning, is essentially left to the user. This clarifies the main limitation of our prototype system: it does not embody one of the most significant expert's capabilities, that is singling out and evaluating symptoms.

3. KNOWLEDGE BASE IMPLEMENTATION INTO A SELECTED FORMAL STRUCTURE

The system build-up started with the Knowledge Base formalization phase in which was useful to separate all the possible diagnoses from all the findings required to their verification. Normally the total number of the findings is much greater than the diagnoses ones: a diagnosis is an univocal entity of the knowledge, while the weight of a fact inside the knowledge generally depends on the number of the diagnoses that it can contribute to verify and on their verification certainty.

Starting from diagnoses classification the problem complexity can be reduced; in fact after that all diagnoses are defined, it is possible to consider at first only the findings able to exclude a great number of possible diagnoses in a preliminar diagnostic search phase. These particular findings are named 'main symptoms' and will be considered at first during the diagnostic search: generally if a considerable number of 'main symptoms' are identified, the partitioning of the KB into 'investigation areas', which include restricted sets of possible diagnoses, is useful in order to implement a forward-chaining or data-driven search strategy able to select the most promising areas in the first search phase. For TVM KB it was found that the number of 'main symptoms' is very low so that the forward-chaining search phase is very fast compared with the backward-chaining one.

The formalized KB consists of a search tree for every possible fault [3]. In fig. 1 is visualized the search tree relevant to the diagnosis of 'bearings misalignment'. To every node of this search tree is associated a 20 character string representing an hypothesis to be verified and in particular:

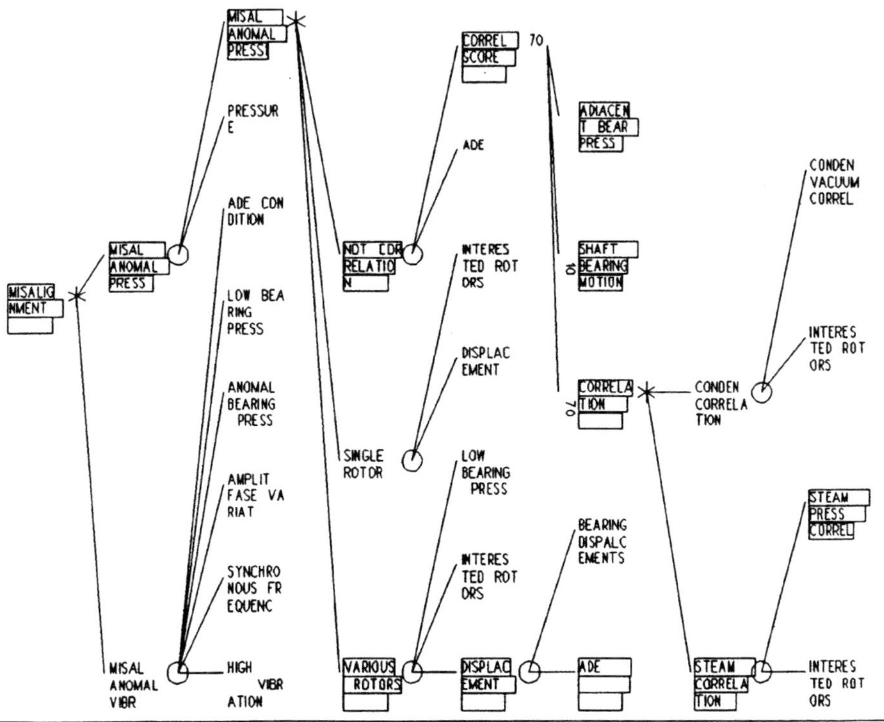

○ identify AND,   ✳ identify OR,   nn identify SCORE node types

Fig. 1 - An example of diagnostic search tree

- the root of the tree represents the diagnostic goal to be achieved;
- the intermediate nodes of the tree are 'dependent hypotheses' generated
  during the inferential search phase as subproblems of a more general
  problem;
- the terminal branches of the tree are 'primitive hypotheses' that can be
  solved by means of an user or Data Base inquiry.

   Either the 'dependent' or the 'primitive' hypotheses are defined as
objects of a class having the following attributes:

```
NAME    --> hypothesis name (20 char.)
VALUE   --> verified/notverified flag
ORIGIN  --> origin node (and/or/score)
TYPE    --> used for multivalued hypotheses
SCORE   --> score of the hypothesis
STATUS  --> hyp.  status (solving/solved)
PARENT  --> parent hypothesis name
GSPLIT  --> grafic output flag
```

The PARENT, ORIGIN, STATUS attibutes are initialized every time, during
the inferential process, a new hypothesis is considered: in particular
the ORIGIN attribute identifies one of the three different node types from
which an hypothesis can arise: an <and> an <or> or a <score> node type.
In the first two cases the dependent hypothesis becomes true if,
respectively, are true all or at least one of its generated hypotheses are
true, otherwise it becomes false; in the last case an initial <score>
value is assigned to the dependent hypothesis and this score increases or
decreases during the verification process of the various generated
hypotheses each one contributing with different weights to the score
value. The dependent hypothesis become true if its score results greater
than a predefined threshold; otherwise it becomes false. For all node
types, after the relative hypothesis verification, the VALUE and STATUS
attributes will be updated.

   The <score> node type allows to consider the possibility that an
hypothesis becomes true if n of its N generated hypotheses (with n<N) are
true; it allows also to associate different certainty factors to a
particular hypothesis according to the obtained score. However with the
adopted formalism,it is not allowed a certainty factor propagation into
the search tree, but only the definition of thresholds establishing a
'reasonable limit' for the hypothesis verification.

4. SYSTEM DEVELOPMENT USING AN OPS5 PROGRAMMING ENVIRONMENT

   Before the selection of a production rules system for the
implementation of the KB acquired from the human expert and structured
with the above described formalism, it was established that the expert
system must respect the following specifications:

1) The representation of knowledge must be 'explicit' inside the
   programming structure and must be 'easy' to add new knowledge or modify
   the existing one. This is a very important feature because in this
   application domain the knowledge is not found out from a single expert
   and there is the necessity to update it more time after the first
   implementation.
2) A reasoning explanation module must be present in the system that, also
   by means of dynamical graphic display of the considered search trees,
   visualizes all the selected and verified hypotheses during the
   inferential process. This function is very useful during the debugging
   and test phase of the system to check the Knowledge Base consistency.
3) The system must use a not very large amount of cpu time during its
   inferential process, and must have the possibility to be also installed
   in a standard hardware environment.

In the OPS5 language structure the hypotheses present in the formalized KB can be easily considered as objects of a GOAL class having the attributes defined in the previous paragraph. To implement a backward-chaining search strategy in the OPS5 production rule system, three principal sets of rules are required:

- the subgoaling rules;
- the rules for handling the immediate soluble goals;
- the control rules for implementing the backward-chaining strategy.

While the first two types of rules depend on the knowledge base contents, the third type depends only on the adopted formalism.

The layout of the total expert system is visualized in fig. 2. The rules part of the system is composed by two different sections:

- the object oriented rule section;
- the control rules section (metarules).

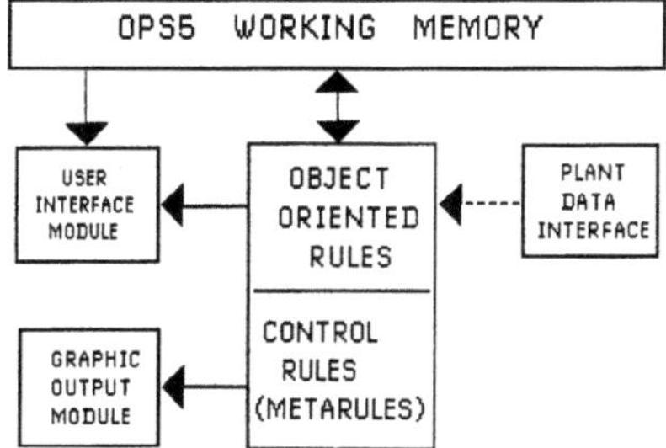

Fig. 2 - Software configuration of TVM expert system

In the first section the KB, formalized into diagnostic search trees, is inserted by means of the first two types of OPS5 rules: the subgoaling rules named 'split' type rules and the 'solve' type rules. The 'split' type rule has the following syntax:

```
(p split:VARIOUS-ROTORS
(goal ^name VARIOUS_ROTORS ^status NIL ^value NIL)
-->
(make goal ^name DISPLACEMENT ^origin AND ^parent VARIOUS_ROTORS)
(make goal ^name LOW___BEARING_PRESS ^origin AND ^parent VARIOUS_ROTORS)
(make goal ^name INTERESTED_ROTOR ^origin AND ^parent VARIOUS_ROTORS)
(make goal ^name INTERESTED_ROTOR ^value OK ^typ 12 ^parent VARIOUS_ROTORS)
(make goal ^name INTERESTED_ROTOR ^value OK ^typ 23 ^parent VARIOUS_ROTORS)
```

This rule concerns the resolution of the dependent hypothesis named

VARIOUS ROTORS in the diagnostic tree of fig. 1. To solve this
hypothesis, three new hypothesis are generated. The hypothesis named
DISPLACEMENT is dependent too and will be solved by means of another
'split' type rule; the other two, named LOW BEARING PRESS and INTERESTED
ROTOR, are primitives and will be solved by means of a 'solve' type rule.
In particular the INTERESTED ROTOR hypothesis will be solved by a 'solve'
type rule having the following syntax:

```
(p solve:INTERESTED_ROTOR
 {<goal> (goal ^name INTERESTED_ROTOR ^value nil)}
 -(goal ^name INTERESTED_ROTOR ^status solved)
 -->
 (write (crlf)
  (crlf) |The pressure variation is present:            | (crlf)
  (crlf) | * on all bearings                 -->Answer ALL | (crlf)
  (crlf) | * on high pressure rotor bearings  -->Answer  1 | (crlf)
  (crlf) | * on low pressure rotor bearings   -->Answer  2 | (crlf)
  (crlf) | * on generator bearings            -->Answer  3 | (crlf)
  (crlf) | * on high and low pressure bearings -->Answer 12 | (crlf)
  (crlf) | * on low pres. and generator bearings-->Answer 23 | (crlf)
  (crlf) |                                    Answer:    |)
  (modify <goal> ^value yes ^typ (accept) ^status solved))
```

It can be noted that the question is asked to the user only if the
INTERESTED ROTOR hypothesis is not just present with STATUS equal to
<solved> into the working memory, and that the goals having status equal
to <OK> generated by the split rule inform the system about the type
values that set the hypothesis as verified. These rules types allow to
implement into the system a KB formalized as diagnostic search trees.

In addition a 'print' type rule can be used containing a call to an
external procedure for printing the final diagnosis reached. This rule,
that fires when the hypothesis relative to the principal node of the tree
is verified, has the following syntax:

```
(p print:MISALIGNMENT
 (goal ^name MISALIGNMENT ^value YES  status SOLVED)
 (goal ^name {<pdia> MISAL_ANOLMAL_PRESS} ^value YES
       ^parent MISALIGNMENT ^status SOLVED)
 -->
 (call printdia <pdia> )
```

The second rule section contains the rules that perform a control of
the inferential process and in particular:

- manages the solution of the hypotheses relative to the various nodes of
  the search trees, depending on the goals currently present in the
  working memory;
- performs a dynamic garbage collection of the working memory;
- contains a reasoning explanation function;
- manages the calling sequence to the graphic output module.

The firing of the rules of the control section occurs, based on the
status of the GOALS present in the working memory, when a node can be
solved, beginning from the more recent one. Since the subgoals recency
guides their verification, the order of generated hypotheses in the split
type rules must be established by means of an heuristic estimate [4].

The user interface module is a collection of external routines which
are generally called at the beginning and at the end of the inferential
process and are used for the 'main symptoms' acquisition and for the
output of the diagnosis reached.

```
                    -  T V M  -

          ** ASSESSMENT OF MAIN SYMPTOMS **

Are you considering a start-up or run-down condition? [yes/no] : no

Is there some bearing presenting either high vibration
amplitude or significant sudden vibration variation? [yes/no]  : yes

Is the oil pressure of some bearings abnormal? [yes/no]        : no

Does the temperature of the oil or of the metal in some
bearing exceed the normal value? [yes/no]                      : no

          ** MALFUNCTION DIAGNOSIS SECTION **

Let's consider the spectral analysis of the abnormal vibration.
Are there peaks at speeds lower than the rotation one? [yes/no]: why

THIS QUESTION IS AIMED AT PROVING THE HYPOTHESIS:
OIL_WHIRL_DUE_TO_STEAM_INSTABILITY_OR_HIT_AND_BOUNCE_RUB

          Type <c> to continue : c

Let's consider the spectral analysis of the abnormal vibration.
Are there peaks at speeds lower than the rotation one? [yes/no]: no

Let's consider the spectral analysis of the abnormal vibration.
Are there peaks at speeds equal than the rotation one? [yes/no]: yes

What is the rotor affected by abnormal vibration:
   HP: high or medium pressure turbine
   LP: low pressure turbine
   GE: generator
                          Answer HP, LP or GE          : HP

What is the age of the rotor? Is it more than 15000 working
hours? [yes/no]                                                : yes

Is there an evident variation of the steam temperature
related to the vibration anomaly? [yes/no]                     : no

What is the trend of the vibration's amplitude and phase over
the period immediatly following the rise of the anomaly?
   V_CONST: it is constant
   V_CONT : it varies in a continuous way
   V_SUDD : it is discontinuous with sudden variation
                          Answer V_CONST,V_CONT,VSUDD : V_SUDD

Let's consider the bearing where the vibration anomaly is more
evident, and the contiguous bearing belonging to the other
rotor. Is the oil pressure in at last one of the two bearings
abnormal? [yes/no]                                             : no

Let's consider the bearing where the vibration anomaly is more
evident and the contiguous bearing in the same rotor. Are the
vibration phases of these bearings almost the same? [yes/no]   : yes


     THE FOLLOWING MALFUNCTION DIAGNOSIS HAS BEEN EVIDENCED:

     A ROTOR IS AFFECTED BY A MASS IMBALANCE OF STATIC TYPE.

     IF THE BEARINGS WHERE THE ABNORMAL VIBRATION LEVEL
     IS PRESENT BELONG TO TWO DIFFERENT ROTORS, THEN THE
     UNBALANCED ROTOR IS THE ONE WHERE THE VIBRATION IN
     THE BEARING IS HIGHEST.

     SUDDEN VARIATIONS IN THE VIBRATION AMPLITUDE COULD
     BE CAUSED BY BREACHES OF INTERNAL PARTS (E.G. BLADES)
     OF THE ROTOR.
```

Fig. 3 - An example of consultation session

The graphic output module dinamically is called by the control rules to update the graphic display of the current diagnostic tree. In fig. 1 the output of a diagnostic tree generated on a Tektronik 4014 graphical display is visualized. The graphic output is available also on a color graphical station type (TEK4100) in which the current state of an hypothesis (true/false/undefined) is remarked by mean of different color codes (blue/red/white) of the relevant string. The frame surrounding an hypothesis, means that the hypothesis was evaluated during the current search tree processing.

Finally the plant data interface module will be developed in the future, in order to allow automated interpretation of the data base created by the data logger during a measurement campaign. It will be a data interpretation expert system, able to provide an estimate of credibility upon the facts which are taken as input of the diagnostic process. This estimate is by now provided by the user of the current version of TVM, by means of a qualitative analysis of the plotted data acquired by the sensors. Experimentation with this version has focused how this estimate requires a great deal of expert skill, so that the system has limited pratical relevance in its current version. The design of the plant data interface, currently under way, is based on the adoption of fuzzy logic [5] for the treatment of the vagueness of the input data.

Current TVM version contains a body of diagnostic knowledge which refers to about 25 different malfunction causes. TVM KB consists of about 200 rules; the system runs on DEC VAX and PDP-11 computers.

5. AN EXAMPLE OF CONSULTATION SESSION

TVM has been tested on a large number of real cases provided by CISE experts. An example of TVM consultation session is reported in fig. 3. The first question of the consultation process is aimed to verify the presence of a set of main symptoms: in the example the result of the first consultation phase determines the sequence of the next questions. When the user answer "WHY" to a TVM question then the system displays the name of the malfunction hypothesis it is trying to verify.

The question-and-answer phase ends up when the malfunction hypothesis under evaluation has reached a defined evidence. Then TVM reports the name of the verified malfunction, along with a justification of the observed phenomena, and, in some case, a list of remedial actions.

ACKNOWLEDGMENTS

The construction of the knowledge base of TVM has been possible thanks to the cooperation of G. Lapini of CISE. His key role in the development of TVM is greatefully aknowledged here.

REFERENCES

1. A. Clapis et al., "Early Diagnosis of Dynamic Umbalancies and of Misalignments in Large Turbogenerators", Proc. IV World Congress on the Theory of Machine and Mechanisms. Newcastle upon Tyne, UK (1975)
2. L. Brownston, R. Farrel, E. Kant, N. Martin, "Programming Expert Systems in OPS5: An Introduction to Rule-Based Programming", Addison-Wesley Series in AI (1985)
3. M.D. Olman, R.B. Worrel, "A Fault Tree Representation Designed for Computer Analysis", Sandia Nat. Lab., Albuquerque, NM, SC-RR-71 0615A, 1972
4. R. E. Korf, "Depth-First Iterative-Deepening: An Optimal Admissible Tree Search", Artificial Intelligence 27 (1985) 97-109
5. L. A. Zadeh, "Inference in Fuzzy Logic", IEEE 10th. Annual Symposium on Multiple-valued Logic, pp124-131 (1980)

EXPERT SYSTEM FOR NUCLEAR POWER PLANT FEEDWATER SYSTEM DIAGNOSIS

R. Meguro[1], Y. Kinoshita[1], T. Sato[1], Y. Yokota[2]
and M. Yokota[3]

[1]Electric Power Control Computer Systems Dept., Fuchu Works,
Toshiba Corp. 1, Toshiba-cho, Fuchu-shi, Tokyo 183, Japan

[2]Nuclear Turbine Plant Engineering Dept., Toshiba Corp. 1-6,
Uchisaiwai-cho 1-chome, Chiyoda-ku, Tokyo 100, Japan

[3]Control & Instrumentation Dept., Keihin Product Operations,
Toshiba Corp. 2-4, Suehiro-cho, Tsurumiku, Yokohama 230,
Japan

ABSTRACT

The Expert System for Nuclear Power Plant Feedwater System Diagnosis
has been developed to assist maintenance engineers in nuclear power
plants. This system adopts our latest process computer "TOSBAC" G8050 and
our expert system developing tool "TDES2", and has a large scale knowledge
base which consists of the expert knowledge and experience of engineers in
many fields. The man-machine system, which has been developed exclusively
for diagnosis, improves the man-machine interface and realizes the graphic
displays of diagnostic process and path, stores diagnostic results and
searches past reference.

[I] INTRODUCTION

The nuclear power plant has been designed and developed to realize
safe operation. Recently, in addition to this, more reliability and
economical efficiency are being required for maintaining safety because of
the growth of percentage of nuclear power among total electric power
generation capacity. In order to realize more reliable and more
economically efficient plant, early detection of abnormality, quick
assessment of its cause, prevention of proliferation of the abnormality,
preventive maintenance and quick recovery from accident are very
important.

If an abnormality happens, knowledge and experiences of engineers in
many field would be necessary for diagnosing and assessing the accident
since the nuclear power plant consists of many systems, equipment, and
parts. The Expert System for Nuclear Power Plant Feedwater System
Diagnosis to be referred hereinafter as "BOP(Balance of Plant) Expert" has
been developed to assist maintenance engineer to detect the cause of an
abnormality. Overview of the system is shown in Figure 1.

Fig. 1. Overview of the system

<u>1.Features</u>

The features of this system are listed below.

- Large knowledge base from many experts in different fields.
- User friendly man-machine interface with mouse and keyboard.
- Graphic display of diagnostic process.
- Record and retrieval of diagnostic results.
- Adoption of our expert system developing tool "TDES2" (<u>T</u>ool for <u>D</u>eveloping an <u>E</u>xpert <u>S</u>ystem 2).
- System configuration and man-machine interface design considering application to online plant process computer.
- Adoption of our latest process computer "TOSBAC" G8050.

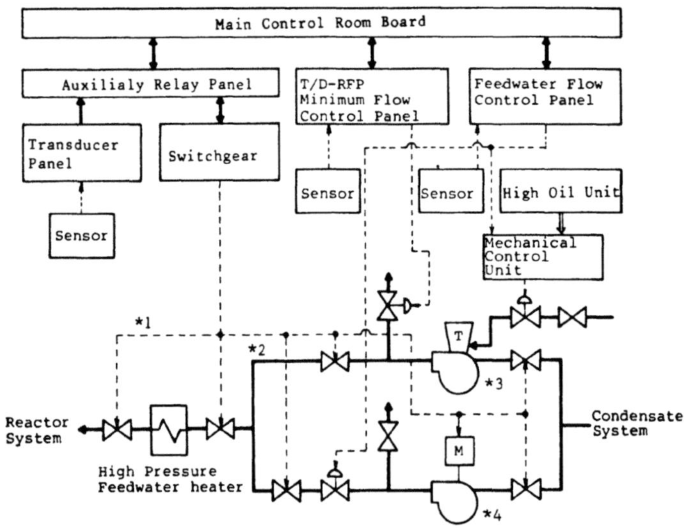

Fig. 2. Objects of diagnosis. *1 Electric power/signal line; *2 Water flow line; *3 Turbine-driven feedwater pump; *4 Motor-driven feedwater pump.

2.Scope of diagnosis

This system covers the diagnosis of turbine/motor driven feedwater pumps, high pressure feedwater heater, piping, valves, control equipments and instruments, consisting of the reactor feedwater system between feedwater pump inlet header and a reactor. Figure 2 shows the objects of diagnosis.

Incidents to be diagnosed by the BOP Expert are (1) Plant trip, (2) Generating power disturbance, (3) Feedwater system abnormality detected by alarm initiating devices, and (4) Feedwater system abnormality which does not lead to plant trip or generating power disturbance, and is not detected by alarm initiating devices.

[II] HARDWARE

BOP Expert hardware system configuration is illustrated in Figure 3. Central processing unit (CPU) is our latest 32bit super mini computer "TOSBAC" G8050 and man-machine interface devices are two full graphic color CRT's with keyboards and mice. As an output device for results, it has a Kanji (Japanese character) typewriter and it has also large storage hard disk memory and console CRT with keyboard and hardcopier for a maintenance.

One of the two full graphic CRTs is dedicated to interactive communication with the operator and the other CRT provides information that aids operator's decision making, such as knowledge tree diagrams and mimics of plant system. Since all input operations can be performed via a mouse and keyboard, an operator who is not necessary familiar with the operation of a computer can use this BOP Expert easily.

[III] SOFTWARE

BOP Expert software system configuration is illustrated in Figure 4. Software system consists of knowledge processing subsystem, man-machine interface subsystem, and interface software between these two subsystems.

1.Knowledge processing subsystem

This subsystem consists of a diagnosis system, a knowledge base and a man-machine interface.

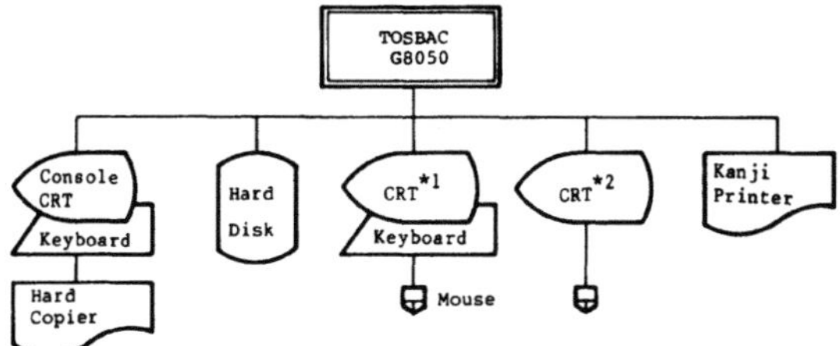

Fig. 3.   Hardware system configuration. *1 Full graphic color
         CRT for interactive communication with operator; *2
         Full graphic color CRT for diagnosis information

The diagnosis system has two functions and is based on our expert system
developing tool "TDES2" which is written in LISP.

One of the two functions of the diagnosis system is the knowledge
base editor. Experts can access the editor through console CRT
man-machine interface and can build and modify the knowledge base.
Description of knowledge in TDES2 is in terms of production rules. The
other function is the inference engine which infers causes of
abnormalities. Inference is performed by forward-reasoning and dialog
with a user in the course of inference is performed through the
man-machine interface subsystem.

2.Man-machine interface subsystem

This subsystem consists of a man-machine system, a data base and a
man-machine interface. The man-machine system has two functions and is
written in a system descriptive program language (PL-7), which has been
extensively used in plant process computer, in order to realize easy
connection with existing plant process computer system.

One of the two functions of the man-machine system is the man-machine
interface function. User can access the function through one of the full
graphic CRT with a keyboard and a mouse, and this function performs dialog
processing between the user and the inference engine in the knowledge
processing subsystem. The other function is the diagnosis supporting
function which edit diagnosed data. Edited data is stored in the data
base and an user can access the data through the other full graphic CRT
with a mouse.

3.Interface software

The interface software links the inference engine in the knowledge
processing subsystem to the dialog function in the man-machine interface
subsystem. This software perform both transmission of data and control of
real-time processing between the above two subsystems.

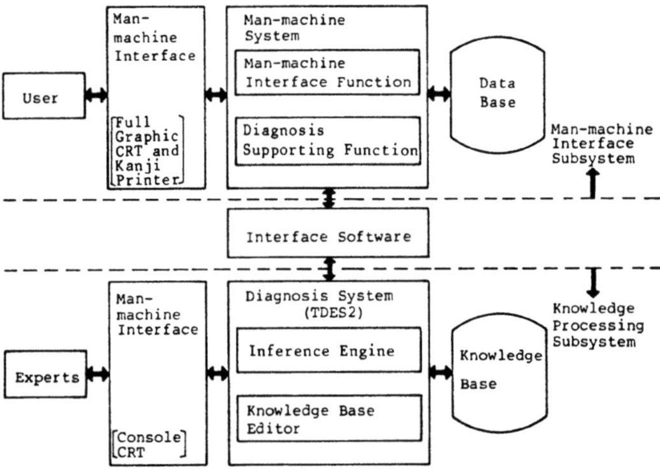

Fig. 4. Software system configuration

[IV] FUNCTION

1.Inference function

This function is included in the knowledge processing subsystem. At
first, the inference engine determines the initial guess of a failure
according to major plant parameters input by an operator. This initial
guess will limit the scope of the subsequent inference to one of the
hardware or the function constituting the reactor feedwater system. Then
the inference engine inferres a cause within this limited scope and
identifies the lacking data necessary to continue the diagnosis. A prompt
message is output on the CRT, and the operator provides the necessary data
according to the message. Because of the following reasons, the operator
may not be able to provide all the data necessary to come up with a single
final cause. Therefore all possible candidates for the cause are detected
by inference.

- Lack of sensor for providing the target data.
- The equipment is located where it is difficult for an operator to
  access and check the status.
- Dismantling of hardware may become necessary to determine the
  final cause.

For the present application the certainty of both a condition part
and an action part of each production rule, and of rule itself are not
taken into consideration, because introduction of certainty makes
inference logic testing very complex. (TDES2 has capability to assign
certainty factors)

2.Man-machine interface function

The man-machine interface functions have been designed to realize
operator friendly man-machine interface. These functions are included in
the man-machine interface subsystem.

(a) Major plant parameters setting function. Before starting
diagnosis, all the major plant parameters, such as a status of motor
driven feedwater pump, before and after occurrence of an abnormal
phenomenon, is set through this function.

(b) Dialog function. This function performs dialog with the operator
in the process of inferring the cause of abnormality. A dialog picture
related to this function is shown in Figure 5. A picture for dialog is
divided into fixed six areas. (1) In the input guide display area, a
question from the system is displayed. (2) In the progress guide display
area, the rule that is now being checked is provided. Owing to the
guidance message, operator can understand what the expert system is
presently inferring. (3) In the record display area, the messages which
operator and the system exchanged in the process of inferring, are
displayed in chronological order. The following items are recorded and
displayed: intermediate results (i.e., candidates for the cause of
abnormality which the system displays in the process of inferring), dialog
records (i.e., system's questions displayed in the input guide display
area and operator's answers) and inferred causes (i.e., causes of the
abnormality which the system inferred). (4) In the abnormality
information display area, the title, date and time of occurence of
abnormal phenomenon which the system is inferring are listed. (5) In the
mimic display area, mimics related to the part of the plant being inferred
now are provided. (6) In the input area, candidates of answers which
operator can give to the system's questions are displayed. An operator
can choose and set his answer using mouse.

(c) WHY function. This function provides an operator with the reason why a certain question is displayed in the input guide display area. An operator can request the function choosing HELP key located in the input area.

3. Diagnosis information edit function

Additional information related to diagnosis is provided to the operator. Dialog is performed on another full graphic color CRT. This function is included in the man-machine interface subsystem.

(a) Module configuration diagram displaying function. Knowledge base in this expert system is divided into several modules as described later. This function displays relationship between each module of knowledge base.

(b) Tree diagram displaying function. Each knowledge module consists of several rules, and the relationship between each rule is displayed by this function.

(c) Report and record function. Results which the expert system finally ascertained as causes of the abnormality and other diagnostic data are summarized, and the summary is output to the printer. The operator can store the summary in a data base if necessary.

(d) Retrieval of record function. Summary of diagnostic results recorded by the above function can be retrieved at any time.

(e) Mimic display function. Several mimic displays are prepared as additional information.

[V] FLOW OF BUILDING KNOWLEDGE BASE

Targets for efficient build-up of knowledge base are (1)To build a large scale knowledge base using more than 10000 rules,

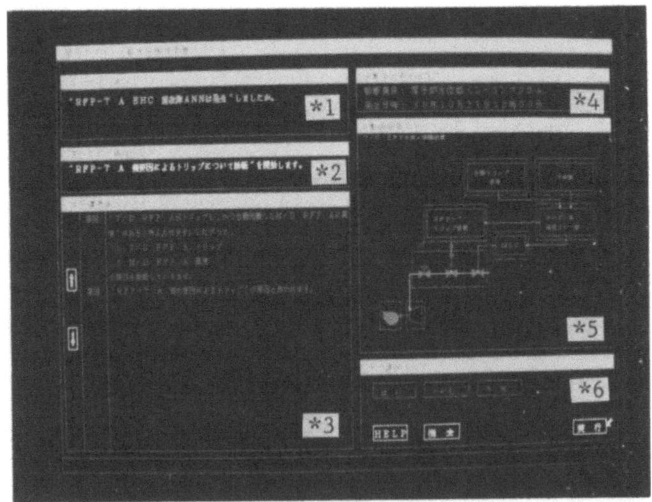

Fig. 5. A sample of dialog pictures. *1 Input guide display area; *2 Progress guide display area; *3 Record display area; *4 Abnormality information display area; *5 Mimic display area; *6 Input area.

(2)Standardization of extracting knowledge from experts in different fields, (3)Speed-up of inference processing in order to apply the expert system to real plants, (4)To improve efficiency of software productivity and (5)To establish procedure of making knowledge base. Procedure of building knowledge base in view of these targets are described in Figure 6.

1.Knowledge acquisition

At first, several experts who design equipments related to the reactor feedwater system were assigned and knowledge acquisition was performed. System design engineers, control comporment design engineers, pump design engineers, piping design engineers, and plant start-up test engineers were assigned as experts.

Assigned engineers extract knowledge in their specific domain from their experience and also from documents such as Design Drawing, Design Specification, Plant Operation Manual, Failure Mode Analysis Report and Plant Accident Case Summary Report.

2.Knowledge arrangement

Next, assigned engineers arranged the extracted knowledge and made AND/OR tree diagrams. AND/OR tree diagrams are diagrams which show relationship between abnormal finding such as "Reactor water level low scram occurd by a drop in flow of total feedwater", observation data such as "Both of the T/D RFP were being operated at the occurence of the plant trip and both of the T/D RFP speed decreased at the occurence of the plant trip", intermediate result such as "Feedwater flow main control module failed or RFP turbine failed", and inferred cause.

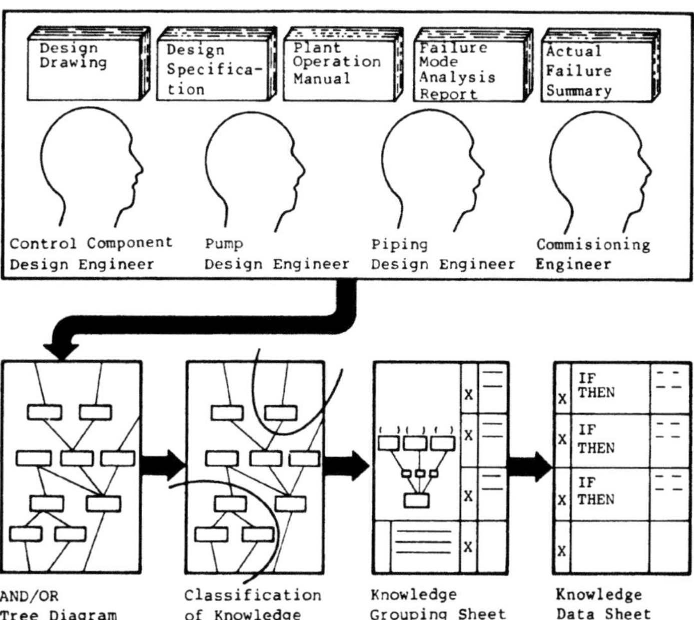

Fig. 6. Flow of building knowledge base

The purpose of making the tree diagram are (1)To arrange knowledge
logically, (2)Deletion of overlapped knowledge, (3)To check consisting of
knowledge, and (4)To unify technical terms. By making AND/OR tree
diagrams according to these purposes, knowledge extracted from experts in
different fields is integrated.

3.Classification of knowledge

After the above stage knowledge arranged by AND/OR tree diagrams are
classified into modules of knowledge related to each equipment and
function. For example "Knowledge for T/D RFP A failure", "Knowledge for
M/D RFP A start-up failure" and so forth. TDES2 can perform inference
based on knowledge modules for the purpose of speed-up of inference
processing.

4.Grouping of knowledge

At this stage, knowledge classified into knowledge modules is divided
into smaller groups. Divided knowledge is documented using a knowledge
grouping sheets. In a knowledge grouping sheet, an abnormal finding and
an intermediate result are used as key indexes, and all of the rules which
are fired from them are described.

5.Description of each rule

At this stage, description of knowledge (rules) documented in the
knowledge grouping sheets are written down on the knowledge data sheets
and implemented into the target computer system.

Following this procedure descrived above, knowledge (rules) extracted
from many experts in different fields are integrated, rearranged, and
documented in order to make speed-up inference processing, to implement
knowledge effectively into the computer and to perform verification test
of knowledge base effectively.

[VI] CONCLUSION

Through our development of the expert system for the nuclear power
plant feedwater system diagnosis, we have following results:

- Prospect for realizing the expert system for nuclear power plant
  feedwater system diagnosis is obtained.

- Diagnosis system man-machine interface being familiar with an
  operator on the plant is realized.

- Accumulation, systematization, computerization, and
  visualization of large knowledge database are realized.

- Basic technology to build large knowledge base from knowledge of
  engineers in many fields is established.

A NEW METHOD OF KNOWLEDGE PROCESSING FOR EQUIPMENT DIAGNOSIS OF NUCLEAR

POWER PLANTS

Makoto Fujii[1], Akira Fukumoto[1], Ichiro Tai[1]
and Toshihiko Morioka[2]

[1]Nuclear Engineering Laboratory, Toshiba Corp. 4-1
Ukishima-cho, Kawasaki-ku, Kawasaki, Kanagawa, Japan

[2]Nuclear Control & Electrical Engineering Dept.
Toshiba Corp. 8, Shinsugita, Isogo-ku, Yokohama
Kanagawa, Japan

1. INTRODUCTION

Regards to the reliable and efficient maintenance work of the nuclear power plants, it is expected to improve the efficiency by introducing an elaborate and intelligent method in the field of the surveillance, inspection and repair work of the plant equipments. Efforts are devoted to the development of the "Maintenance and Diagnosis Support System" to support the personnel who works in the maintenance fields and led to the development of the system based on the artificial intelligence.

For this system a new method of knowledge processing technique for equipment diagnosis has been devised. By applying this technique, flexible man-machine interface and highly powerful knowledge editing function have been realized.

In this paper, the outline of the new method of knowledge processing and the evaluation result of the system functions by applying to the "Reacter Instrumentation Maintenance and Diagnosis Support system" are described

2. SYSTEM FUNCTIONS

The "Maintenance and Diagnosis Support System" developed in this work has functions to assist operators or engineers to observe and evaluate the conditions of the plant equipments based on the experts' knowledge. This system supports the users to evaluate some phenomena observed around a certain equipment to investigate its status, and to identify the cause of the phenomena.

Fig.1 shows the conceptual figure of the system functions. This system can be categorized to a kind of consultation system. Users start up the system by introducing an input status of some anormaly occured in the plant on this system. The system can give guidance of additional specific tests and its procedure for users to proceed the diagnosis. Required plant data should be collected and inputted to the system by users according to the guidance.

This system consists of two subsystems: diagnosis subsystem and countermeasure decision subsystem. The diagnosis subsystem performs the functions to give guidance for diagnosis procedure and to infer the status of the objective equipments based on the obtained data. On the other hand, countermeasure decision subsystem performs the functions to identify the cause of the anormaly and finally to decide the countermeasures against it.

In addition, this system has function to determine what kind of operational action should be taken. If it appears that some operations are necessary at any time in the middle or the last stage of the diagnosis process, the system gives the pertinent guidance to users.

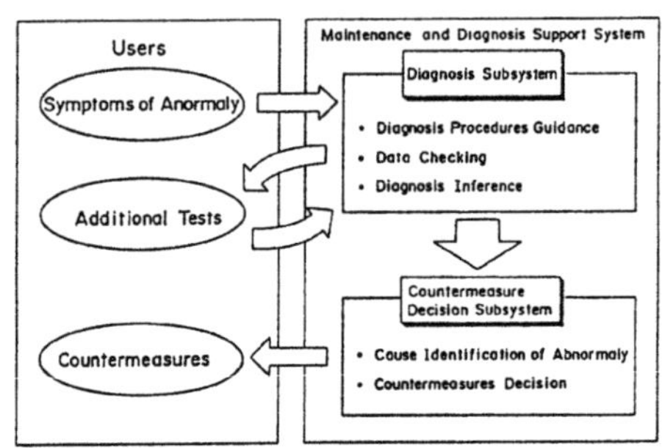

Fig.1 System Functions

3. KNOWLEDGE PROCESSING METHOD

To easily manage the knowledge base, the authors have developed a new method of knowledge processing for equipment diagnosis founded on the analysis of the diagnosis procedure and thinking process of experts.

In the thinking process of experts for fault diagnosis, possible causes are enumerated for some symptoms at first, and then related data are checked to explore their possibility of occurence. By repeating the data check, the number of enumerated causes decreases and the most probable cause is finally inquired. The concept of this diagnosis process is shown in fig.2. This concept of thinking process for fault diagnosis, "Cause Generation and Checking" concept, is realized by the combination with modularized and hierarchically structured knowledge base.

When experts perform equipments diagnosis in the actual plant, the heuristic know-hows have much significance for its efficiency. For example, the sequence of data checking execution is one of such know-hows for rapidity of diagnosis process. Another requirement for the efficiency is the correctness of diagnosis. The authors have realized the processing method to be able to satisfy these requirements

Details about the processing method are discussed in the following sections.

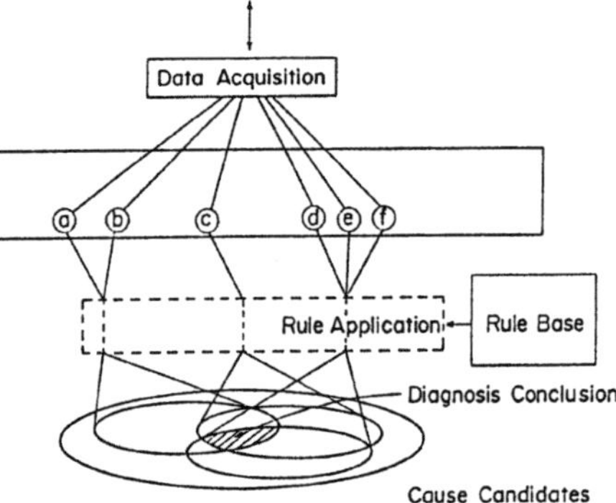

Fig.2 Concept of Fault Diagnosis

3-1 Knowledge Representation

The knowledge for equipment diagnosis acquired through the analysis of the diagnosis procedure and thinking process of the experts is represented by production rules. Fig.3 shows the knowledge representation format adopted in this system. Production rules are modularized for one piece of dialogue steps of diagnosis and each module consists of four different hierarchically structuted rules;

582

(1) module activation rule:
> to make judgements if the rule module should be activated at each stage of the diagnosis.

(2) man-machine interface rule:
> to give a diagnosis guidance message for users and to obtain corresponding data from users.

(3) data checking rule:
> to evaluate the input data based on the level check or the status check base relating to the current status.

(4) diagnosis processing rule:
> to proceed diagnosis by eliminating enumerated causes which are estimated not to occur in reality.

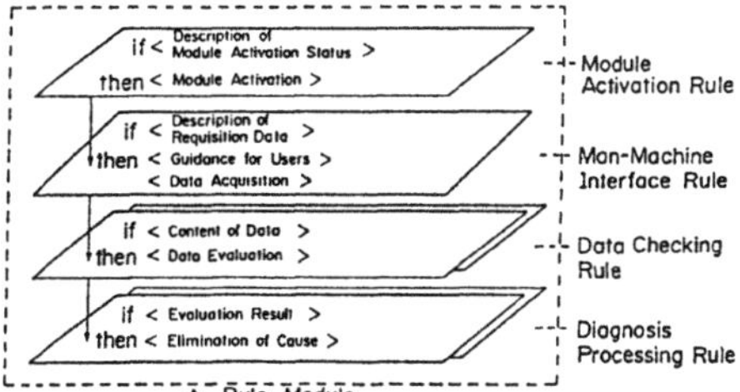

Fig.3 Knowledge Representation

These four kinds of rules are the basic rules for the inference. Other than these rules, following rules are provided to achieve the system functions.

(5) countemeasure decision rule:
> to judge countermeasures to be taken and to give the message on it.

(6) user supporting rule:
> to realize the user supporting functions (to be mentioned later).

3-2 Inference Function

Fig.4 shows the conceptual figure of the inference process of this knowledge processing method. Rule base is a storage place to preserve knowledge for maintenance and diagnosis, working area is the memory area of temporary data necessary for diagnosis process, and the man-machine interface is the functional part to manage dialogue between the system and users.

The inference for diagnosis is proceeded in the following process.

Step1: When users input to the system information of what abnormal phenomenon is found in the plant, all the possible cause candidates regards to this phenomenon are enumerated in the working area. At this time, each candidate is given the priority value which is decided on the basis of the knowledge (know-hows) derived from experts.

Step2: The "If-clause" of module activating rules are checked and are searched if they match to the present status of the cause candidates group stored in the working area.

Step3: The rule module which fits best to the contents of the working area is activated, and the hierarchical rules are fired sequentially.

Step4: The dialogue between users and the system is performed and the necessary data is acquired. The data obtained in this step is stored in the working area to be referred on users' demand.

Step5: The data got from users is evaluated and the status of the objective equipment is investigated.

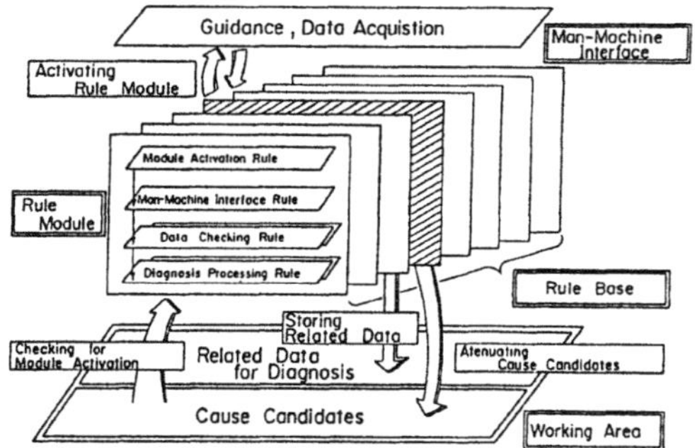

Fig.4 Concept of Inference Process

Step6:Among the cause candidates enumerated in the working area, some
       ones are eliminated if they proved not to take place in the plant
       by the results of data evaluation.

   By repeating these steps (Step2-Step6), diagnosis is proceeded and
the causes remained in the working area at the final stage are decided
to be the most probable ones.

4.USER SUPPORTING FUNCTIONS

   By applying this knowledge processing method, the following
remarkable user supporting functions are equipped to the system. These
functions are activated by users' request according to the situation of
diagnosis process.

(1)Flexible Inference
   This system can flexibly infer the procedure of the diagnosis. In
case the required information cannot be obtained from users during the
diagnosis process, a substitutive measurement parameter for it is newly
decided. This system can manage its inference according to the outer
status.

(2)Intermediate Processing Status Display Fuction
   All the possible causes can be displayed at any time in the
course of consultation by users' request. By this function, the
intermediate processing status of diagnosis can be displayed and the
users can understand the processing status clearly.

(3)Precise Explanation Function
   The precise explanation for a guidance can be shown on users'
demand. Users can use this information to find how required data can be
obtained in the plant, what can be proved with this data, and so on.

(4)Diagnosis Step Reverse Function
   This function is very important for the consultation system to
prepare for the users' mistake. Whenever users input a wrong data to
the system, the dialogue step can be reversed and the stage of
diagnosis can be restored to the former one, from which users should
resume the diagnosis again.

(5)Dialogue Process Display Function
   This system can show the logging of the dialogue performed between
the system and users. This function is very important for the
confirmation of the diagnosis and of the correctness of the results.

   With all these user supporting functions, the dialogue between the
system and users can be carried out smoothly and the diagnosis process
is proceeded exactly.

5.KNOWLEDGE EDITING FUNCTION

   Knowledge editing function is the key technology for the expert
system. If any new information which have never been experienced before
is found, it is necessary for the experience to be added into the
knowledge base of the expert system. Concerning to the knowledge
modification of this system, the contents of knowledge base can be
easily edited because of the modularity of the rule based knowledge and
the independence of each rule module.
   As mentioned before, modularized and hierarchically structured
production rules are applied for the knowledge representation of the

system, and each rule module corresponds to the consultation step of
the diagnosis. By this processing method, one-to-one relationship
between a diagnosis step and a rule module is established. From the
characteristics of the processing method, knowledge editing function
can be easily achieved.

Using this feature, "Graphical Rule Editor" can be realized, which
has functions to display diagnosis procedure with tree structures on
graphic CRT display and to edit the contents of knowledge base by
handling them on graphic information.

6.EVALUATION OF SYSTEM FUNCTIONS

In order to evaluate the system functions, the authors have
developed the "Reactor Instrumentation Maintenance and Diagnosis
Support System" based on this knowledge processing method.

6-1.Requirements

For BWR nuclear power plants,
reactor instrumentation is one of
the most important systems because
it performs the functions of monitoring of the reactor power in the
safety grade.

Simplified configuration of the
reactor instrumentation system is
shown in fig.5. The system consists
of three subsystems: source range
monitor (SRM), intermediate range
monitor (IRM) and local power range
monitor (LPRM). They are availed to
measure the neutron flux level in
the reactor core according to its
measurement range.

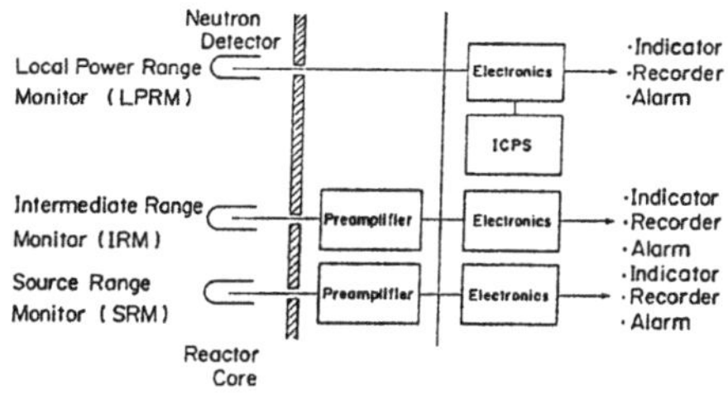

When some anormalies are found
in the reactor instrumentation sys-

Fig.5   Configuration of
Reactor Instrumentation

tem during its operation or maintenance, operators and engineers must
diagnose the system status to take appropriate counteractions as soon
as possible. However, the information on the reactor instrumentation
can be obtained only through its signal line so that plant engineers
must often execute additional special experiments and tests for system
diagnosis. Thus, in order to perform these tasks, much knowledge and
experience about its design, manufacturing and treatment of the system
enough to accomplish these works are necessary.

To realize a successful and practical system, know-hows for the
reactor instrumentation diagnosis are gathered energetically and
represented in the knowledge base based on the newly developed
knowledge representation method.

6-2. System Functions

Fig.6 shows the examples of abnormal status of reactor
instrumentation system treated in this application. They are all the
abnormal phenomena which can be found in the control room of the
nuclear power plant. Users start up the diagnosis process by inputting
the information about what alarm is found on the reactor
instrumentation panel. Then this consultation system starts and
investigates the cause and decides the countermeasures through dialogue
with users.

Fig.7 shows the outline of the "Reactor Instrumentation
Maintenance and Diagnosis System". This system executes its functions
through two display terminals as shown in the photograph: one is used
to perform the dialogue between system and users, and the other is used

to show the colored graphical information for suplementary use such as
the characteristic curve data of
detectors. By using such graphical
information   exchange   interface,
total system has attained its flexible   and   efficient   diagnosis
function.

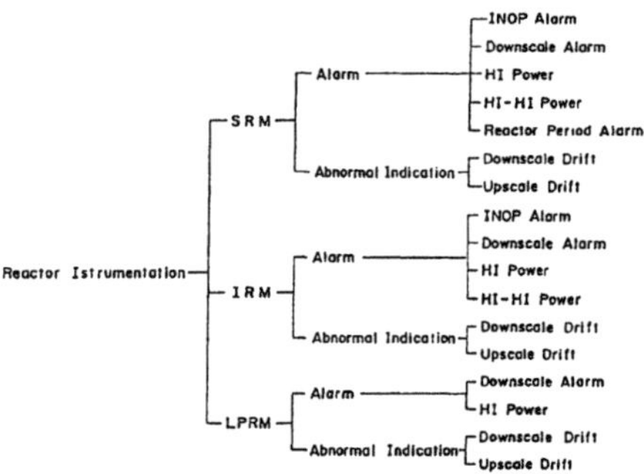

Fig.6   Abnormal Status of
Reactor Instrumentation

Fig.7   The outline of "Reactor Instrumentation
Maintenance and Diagnosis support System"

6-3 Function Evaluation

To evaluate system functions, simulation of diagnosis about
abnormal phenomena of reactor instrumentations has been carried out.
The simulation began with the above mentioned abnormal alarm detection.

From the simulation test, the following results have been
confirmed.

(1)By consulting with this system, incidents experienced in the reactor
instrumentation system can be successfully traced even by
non-experts. And guidance regarding countermeasures against anomaly
is accurate.

(2)The diagnosis process carried out with this system is almost same
as that executed by experts.

(3)User supporting functions equipped with this system are very
inevitable and useful for the smooth diagnosis procedures.

(4)With man-machine interface devices, such as a color graphic display
   and a japanese character display, dialogue between users and the
   system is easy to be understood, and the diagnosis can be carried
   out efficiently.

7.CONCLUSIONS

   In this work, the authors have completed the development of a new
knowledge processing method and representation for equipment diagnosis
of nuclear power plants and have eveluated its functions by applying to
the maintenance and diagnosis support system of the reactor
instrumentation.
   This knowledge processing method system is based on the "Cause
Generation and Checking" concept and has sufficient performance not
only in the diagnosis function but also in the man-machine interfacing
function. The maintenance and diagnosis support system based on this
method leads to the possibility for users to diagnose various phenomena
occured in an objective equipment to the considerable extent by
consulting with the system, even if they don't have enough knowledge.
With this system, it becomes easy for operators or plant engineers to
take immediate actions to counteract against the anormaly. So that, the
maintenability of the equipments is improved and MTTR (Mean Time To
Repair) is expected to be shorter.
   This new knowledge processing method is proved to be suited for
fault diagnosis of the equipments of nuclear power plants.

CHAPTER 10

OPERATION ANALYSIS AIDS

THEXSYST - AN INTELLIGENT MONITOR FOR REACTOR SAFETY CALCULATIONS

B. Burger[1] and F. Schmidt

IKE, University of Stuttgart, Pfaffenwaldring 31,
D-7000 Stuttgart 80, West Germany, Tel. 0711/685-2116

Abstract

A concept of a knowledge based system for the support of thermal hydraulic calculations *(THEXSYST)* is presented. Potential support for the thermal hydraulic calculations by knowledge based systems and prerequisites of a tool to control thermal hydraulic calculations are discussed first. *THEX* - the shell of a knowledge based system for monitoring time rows is described next. To demonstrate the capabilities of *THEX*, a system including *THEX*, *RELAP* and techniques available in RSYST was prototyped. *RELAP5* calculations were performed and analyzed to gather experience for the knowledge base and to verify existing rule sets. For building up the knowledge base we developed the Pascal oriented language *MONILA*. In the frame of this system results from a *RELAP5* calculation were examined to identify irregularities like undesired peaks and fluctuations.

1 Introduction

The use of thermal hydraulic codes like *RELAP* [1] or *TRAC* [2] requires experience in preparing calculational models and input data, in running and controlling the code, and in analyzing the calculational results. Usually these tasks require much more time than it needs to submit the codes to a computer and get the results back. These problems become even more severe if the codes are run on modern supercomputers like the *CRAY-2* where simulations faster than real time are possible. Furthermore if the codes are included into a plant analyzer which is intended to support accident management and therefore requires several runs and analyzes during the course of the actual transient, the solutions of the above mentioned problems determine the success of the whole simulation.

The methods of artificial intelligence and especially those of expert systems offer one way to give the necessary support to the analyst. It is the purpose of our work to explore the potential of these methods in supporting complex simulations as they may be performed with the help of thermal hydraulic codes.

2 Potential Support for Thermal Hydraulic Calculations by Knowledge Based Systems

The characteristics of knowledge based systems are described in [3] and [4]. In general, thermal hydraulic programs are very complex codes written in *FORTRAN*. Much effort was put into the validation of the codes. Therefore the installation of thermohydraulic programs on *LISP machines* and their integration in the AI-world is inproper. Instead one has to make AI-systems accessible to thermal hydraulic programs.

---

[1] On leave from RWTÜV Essen

The various steps of a thermo hydraulic calculation and their resulting problems are described in [3]. From there we take that AI-methods may support the following steps of thermal hydraulic calculations performed by already existing programs:

- Validating of input data,

- analysis of output data,

- control of the calculational flow.

In this paper we concentrate on intelligent control of calculational flow. To control the flow of a thermal hydraulic calculation we have to be able to interrupt the calculation, analyze the results, modify restart or input data, validate the new input data, and to continue the calculation. Thus the prerequisites of a tool for the control of thermal hydraulic calculations are:

1. identification of irregularities in the transients,

2. identification of the causes of the irregularities,

3. intervention at the calculational flow regarding to the causes of errors, that is to stop the calculations and repetition of the calculation with the modified input.

In the following we will present the concept of the expert system *THEXSYST* which will satisfy the demands, listed above.

3 THEXSYST

3.1 Concept

In Fig. 1 we show our basic concept of supporting thermal hydraulic programs with a knowledge based system. The concept utilizes essential features of the data and software management system *RSYST* [5,6,7]. *RSYST* supports the engineer in the areas manipulation, management and graphical representation of data, provides tools to connect data and user written applications and it offers a unified dialog system.

The *RSYST* dialog system supports the communication between user, *RSYST* and in *RSYST* integrated user written programs. Various input media are accessible. Input sequences generated with the system may be stored in the *RSYST* data base. Applications are programmed in *RSYST* in the form of functional modules. Each module has interfaces in the form of abstract data types (ADT). *RSYST* provides for standard data types ADT modules which perform various tasks. ADT modules to treat knowledge are developed in the frame of *THEXSYST*.

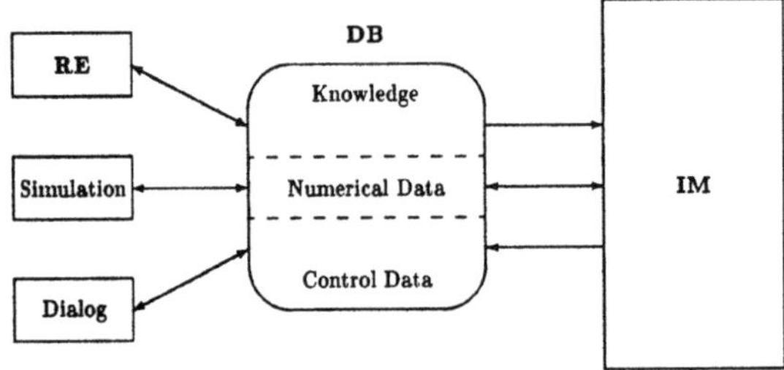

Figure 1. Basic concept of THEXSYST

Data are stored in the data base of *RSYST* as objects. Objects of well defined type (e.g. vector, matrix, picture or rule) are characterized by a data description part which allows to interpret structure and semantic of objects independent from the program which generates or uses the data. By this technique it becomes possible to exchange data between programs freely. Besides the standard data types *RSYST* allows to define general data types by using a class scheme. For such data types the description part is stored in the data base itself. To represent complex objects *RSYST* allows to connect data object into hierarchical tree structures. Each data object may belong to different trees. Relational operations on the tree are under development.

Using these features we now desire the following structure of the expert systems. The data base is clearly separated from the application modules. It contains common and program specific knowledge, principals of connecting common and program specific knowledge, simulation data, and flow of control data. Various completely independent programs (module sequences) operate on these data:

- a rule editor (RE) generates knowledge in the form of rules in a structure described in chapter 4,

- a simulation component produces numerical data like time series or matrices,

- a monitoring system controls the flow of calculation using standard input parameters.

In addition to this classical *RSYST* structure an inference machine (IM) is developed. It operates with both rules and numerical data and generates changes in the control data. This architecture enables an intelligent control of the system. The task of the data base as well defined interface makes a modular design of the system possible. The inclusion of the simulation program into this monitoring system requires no changes of the code, because already existing interfaces can be used with the aid of the data base . Therefore this system is problem independent.

The realisation of this concept for the subject *thermal hydraulic calculations* is shown in Fig. 2. This System *THEXSYST* (Thermal Hydraulic Expert System) consists of an expert shell *THEX*, the *RSYST*-data base, and the thermal hydraulic code *RELAP5*. The whole system is realized in the frame of *RSYST* as a module sequence.

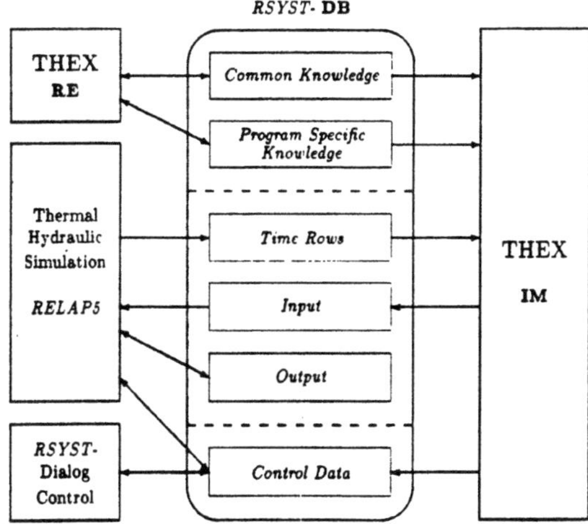

Figure 2. Realisation of *THEXSYST*

3.2 The Shell THEX

3.2.1 Architecture

Presently *THEX* consists of the two components: inference machine and rule editor. The rule editor supports the creation of the knowledge base and enables changes of rules and facts. It is planned to use the *language sensitive editor (LSEDIT)* [8]. The *LSEDIT* is capable to support the input of language elements of a user defined rule syntax and to check on syntactical correctness.

The inference machine can access the three parts of the data base. The inference machine checks the time rows of the simulation calculation regarding the common rules, allows interventions into the calculational flow through the control of the *RSYST*-module sequence with the aid of program specific rules, and enables a modification of the input of the simulation code regarding the results from the reasoning process.

3.2.2 Program Flow

The input of rules and goals ensues with the aid of the rule editor. This also enables changes of previously written rule sets. The rule editor conducts a syntactical checking. Subseqently the rules and goals will be written into the *RSYST*-database.

The thermal hydraulic code *RELAP5* will be started from the *RSYST*-environment. Restart dumps will be written on the *RSYST*-data base at determined points of time. This restart dumps can be used for the repetition of a part of a macro time step. Information about the restart points are stored in the control data part. Furthermore *RELAP5* produces plot data which also will be written in the *RSYST*-data base.

At the end of a macro time step the thermal hydraulic calculation will be interrupted and the monitoring system will be started. Rules, facts, goals and time rows will be called from the data base and pass a lexical analyzer *(scanner)* and syntactical analyzer *(parser)*. With the aid of a code generator a pseudo code will be produced which the interpreter is using.

The extraction of knowledge from time rows is called monitoring. The monitoring is performed with the aid of an internal window technique. A window is pushed along the macro time step interval (Fig. 3). The window includes all time rows which have to be investigated. The length of the window depends on the formulation of the problem of monitoring. The examination of the excess of limits can be performed with only one time point, whereas for the examination of peaks at least three points are required within the window.

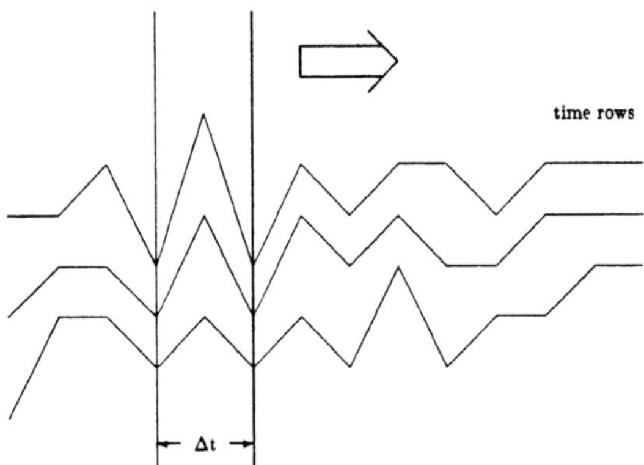

Figure 3: Monitoring of time rows

If no irregularities will be detected after working up of rules and goals the thermal hydraulic calculation will be continued with the aid of the *RSYST*-dialog control. In the other case necessary interventions into the program flow will be reasoned from program specific rules. The results is either a breaking off of program execution or a repetition of the calculation with modified input. As starting point the last restart dump which was written just before the irregularities have occurred will be used. Also, the update of the input deck will be deduced by means of program specific rules.

4 Present State of THEX

A first prototype of the thermohydraulic expert *THEX* has been finished at the end of 1986. This prototype was installed on a *VAX-780* and afterwards on a *CRAY-2*. This system constisting of an inference machine, which is coupled with the *RSYST*-data base, is capable to inform the user about critical values.

The inference machine is constructed by a compiler-interpreter system written in *Pascal*. The syntactical analysis by the parser requires an involvement of procedures for symbol identification corresponding to the priorities in logic and mathematics. Symbols with equal priorities will be worked up by the same procedure. Therefore a possibility for recursive programming is necessary to realize the parser. Such a possibility can be implemented in *FORTRAN* only with difficulties [12]. *Pascal* however is well suitable for recursive programming, because *Pascal* supports dynamic heap allocation. Typical AI-languages like *Prolog* or *LISP* fulfil these requirements too but lack of portability. In addition the interfaces between these AI-languages and *FORTRAN* are unsatisfactory. Therefore we felt that the use of these languages would not be adequate for our purposes.

For the programming of the inference machine we used techniques of a table controlled compiler [11]. The advantages of these techniques are good clearness in structure and easy capability of syntax extension. The inference machine consists of two parts, a compiler and an interpreter. The compiler includes scanner, parser, and code generator. The code generator produces a pseudo code which will be worked up from the interpreter with the *backward chaining method* [10]. For reasoning the *depth first searching method* [13] is used.

To be processed by *THEX*, knowledge must be presented in a problem oriented input language. Therefore we developed the input language *MONILA* (**moni**toring **la**nguage). This language is largely *Pascal* oriented. Due to its procedural character *MONILA* has some differences compared to other common input languages for expert systems. But for the kinds of problems described above, *MONILA* is advantageous in comparison to descriptive languages, because it is a very powerful tool for monitoring of time rows with the window techniques (3.2.2). A detailed description of the *MONILA* syntax is given in [9].

5 Monitoring of RELAP5 Calculations by THEXSYST

Our current work concentrates on building a knowledge base for monitoring *RELAP5* calculations. The knowledge is extracted from *RELAP5* calculations in which the nodalisation includes the whole primary loop and parts of a secondary loop of a PWR. The analyses refer to different cases of leaks (leak in hot leg, cold leg, vessel or rupture of surge line), different leak sizes (small and large leak) and various cases of high pressure injection (HPI in hot leg and cold leg). The objective of these calculations is to find rules for the discriminations of numerical fluctuations from physical ones.

An example of monitoring pressure and temperature transients in case of a small pressurizer leak (20 cm^2) can demonstrate the capabilities of *THEX*. In Figs. 4 and 5 we show the fluid temperatures and pressures of the primary and the secondary loop. After beginning of the leakage an equalization of the temperature of the primary and the secondary loop ensues (Fig. 4). As expected, a decrease of both temperatures follows nearly on the same line in consequence of high pressure injection and heat removal by the leak. The secondary loop temperature

subsequently persists at the saturation level whereas the temperature of the primary loop changes
to the subcooled range. Also after beginning of the shut down of the reactor with 100 K/h at
1800 sec, the primary loop is acting totally decoupled from the secondary loop. The extremely
strong fluctuations in the pressure of the primary loop (Fig. 5) give a possible explanation for
this behaviour. These fluctuations lead to an overestimation of heat removal from the primary
circuit by the leak. Corresponding to the nodalisation, the flow reversal at the leak causes an
additional subcooling of the primary cirquit. Fig. 6 give rules which were used to analyze these
curves. All irregularities in the curves in Figs. 4 and 5 could be detected by the rule based
system.

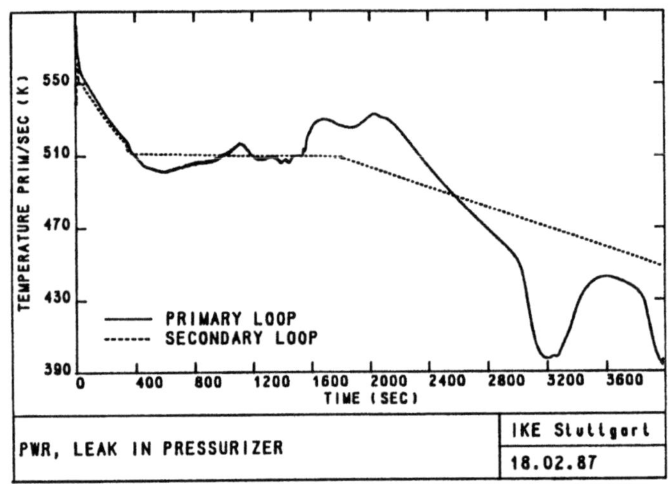

Figure 4. Temperatures of primary and secondary loop of a PWR in case of a pressurizer leak
calculated with RELAP5

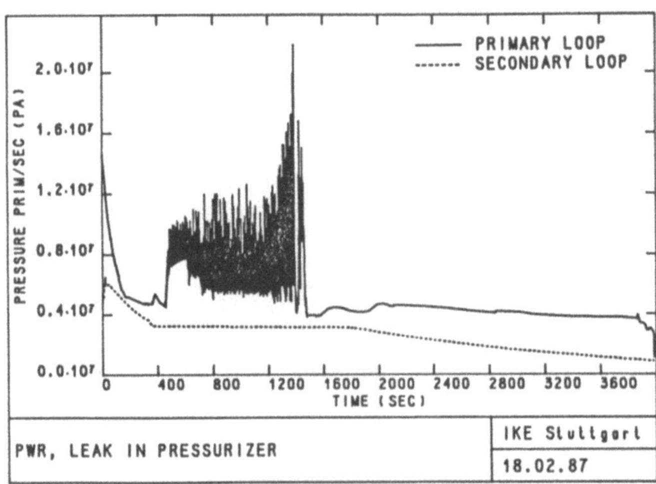

Figure 5. Pressures of primary and secondary loop of a PWR in case of a pressurizer leak
calculated with RELAP5

```
module steamgen;

        (*set of rules to detect irregularities in specific
           time rows of a RELAP5 calculation*)

type temp_prloop:       $TEMP_310010000,(*temperatures primary loop*)
                        $TEMP_320010000,
                        $TEMP_330010000,
type p_prloop:          $P_310010000,   (*pressures primary loop*)
                        $P_320010000,
                        $P_330010000,
type temp_sloop:        $TEMP_600010000,(*temperatures secondary loop*)
                        $TEMP_600020000,
type p_sloop:           $P_600010000,   (*pressures secondary loop*)
                        $P_600020000,

const:   prloopmax=3, (*number of checked volumes in primary loop*)
         sloopmax=3,  (*number of checked volumes in secondary loop*)
counter: i,j,k,l,mts;
boolean: fluctuation;
const:   klt=100, knt=0.25, klp=10.E5, knp=0.25;

procedure prpeak(kp1,kp2,kp3,t1,t2,t3,kl,kn,peak);
(*standart procedure to detect peaks*)
boolean: fk1_fx1,nivrel_1,k_2_extr,inc_1_2_large,peak;
begin
  fk1_fx1:=(kp2_kp1)/(t2_t1);
  nivrel_1:=abs((kp2-kp1)/(kp2-kp1)>kn;
  k_2_extr:=(kp2>kp1)and(kp2>kp3)or(kp2<kp1)and(kp2<kp3);
  inc_1_2_large:=abs(fk1_fx1)>kl;
  if inc_1_2_large and k_2_extr and nivrel_1 then certain peak;
end;

procedure prfluctuation(tvek,vek,mts,kl,kn,fluctuation);
(*standart procedure to detect fluctuations*)
boolean: fluctuation;
counter: mts,k;
rarray: tvek[length],vek[length];
larray: peak[5];
begin
  for k:=1 to 5 do
  begin
    t1:=tvek[mts+k-1]; t2:=tvek[mts+k]; t3:=tvek[mts+k+1];
    kp1:=vek[mts+k-1]; kp2:=vek[mts+k]; kp3:=vek[mts+k+1];
    call prpeak(kp1,kp2,kp3,t1,t2,t3,kl,kn,peak[k]);
  end;
  if(peak[1] or peak[2]) and (peak[3] or peak[4]) and (peak[5] or peak[6])
  then almost_certain fluctuation;
end;

(*main part*)
begin
  for mts:=1 to length do
  begin (*loop for window*)
    for j:=1 to prloopmax do
    begin (*primary loop*)
      call prfluctuation($TIME,temp_prloop.j,mts,klt,knt,fluctuation);
      if fluctuation then result('fluctuation',$TIME[mts],temp_prloop.j);
      call prfluctuation($TIME,p_prloop.j,mts,klp,knp,fluctuation);
      if fluctuation then result('fluctuation',$TIME[mts],p_prloop.j);
    end;
    for j:=1 to sloopmax do
    begin (*secondary loop*)
      call prfluctuation($TIME,temp_sloop.j,mts,klt,knt,fluctuation);
      if fluctuation then result('fluctuation',$TIME[mts],temp_sloop.j);
      call prfluctuation($TIME,p_sloop.j,mts,klp,knp,fluctuation);
      if fluctuation then result('fluctuation',$TIME[mts],p_sloop.j);
    end;
  end;
end.
```

Figure 6: Set of rules to detect irregularities in the transients of pressure and temperatures of primary and secondary loop of a PWR

6　Conclusion and Perspective

The aim of the present paper was to show that the use of expert systems can be a sensible alternative to the manual control of thermal hydraulic calculations. The concept of *THEXSYST* which consists of the coupling of thermal hydraulic code, an inference machine, and a rule editor through a data base, is very advantageous, because the particular components of the system are exchangeable. Therefore this concept could be used for other applications.

The next step of our work will be a coupling of *THEX, RELAP,* and the *RSYST*-data base by an *RSYST*-module sequence on the *CRAY-2.* It is planned to devide *THEXSYST* into two parallel processes. While *THEX* is monitoring the data of one macro time step, *RELAP* will execute the calculation of the next macro time step. Simultaneously to this work, we are running some test calculations with *RELAP5/MOD2* to gather more experience for our knowledge base.

References

[1] V.H. Ransom et al., "RELAP5/MOD2 Code Manual", Idaho National Engineering Laboratory, 1985.

[2] D.R. Liles et al., "TRAC-PF1/MOD1 an Advanced Best-Estimated Computer Program for Pressurized Water Reactor Thermal Hydraulic Analysis", Los Alamos National Laboratory, 1984.

[3] B. Burger, F. Schmidt, "Use of Rule Based Methods for the Control of Thermal Hydraulic Calculations", International Topical Meeting on Advances in Reactor Physics, Mathematics and Computation, Paris, 1987.

[4] D. Cain, F. Schmidt, "Expert Systems / Basic Principles and Possible Applications", International Topical Meeting on Advances in Reactor Physics, Mathematics and Computation, Paris, 1987.

[5] R. Rühle, "RSYST I-III, Experience and Further Development", Atomkernenergie, 26,3, 1975.

[6] I. Löbich, L. Ehnis, U. Lang, R. Rühle, "DFN-RSYST", Rechenzentrum Universität Stuttgart, Abt. Anwendungssysteme, 1986.

[7] R. Rühle, "RSYST an Integrated Modular System for Reactor and Shielding Calculations", Conf. on Mathematical Models and Computational Techniques for Analysis of Nuclear Systems, Ann Arbor, USAEC Conf-730414-P2, 1973

[8] Digital Equipment Corporation, "VAX Language-Sensitive Editor (LSE), Users Guide", Digital Equipment Corporation, Maynard, Massachusetts, AA-FY24A-TE, 1985.

[9] B. Burger, "THEXSYST, ein wissensbasiertes System zur Überwachung von Thermohydraulik-Rechnungen, 1. Zwischenbericht", RW-TÜV Essen/IKE Stuttgart, 1987.

[10] A. Barr, E.A. Feigenbaum, "Handbook of Artificial Intelligence Volume 1", William Kaufmann, Inc., Los Altos, California, 1981.

[11] N. Wirth, "Compilerbau", B.G. Teubner Stuttgart, 1981.

[12] S. Sachs, "REASON II, Ein System zur Generierung und Bewertung von Handlungsalternativen", Anlage zum Abschlußbericht DBF Anwendermodell (Be 156/6), 1986.

[13] N.J. Nilsson, "Principles of Artificial Intelligence", Spinger Verlag, New York, Heidelberg, Berlin, 1982.

DEVELOPMENT OF "ADRA" - AN ALARA DESIGN REVIEW ASSISTANT

John S. Brtis

Sargent & Lundy
Nuclear Safeguards & Licensing Division
55 E. Monroe, Chicago, IL   60603

INTRODUCTION

All nuclear power plant modifications must be designed in a way
that will keep the radiation dose to radiation workers and the public As
Low As is Reasonably Achievable (ALARA). (This is not only a good
design practice, it is required by the NRC in accord with Title 10 of
the Code of Federal Regulations, Part 20.) As a result, millions of
dollars are spent each year developing, analyzing, and justifying the
ALARA design of plant modifications. Providing the necessary design
reviews is a major challenge to the industry. In response, Sargent &
Lundy has developed ADRA (the ALARA Design Review Assistant), a
microcomputer-based system that will aid in developing, reviewing, and
documenting good ALARA designs. ADRA is now in an early operational
stage, undergoing testing at Sargent & Lundy. This paper reviews the
findings and results of the ADRA development effort.

THE CHALLENGES OF PERFORMING ALARA DESIGN REVIEWS

Performing ALARA design reviews is as much an art as a it is a
science. Characterizing and understanding a design modification and
evaluating the modification's impact on worker doses requires both broad
and detailed knowledge on the part of the reviewer. It requires a broad
understanding of the radiological impact of design modifications. The
reviewer must also be able to envision whether the modification will
have a detrimental impact on the radiological characteristics of the
area where it is being installed, and in doing so must address the
radiological impact of all aspects of the design during installation,
normal operation, failure conditions, maintenance, and decommissioning.
The organization responsible for performing ALARA design reviews faces
four major challenges:

- performing quality reviews,

- performing consistent reviews,

- documenting the reviews, and

- maintaining a staff of experts.

Each of these challenges is made more difficult by the subjective nature
of ALARA design reviews.

The challenge to a utility of providing adequate manpower and
assuring technical quality has two evident solutions: 1) the maintenance
of a pool of veteran reviewers or 2) a dependence on outside consultants
to perform ALARA reviews on call. There are difficulties in
implementing either of these solutions. But unless this challenge is
met, the possibility exists that ALARA reviews may be performed in a
cursory, pro forma, and potentially inadequate manner.

The challenge of providing consistent and well documented reviews
has most often been addressed through the adoption of standardized ALARA
review checklists and reporting formats. One problem with checklists is
that they are always a compromise between the need to be concise and the
need to be comprehensive.

An attempt to make a checklist both concise and comprehensive leads
to inevitable conflicts. To keep an ALARA checklist from becoming so
large that it is unusable, the questions about good ALARA design are
often phrased in broad terms. This leaves the checklist open to varied
interpretations (the very problem that the checklist was called on to
solve). Also, such a checklist is of little use to the inexperienced
reviewer. On the other hand, if the checklist is too comprehensive, the
user will be faced with a huge number of issues, many of which may have
nothing to do with the modification being reviewed. Since a checklist
cannot "decide" what is important and what is not, the user is required
to address all issues.

What is needed is a checklist that can adapt to both the user and
the modification. What is needed is a checklist that is both
comprehensive and concise. This is impossible for a static evaluation
methodology, such as a written checklist, because it cannot adapt to the
design under review or to the experience level of the user. But through
the application of artificial intelligence techniques, a computer-based
adaptive checklist is possible.

Artificial intelligence (AI) offers tools and techniques that are
well suited to addressing the challenges of providing ALARA design
reviews. Artificial intelligence based expert systems can adapt
dynamically to both the type of modification and the experience level of
the user. In this way, they can maximize the utilization of manpower;
minimize the effort on the part of the reviewer; and provide a concise,
comprehensive, and fully documented review.

ADRA: AN EXPERT SYSTEM FOR ALARA DESIGN REVIEWS

ADRA (ALARA Design Review Assistant) is the result of Sargent &
Lundy's effort to develop an adaptive ALARA design review aid through
the use of expert system techniques. Although ADRA is in an early
operational stage. The program structure, strategies, and I/O have been
developed and demonstrated to be workable. Its rule base is about 40%
of its final size. Those rules that do exist are fully operational.

ADRA performs three main functions during a review. First, ADRA
guides the user in properly characterizing the design. Second, ADRA
identifies the ALARA concerns that are germane to the design and
presents the user with a discussion of what a good design would be for
this ALARA concern. Finally, ADRA documents the review in the form of a
hard-copy report. The first two of these functions make use of AI
techniques.

ADRA'S ALARA REVIEW STRATEGY

     In an ALARA design review, the adequacy of design can be measured
by the amount of avoidable radiation dose that has been saved. The goal
is to reduce the radiation dose to the workers and the public to As Low
As is Reasonably Achievable. During an ALARA design review, it is the
reviewers job to identify what design improvements, if any, could
significantly reduce the doses associated with the modification, and to
determine the practicality of implementing any such improvements.

     ADRA helps the user identify what design improvements could result
in the greatest reduction of dose. It has two techniques for doing
this: one qualitative and the other quantitative.

     At the start of a review, ADRA prompts the user for information
that will characterize the design. From this information, ADRA scans
its rule base to identify pertinent ALARA concerns. The rule base
contains information on good ALARA design. ADRA presents the user with
this information, and asks him to rate the design against each ALARA
concern. This rating can be either qualitative or quantitative. The
qualitative rating results in a prioritized list of options for design
improvement. The quantitative rating additionally provides an estimate
of the dose that could be avoided by implementing each design
improvement.

ADRA'S SOFTWARE BASE

     ADRA was developed using an expert system shell, capable of forward
chaining, backward chaining, confidence factors, and fuzzy variables.
The shell also includes procedural programming, data base management,
and screen control (forms and menus).

ADRA'S INFERENCE ENGINE

     ADRA is largely driven by backward chaining. The goal of the
backward chaining is to complete the review. The rule selection
strategy is one of choosing the rules that have the fewest number of
unknowns for first consideration, unless an alternative priority
strategy has been established by the user. Both backward chaining and
the rule selection strategy reflect the desire to minimize user input.
In addition to pure backward chaining, ADRA takes advantage of mixed
forward chaining to propagate any newly identified facts through the
rule base. By doing this, new paths of minimum effort to completion are
uncovered.

     Neither certainty factor algebras nor fuzzy variables were found to
be useful in developing ADRA. ADRA's evaluation is based on avoidable
dose, a quantity that is linear and additive. This cannot be
represented by the common certainty factor algebras which are based on
statistics. There is no need to account for ambiguity of set
membership. As a result, fuzzy variables were not used.

ADRA'S RULE BASE STRUCTURE

     ADRA's knowledge base is made up of IF-THEN rules. These rules are
of two types: those that characterize the design and those that
identify ALARA concerns.

     The rules that categorize the design drive the interrogation of the
user in an effort to determine those modification design considerations
that could be germane to ALARA decision making.

The rules for identifying ALARA concerns make use of the characterization information. When such a rule fires, it informs the user of the ALARA concern, provides him with a discussion of what a good design should include, and requests that he perform either a quantitative or qualitative evaluation of the design vis-a-vis the ALARA concern.

ADRA's rule base is modular. A controlling rule base contains rules that can call on any of a group of sub-rule bases. With few exceptions, each sub-rule base corresponds to the ALARA design considerations that are unique to a particular type of equipment or design. This modular structure of ADRA's rule bases keeps the number of rules in a single rule base small, usually fewer than 25. This has several advantages. It improves execution speed, simplifies the debugging of the rule base, and will also simplify future rule base expansion. This modular approach follows the conceptual grouping of the knowledge source document.

ADRA's ALARA decision making is based on the recommendations found in Sargent & Lundy's "ALARA Design Manual." Sargent & Lundy has extensive experience in the design and review of modifications for both BWR and PWR nuclear power stations. The "ALARA Design Manual" represents more than 100 man-years of ALARA design experience.

USER INTERFACE

A major effort has been made to assure that ADRA is user friendly. It cannot be assumed that the user is adept in the use of computers or trained in typing. Clear and concise screen presentations and the use of menus and other techniques to minimize user input (and the resulting chance for typographical errors) are used throughout.

ADRA's question and answer interface has been designed to eliminate ambiguity and to minimize input errors. All questions are presented with a discussion of their meaning. Most questions are presented on the screen with "point and shoot" menus to facilitate user response. These questions can be answered by using the keyboard cursor to select the desired response. The user is required to type answers only when necessary.

Some of the ADRA's users may use it infrequently. Since cumulative-training-through repetition cannot be assumed, it is important that the learning curve be close to zero. For this reason, ADRA is being developed as a self-directing and self-explanatory program.

ADRA produces a hard-copy report that documents the information flow of the design review. The report is a stand-alone document, ready to be signed by the user and submitted to an independent reviewer. In this report, ADRA provides a detailed chronology of the user's entries and any deductions made by the rule base. It lists the aspects of the design that contribute to the potentially avoidable dose and makes recommendations of good design for each deficiency. In the case of a quantitative review, it lists estimates of the total (potentially) avoidable dose.

In more advanced, customized versions, ADRA will be able to further minimize user input by drawing directly on data bases that contain information on the plant and its systems. For example, a data base could contain the dose rates and contamination levels in the various plant areas.

A DESCRIPTION OF AN ADRA CONSULTATION

At the beginning of each consultation, ADRA asks the user to
characterize the situation. Specifically, the user is asked to enter
information about himself, the station where the modification will be
implemented, and the modification itself. Then ADRA enters its "expert
system" mode. It presents the user with a series of questions about the
modification: some that help to characterize the modification, others
that ask the user to rate the modification for the ALARA concerns
identified by ADRA.

ADRA is designed to select questions that allow the review to be
completed with the least amount of effort on the part of the reviewer.

ADRA automatically adapts its line of questioning to the
characteristics of the modification and the knowledge level of the
user. ADRA's "inference engine" works to achieve two important goals:
1) that no pertinent questions go unasked and 2) that no irrelevant
questions are asked.

When the review is complete, ADRA asks the user whether a printout
of the consultation and its findings is desired. ADRA will print a
report that includes a full chronology of the review (documenting all
the expert system's questions, all the user's answers, and all the
expert system's deductions), an ALARA rating of the proposed
modification, and a list of recommendations for improving the ALARA
design of the modification in order of potential dose impact. This
report is complete and ready for the user's signature.

ADRA SPECIFICATIONS

Hardware Requirements:

* IBM AT or Compatible,
* 640-K Memory,
* Color Graphics,
* Fixed Disk, and
* Printer.

Software Requirements:

* MS DOS 2.0 or Higher

Software Base:

* In its end user form, ADRA is delivered as a run-time module,
  ADRA was developed using a commercially available expert system
  shell that supports procedural programming, mixed forward and
  backward chaining, confidence factors, and fuzzy variables.

Rule base Size:

* Roughly 300 IF-THEN Rules.

EXPERT SYSTEM SHELL CHARACTERISTICS NEEDED TO DEVELOP ADRA

Functional characteristics that were vital to developing ADRA:

* Backward Chaining: with control over the rigor of search,
* Math Capabilities: for dose analysis,
* Menu Interface: to minimize user typing,

- Report Generating Tools: to create a report for sign-off,
- Procedural Programming: for activities outside the inference engine,
- Data Base Management: to accumulate and manipulate information obtained during the review,
- Good Screen Graphics: to maximize user friendliness, and
- Knowledge Base Security.

Functional characteristics that were useful to developing ADRA:

- Windowing,
- Color Capability,
- "How" and "Why,"
- Business Graphics: for graphic display of the result,
- Run-time Modules: for security,
- Graphic View of The Rule Base: for knowledge base development, and
- Forward Chaining.

USE WITHIN AN ORGANIZATION

ADRA is designed for use by an engineer or health physicist who has a good understanding of the plant systems and areas and a working understanding of radiation sources and their effects. In the event that the user's knowledge level is not sufficient to properly complete a review, ADRA will recommend that an ALARA expert be consulted and will document the review up to that point.

ADRA should not be considered to be a final decision-maker. It should be seen as an assistant to the user. It can help the user gather data, reduce the chance that concerns will be overlooked, help the user avoid reviewing impertinent items, and prepare documentation of the review. However, the user must remain the decision-maker and must review ADRA's recommendations.

VALIDATION OF ADRA

ALARA design reviews are non-safety-related. Moreover, it is unlikely that a subtle error in an ALARA design review could result in a significant radiation dose. Still, good engineering practice dictates that there be confidence in the results of the program.

ADRA's integrity can be shown in the same way as is done for the design review checklists that are now in use. The "knowledge base" can be scrutinized by experts. Then, at the time of use, the report generated by ADRA can be subjected to review and approval.

Since the method of validation for ADRA is much the same as that for checklists, and since checklists are more ambiguous and less transparent in their documentation, it is safe to say that validation is feasible and that ADRA can provide results that are superior to those of using checklists.

Control of the ADRA software is necessary to assure the integrity of the knowledge base. By using run-time modules and an encrypted rule base, ADRA is secured from modification by unauthorized personnel. The version of the knowledge base is displayed on the monitor during use and is included on printed and mass-storage output for accountability.

CONCLUSION

Expert systems offer tools and techniques that are well suited for
application to the problems associated with performing ALARA design
reviews. An expert system can aid both the expert and the novice in
performing and documenting high-quality, consistent and comprehensive
reviews. Sargent & Lundy's development of ADRA demonstrates the
feasibility of creating such an expert system.

DOCUMENTATION OF KNOWLEDGE IN THE DEVELOPMENT OF CLEO,

A REFUELING ASSISTANT FOR FFTF

Devin E. Smith

Department of Computer Science
SUNY at Stony Brook
Stony Brook, NY 11794-4400

INTRODUCTION

The developer of an expert based system is a translator
between the world of a specialist and the world of computer
models and algorithms. In the process of translating, the
developer must make explicit the implicit assumptions and
actions of the specialist. The process is akin in many ways
to that of developing a two way compiler which takes the
input of the specialist, parses it for semantic content, and
develops the action routines to be performed based on the
semantic contents. The process is complicated by the need
for it to be two way, for the specialist to be able to verify
that the translation is correct. The developer must act as
the inverse function to decompile the action routines into
the specialist's language. As the process continues, iteration after iteration, the development of the "compiler"
documents the knowledge of the specialist in a more regular
form.

A DESCRIPTION OF CLEO

CLEO is an expert system to assist in the planning of the
refueling process at the Fast Flux Test Facility (FFTF) in
Richland, WA. The reactor is composed of a hexagonal lattice
of subassemblies within a reactor vessel. The reactor vessel
is divided into three sectors, each with its own subassembly
in-vessel handling machine (IVHM), in-vessel storage (IVS),
and fuel transfer port (FTP). Additional subassembly storage
is provided external to the reactor vessel in the interim-
decay storage (IDS). The refueling process modeled by CLEO
focuses on the steps required to transform the old core configuration to the new configuration through the shuffling of
subassemblies within a core sector, between core sectors, and
between IDS and the core sectors.

THE DEVELOPMENT OF CLEO

The development of CLEO has gone through four iterations,
going from a concept, to a design for a production version.

Several independent versions of CLEO were developed in the
process. The four iterations were feasibility study, small
prototype, large prototype, and production version design.

The feasibility study focused on determining exactly
what was the process to be modeled. This step focused on the
expert as a system and the inputs and outputs from that system.
The local store of reference material and the rules used by
the expert were given only cursory examination. The intent
was more to determine whether the problem could be assisted by
a computer system and if it was worth doing. The result of
this step was a high level data flow diagram model of the
system and a concept of how the planning function could be
modeled by the computer.

The small prototype was developed to test the model for
the smallest possible refueling case involving only a few
types of subassembly moves. Discussions with the expert
focused on obtaining exact descriptions of the input and output used and the rules that pertained to this limited problem.
As other rules were discussed, they were recorded but not
implemented. This prototype later formed the kernel of the
lowest layer of CLEO. Three versions were written at this
stage: the first in Pascal, the second in LISP, and the third
in Pascal modeling a PROLOG solution.

The large prototype focused on the complete problem ex-
cluding I/O and a user-friendly interface. Since the first
refueling cycle after the start of the development of CLEO,
the expert had sought to standardize and refine his approach
to refueling. The rules to be modeled were in some respects
a moving target. Both a set of rules and a strategy for
applying the rules were identified and implemented in the
large prototype. Special conditions and exception handling
were noted and not implemented at this stage. Comparisons
between CLEO and expert planning identified rules used by the
expert outside the agreed set. The refinement process had
the form of a set of case studies identifying special conditions and the reasoning and rules used in such cases.

The design of the production version formalized the
discussions, notes, and documentation of the earlier phases.
The requirements for an acceptable production version were
documented. The rules for each type of subassembly move to
be included were detailed along with the conditions on when
such moves were permitted. Higher level rules about move
rules and overall order of positions, sectors, and facilities
were also detailed. The rules were detailed by categories
such as physical limitation, technical specifications, and
engineering judgement among others. The first two were to be
considered absolute rules while the latter was flexible but
required justification for violating. Even at this stage new
rules were uncovered as supporting documentation for the
rules were discussed with the expert. The major focus at
this stage was the user interface. Acceptance depended on
ease of use. A menu driven structure was developed to create
and edit input and output, control the overall planning
strategy, and step through the planning process. Though not
yet implemented, the I/O requirements will greatly increase
the size of the code and will form the top level of CLEO.

CONCLUSIONS

Documenting of the requirements for the production version
of CLEO was the last step in documenting the reasoning used in
planning refueling at FFTF. While the requirements for the
production version are not written in a tutorial form, they
could be used along with existing documentation as a reference
source for the refueling planning process. The requirements
are helpful in that they also specify what subjects are not
covered in the rules. Knowing the holes in the rules is almost as important as the rules themselves. The interaction
between the expert and the specialist was two way. As a
result of the discussions, other supporting documents for the
refueling planning process were updated. The planning process
as a whole was refined and standardized. The written requirements presents the data sources, rules, and strategies used
by the expert while the large prototype version of CLEO
presents an instance of an implementation of the data, rules,
and strategies. Together they document a model of the
refueling planning process at FFTF.

ATHENA AIDE - AN EXPERT SYSTEM FOR ATHENA CODE INPUT MODEL PREPARATION[a]

R. K. Fink, R. A. Callow, T. K Larson, and V. H. Ransom

Idaho National Engineering Laboratory
EG&G Idaho, Inc.
P. O. Box 1625
Idaho Falls, Idaho 83415

ABSTRACT

An expert system called the ATHENA AIDE that assists in the preparation of input models for the ATHENA thermal-hydraulics code has been developed by researchers at the Idaho National Engineering Laboratory. The ATHENA AIDE uses a menu driven graphics interface and rule-based and object-oriented programming techniques to assist users of the ATHENA code in performing the tasks involved in preparing the card image input files required to run ATHENA calculations. The ATHENA AIDE was developed and currently runs on single-user Xerox artificial intelligence workstations. Experience has shown that the intelligent modeling environment provided by the ATHENA AIDE expert system helps ease the modeling task by relieving the analyst of many mundane, repetitive, and error prone procedures involved in the construction of an input model. This reduces errors in the resulting models, helps promote standardized modeling practices, and allows models to be constructed more quickly than was previously possible.

INTRODUCTION

Preparation of models using a large engineering analysis code is a demanding task in terms of both time and expertise required. The cost of performing analyses is frequently dominated by the engineering labor and time associated with accumulating system data and developing the system models, as opposed to the computer costs to generate the final result. These time-consuming tasks are made easier with experience, and the quality of the model produced is strongly influenced by the expertise of the analyst. The use of expert systems technology is one possible way to reduce the effort and cost associated with computer-aided engineering analysis.

This paper provides a brief description of the ATHENA AIDE (Fink et al. 1987), an expert system developed over the past year and a half at the Idaho National Engineering Laboratory. The ATHENA AIDE assists in

---

a. This research is sponsored by the U.S. Department of Energy, Office of Basic Energy Science, under DOE Contract No. DE-AC07-76ID01570.

the preparation of models for the Advanced Thermal Hydraulic Energy Network Analyzer (ATHENA) thermal-hydraulics computer code (Chow et al. 1985).

The AIDE is a prototype developed as part of a research program investigating the application of expert systems technology to the use of large engineering analysis codes. Although developed as a prototype, the ATHENA AIDE is substantially complete and has been used in several working applications.

To provide some context for the description of the AIDE, we first briefly describe the ATHENA code itself. We then describe the capabilities of the ATHENA AIDE, and its usage in developing an ATHENA model.

ATHENA DESCRIPTION

The ATHENA computer code was developed at the Idaho National Engineering Laboratory (INEL) and is a direct descendant of the RELAP5 (Ransom et al. 1985) series of codes. ATHENA is general and can be used in the steady-state and transient simulation of a variety of nuclear and non-nuclear systems involving various working fluids and two-phase mixtures. The code currently can be run on CYBER and CRAY mainframe computers.

ATHENA is based on nonhomogeneous and nonequilibrium models for two-phase flow processes. A six equation model of the process is solved numerically using efficient finite difference methodologies to permit economical calculation of steady-state and transient simulations. An inventory of working fluids is available and includes water, lithium, freon, and helium.

Physical systems are modeled using ATHENA components. These components include generic items such as pipes, pumps, valves, heat structures, heat pipes, and control system components. Special process models are used to describe form losses in hydraulic networks, choked flow, branched flows, noncondensible gas effects, etc. These components and process models are arranged and utilized to form hydraulic and energy networks that represent the geometry and thermodynamic state of the physical system being modeled.

Development of an ATHENA model for a large complex system such as a commercial pressurized water reactor is a time-consuming task, requiring as much as a man-year of effort. The input file for such a model may contain upwards of 5000 lines of input records.

ATHENA AIDE DESCRIPTION

The ATHENA AIDE is an expert system that assists in the construction and manipulation of models for the ATHENA code. The system currently operates on Xerox 1100 series workstations. An object-oriented programming approach using LOOPS and Interlisp-D (Stefik and Bobrow 1986, Xerox 1985) is used for representation of the knowledge and information required to construct an ATHENA model. The ATHENA AIDE is user friendly and utilizes a menu driven, graphics interface with the user. Figure 1 illustrates a typical screen display from the ATHENA AIDE.

The goals of the ATHENA AIDE system are twofold: first, to provide an intelligent environment, capable of handling most of the mundane data management tasks for the analyst (to reduce the development time for the model); and second, to provide a platform for expert system assistance to the analyst (in the hope of improving the quality of the model).

612

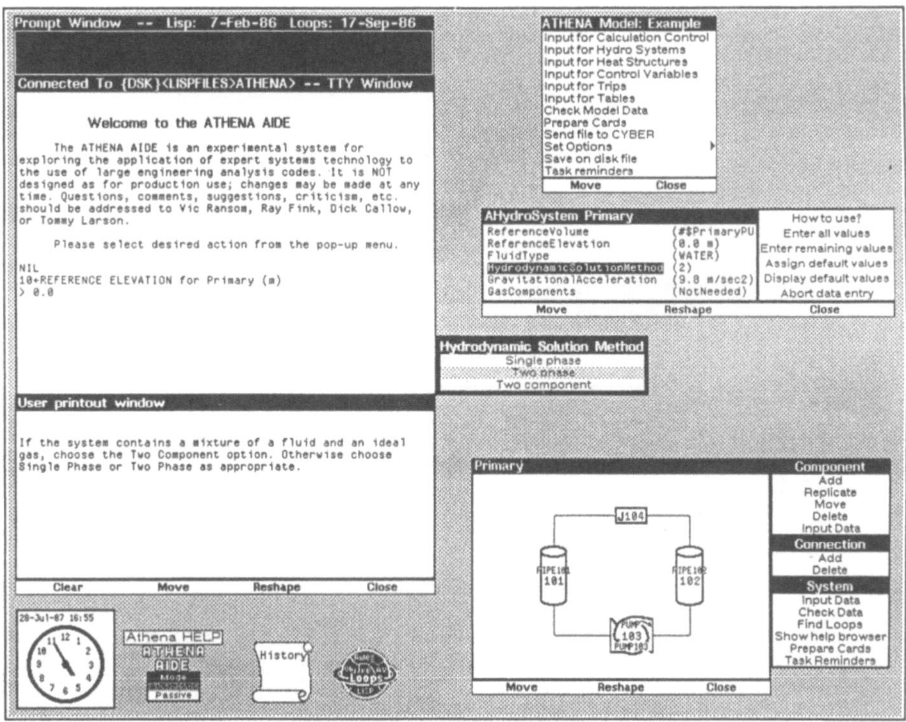

Fig. 1. ATHENA AIDE screen display.

An ATHENA model of the physical system of interest is constructed using the graphical interface features of the ATHENA AIDE. The analyst performs the initial abstraction of the physical system by arranging and connecting iconic representations of ATHENA components such as pipes, pumps, valves, etc. on the graphical interface. Figure 2 illustrates a typical hydrodynamic system layout on the ATHENA AIDE screen. If the analyst wishes to work with an existing model (developed outside of the AIDE), the ATHENA input file for the model can be read into the AIDE.

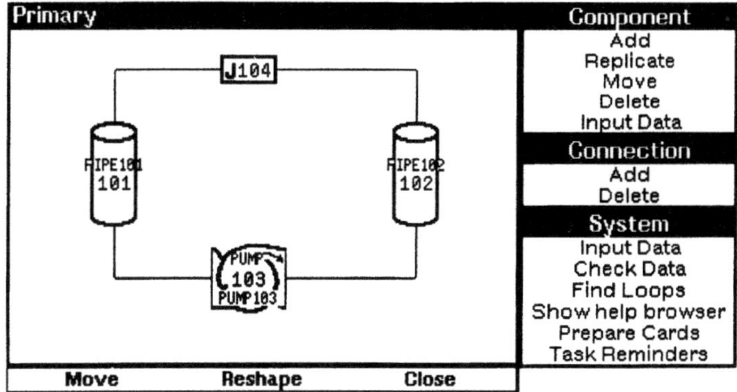

Fig. 2. Sample hydrodynamic system layout.

Rule-based consultations based on heuristic knowledge for specific
physical systems are available to assist in the component selection and
to prevent unacceptable modeling practices; Figure 3 shows the screen
display for one such advisory system. These expert system advisors are
invoked in response to help requests from the analyst, or they may be
consulted on a stand-alone basis. If appropriate, the result from the
expert system consultation may be incorporated directly into the model
under consideration (subject to approval by the user).

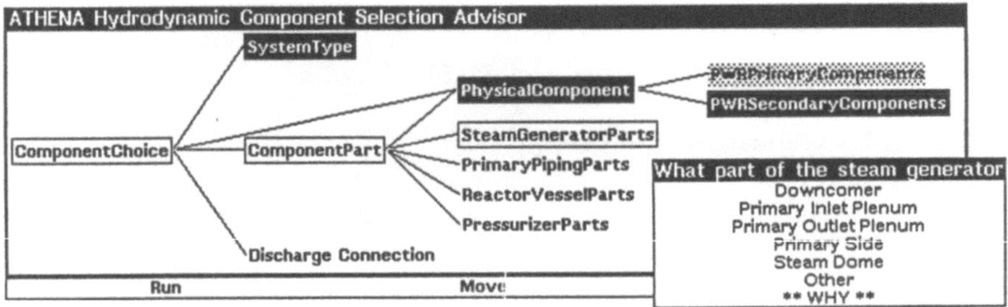

Fig. 3. Expert system advisor for component selection.

Parameter input for each component is handled through an edit window
as shown in Figure 4. Each component has knowledge of what parameters are
required, what default values may be suitable, and what requirements apply
to any parameter value. The expert system prompts the user for required
geometric and thermodynamic state information for each selected component.
Error checking is provided to prohibit incorrect input and help minimize
modeling errors. Help functions can be activated at nearly any stage of
the model construction process to provide detailed information on input
requirements. Engineering units conversions are handled automatically so
that the analyst can work in any system of units desired and change units
at will during the model construction process.

ASingleVolume PrimaryLowerHead		
Length	(27.2 in)	How to use?
FlowArea	(12.4 in2)	
InclinationAngle	(0.0 deg)	
WallRoughness	(.00005 m)	
HydraulicDiameter	(3.97355 in)	Enter all values
VCFWallFriction	(0)	
VCFEquilibCalc	(0)	
HasBoron?	(0)	
IVCThermoState	(3)	Enter remaining values
EquilibriumPress	(1.1 atm)	
EquilibriumQuality	(NotNeeded)	
EquilibriumTemp	(40.0 C)	
LiquidSpecificInternalEnergy	(NotNeeded)	Assign default values
LiquidTemp	(NotNeeded)	
NoncondensibleQuality	(NotNeeded)	
Pressure	(NotNeeded)	Display default values
SteamPartialPressureRatio	(NotNeeded)	
Temperature	(NotNeeded)	
VaporSpecificInternalEnergy	(NotNeeded)	
VaporTemp	(NotNeeded)	
VaporVoidFraction	(NotNeeded)	Abort data entry
InitBoronConc	(NotNeeded)	
Move	Reshape	Close

Figure 4. Sample input edit window.

The expert system automatically maintains a list of tasks remaining
in the construction process to remove such burden from the analyst. Each
component in the model has knowledge about its current state of completion, and what sequence of additional tasks is necessary to complete its
specification. This task list is used to provide an agenda for the analyst, allowing the analyst to resume threads of model development that
were interrupted by delays to obtain additional data, meetings, weekends,
or other impediments to engineering progress.

An audit capability in the ATHENA AIDE maintains the creation date
and dates on which components in a model are altered. For historical
documentation or quality control purposes, the analyst may also enter comments regarding decisions and assumptions made during the modeling process
into the audit record. This audit trail is maintained as an integral part
of the model, and can be sorted by date or by component as required.

The ATHENA AIDE has two modes of operation: automatic and passive.
In the automatic mode the system will interact closely with the user and
lead the user through the model construction process. In the passive
mode, the system interacts in a less obvious, nonintrusive manner; the
user retains control over choice of "what to do next." The passive mode
is intended for the more experienced analyst, who is more likely to have
a definite opinion on what order should be used to develop parts of the
model. The mode may be changed at any time via a menu selection.

After development of the model is completed, the ATHENA card image
input file is automatically generated in the required format. This file
can then be stored on hard disk, floppy disk, or routed via standard file
transfer packages to a mainframe computer.

BENEFITS OF THE ATHENA AIDE

Preparation of ATHENA models for complex physical systems is a difficult and time consuming task. Efficient, successful use of the code is
an art for which guidelines addressing use of the ATHENA code and proper
model construction practices have been developed from practical applications experience with the code. The ATHENA AIDE utilizes this heuristic
knowledge to promote adherence to standard modeling practices.

The extensive error checking, units conversions, and help functions
all assist in substantially reducing the time required to develop a correct and complete model. Although the AIDE has not been been placed in
production use, it has been used by novice ATHENA analysts in several
small-to-medium sized modeling applications. In these applications,
model construction time has been reduced by fifty to eighty percent of
that typically required.

LESSONS LEARNED

Development of the ATHENA AIDE prototype has led to several conclusions regarding the use of expert systems for engineering analysis
computer codes. Perhaps most important is the need for an intelligent
modeling environment. A stand-alone consultant expert system can only be
used in very limited contexts, and does not address the enormous data
management needs of the analyst. The intelligent environment can substantially ease the analyst's data management problems, as well as
provide the necessary expert system support.

Much of the value in an intelligent modeling environment is obtained
from relatively mundane support capabilities. Engineering units conver-
sion, help messages, and immediate error checking are examples of these
kinds of necessary capabilities.

The characteristics of the user interface affects the users' perceptions of a system. In the ATHENA AIDE, it was necessary to use graphical
symbols for the hydrodynamic system that were very similar to the diagrams typically used for analysis reports, in order to gain acceptance of
the system. Analysts who were unaccustomed to a large bandwidth interface (multiple windowing, pop-up menus, mouse-sensitive screen regions,
etc.) needed more time adjusting to the system than we had originally
expected. To solve this problem, we found it necessary to "tone down"
the interface: all text input constrained to one window, all pop-up
menus occurring at the same screen location (and strongly highlighted),
and explicit feedback on system status (e.g., working, waiting for input,
or idle). Although these efforts did little to improve the capabilities
of the system, they strongly influenced the acceptance of the system.

Finally, the choice of delivery system is important. Although the
Xerox AI workstation was an excellent platform for development, most analysts expressed a preference for having the system on more widely available
computers such as PC's or the time-shared mainframes. The analysts preferred having such a system in their office (rather than down the hall or
across the building) and were unlikely to purchase a special purpose workstation solely for modeling work.

CONCLUSIONS

The ATHENA AIDE modeling environment provides the capability for more
rapid construction of complex models, promotes the use of standard modeling practices in these models, helps to reduce errors in the input, and
provides a convenient historical documentation of model development and
evolution. Much of the utility of the system derives from support for
the analyst's data management problems, rather than the traditional expert
system advisory capabilities. The quality of the user interface was
found to have a substantial impact on the perceived utility of the system.

REFERENCES

Chow, H., et al. (1985) ATHENA Code Manual, EGG-RST-7034, EG&G Idaho,
Inc., Idaho Falls, ID, September 1985.

Fink, R. K., et al. (1987) "Expert Systems Approach to Modeling Systems
with the ATHENA Thermal-Hydraulics Code," Transactions of the Fourth
Symposium on Space Nuclear Power Systems, Albuquerque, NM, January 1987.

Ransom, V. H., et al. (1985) RELAP5/MOD2 Code Manual, NUREG/CR-4312,
EGG-2396, EG&G Idaho, Inc., Idaho Falls, ID, August 1985.

Stefik, M. and D. G. Bobrow (1986) "Object-Oriented Programming: Themes
and Variations," AI Magazine, VI(4):40-62.

Xerox Corporation (1985) Interlisp-D Reference Manual, October 1985.

A BWR FUEL CHANNEL TRACKING SYSTEM

R. S. Reynolds

Professor of Nuclear Engineering
Mechanical and Nuclear Engineering Department
Mississippi State University

A relational database management system with a query language, Reference 1, has been used to develop a Boiling Water Reactor (BWR) fuel channel tracking system on a microcomputer. The software system developed implements channel vendor and Nuclear Regulatory Commission recommendations for in-core channel movements between reactor operating cycles.

A BWR Fuel channel encloses the fuel bundle and is typically fabricated using Zircaloy-4. The channel serves three functions: (1) it provides a barrier to separate two parallel flow paths, one inside the fuel assembly and the other in the bypass region outside the fuel assembly and between channels; (2) it guides the control rod as it moves between fuel assemblies and provides a bearing surface for the blades; and (3) it provides rigidity for the fuel bundle. All of these functions are necessary in typical BWR core designs. Fuel channels are not part of typical Pressurized Water Reactor (PWR) core designs.

Axial fuel channel bowing may result when a channel is irradiated in a region of the core having a relatively steep gradient in the fast neutron flux. This condition exists primarily in the peripheral region of the core where the fast neutron flux level on one side of the channel is significantly greater than the fast neutron flux level on the opposite side of the channel over an extended period of time. Under these conditions a difference in channel side elongation occurs due to irradiation growth and results in bowing of the channel, i.e., the highly irradiated side of the channel becomes longer than the opposite, less irradiated, side. The degree of channel bowing is directly proportional to the difference in elongation of opposite channel sides. Further, the bowing will tend to increase with continued exposure to fast neutron flux gradients if the same channel sides remain facing the core center from cycle to cycle.

Irradiation-induced channel bowing, when oriented toward the control blade, will affect the size of the control blade gap between fuel assemblies. A reduction in the control blade gap size can increase friction between the control rod and the fuel channel and in extreme cases could affect the scram function of the control rod.

In order to reduce fuel channel bowing and the subsequent effects on control rod functions, the fuel channel vendor has developed a series of recommendations for channel placement in the core as a function of previous

channel location and exposure history. The General Electric recommendations are given in Reference 2, and the less stringent guidelines in the
utility Final Safety Analysis Report (FSAR) are listed in Reference 3.
These recommendations are summarized below in order to clarify the purpose
of the Channel Tracking System:

1. Records should be kept of channel location and exposure for each
    cycle of fuel operation.

2. When possible, channels should not reside in the outer row of the
    core for more than two operating cycles.

3. Channels that reside in the outer row for more than one cycle
    should be positioned in core locations that permit different
    channel sides to face the core edge on successive cycles.

4. Channels that reside in the outer row of the core for three or
    more cycles should not be shuffled inward.

5. At the beginning of each cycle, the combined sum of outer row
    residence times for any two channels in any control rod cell should
    not exceed four peripheral cycles.

6. Channels which reside in the outer row of the core during their
    initial cycle of operation should not be positioned with their
    highest fluence sides facing each other in the same control rod
    cell during the next cycle of operation.

These recommendations bound the guidelines given in the FSAR with the
exception that the FSAR requires that control rod drive friction testing
be performed for those core cells not conforming to the recommendations or
for those core cells containing fuel channels with a cumulative exposure
greater than 30,000 MWd/T. The FSAR allows fuel channel deflection measurements to be substituted for friction testing.

The Channel Tracking System is designed to provide the record-keeping
functions suggested in the first recommendation and to test proposed core
loadings against the other recommendations to determine if friction testing
or deflection measurements are required. The results of the analyses may
be used as a decision aide in determining whether it is more economical to
move the channel to a new acceptable location or to test it. It should be
noted that the system as developed is not an expert system in the usual
sense because there is no inferential reasoning applied to perform the
channel management task. The Channel Tracking System is driven by both data
and procedures while an expert system, by most definitions, is driven only
by data. The Channel Tracking System accepts data that describes a proposed core loading, which includes channel locations, and systematically
checks each channel serial number and proposed location against previous
locations for the same channel and the total accumulated channel exposure
in order to determine if any of the channel movement recommendations have
been violated. The process is based on well-defined algorithms and no
inference is used.

The system developed may be described as intelligent because it implements and executes a complex set of rules and alerts the operator if a
proposed core loading is unacceptable. Each channel movement that is not
allowable is flagged so that adjustments can be made. It should be noted
that the system does not make recommendations about where a channel should
or could be placed, but rather, once a proposed location is provided, will
indicate whether or not that location is acceptable based upon the past
history of the channel in question.

The system currently consists of eighteen data forms and twenty-two
reports. Six of the forms are defined by the <u>DataEase</u> software and the
remaining twelve are defined for use by the Channel Tracking System.
These forms and reports are supported by seven menus and twenty-two relationships between data forms. The Channel Tracking System can be accessed
in one of two ways: through the <u>DataEase</u> software menu or through a menu
developed specifically for the Channel Tracking System. The operator menu
for the system, the seven menus mentioned above, allows the user to perform
certain prescribed tasks and print reports without any special knowledge
of the system or software, as long as the databases are current.

When a new fuel channel is received, it is logged into the <u>New Channel
Data</u> form, which records channel serial number, receipt data, storage location, whether or not the channel is damaged, the data entry date, channel
relocation data, and a comment field. The channel serial number is very
important since it is used to define the basic relationships between all
primary data forms in the database. Other primary data forms include <u>NFV
Channel Data</u>, for the new fuel vault; <u>Core Channel Data</u>, for channels
currently located in the core; <u>SFP Channel Data</u>, for the spent fuel pool;
and <u>Channel Disp Data</u>, which logs the channel off-site for permanent disposal. Each of these forms, or databases, is related to the others through
a common field, the channel serial number, which appears in each. Consequently any record in a given database form has access to all the data in
a matching record in a different database form. The match, in these cases,
is based on the channel serial number, which is unique. An additional
record field, typically cycle number, is also defined as unique so that the
combination of these two fields will unequivocally identify a specific
record in a database form.

There are several supporting forms that provide data to the primary
forms in terms of look-up formulas to prevent data entry errors and to
speed the process of data entry by automatically providing field values.
The <u>Core Analysis Map</u> form provides fuel bundle ID, core location, control
rod index, and channel serial number, for the purpose of analyzing the proposed core map for channel movement violations. The <u>Core Description</u> form
provides core location information including control cell blade number,
fuel assembly orientation, control rod location, and outer row information.
The <u>Core/Bundles</u> form provides the isotopic and burn-up data necessary for
the analysis. The isotopic and burn-up data are maintained by the plant
process computer and can be retrieved electronically, automatically re-
formatted, and imported into a database form. A single report may then
be run from the main menu which will update the necessary forms in the database using the newly imported isotopic and burn-up data. Consequently,
very little keyboard activity is required to routinely maintain the system
during the operating cycle.

Updating the various forms in the Channel Tracking System is more time-
consuming than the routine monthly maintenance mentioned above. Prior to
and during a refueling outage most of the primary data forms will require
updates. The movements from channel storage to the new fuel vault, new
fuel vault to the core and the core to the spent fuel pool will have to be
entered. For a large BWR this means the addition of as many as 800 records
in the <u>Core Channel Data</u> form, and 200 to 300 additions to both the <u>NFV
Channel Data</u> form and the <u>SFP Channel Data</u> form. Less frequently, new data
will be required for the <u>New Channel Data</u> form as new channels are pur-
chased; and even less frequently, the channel disposal file, <u>Channel Disp
Data</u>, will require updating as spent fuel is shipped off site or damaged
channels are disposed of. Note that a single channel may appear in a given
form, e.g. the <u>Core Channel Data</u> form, three or four times depending on the
fuel management scheme. These are not duplicate records since either the
location or the cycle number will be different. This approach is necesary

since a channel will more than likely occupy more than one location in the
core during its life, and may well be moved about in the spent fuel pool or
in other locations described by the data forms. This means that the <u>Core
Channel Data</u> form may have twenty-four hundred records in it at any given
time. As channels are disposed of or reach the end of their useful life,
their corresponding records may be deleted from the core data file.

The Channel Tracking System reports are generally organized into three
categories: history or tracking reports, channel management reports, and
miscellaneous reports. The tracking reports will provide a history for the
channels in individual locations, such as the core, or track the channel
for its lifetime through all of the possible channel locations. The channel management reports implement the channel management rules by checking
the current condition of each channel in a proposed reload core and determining whether it satisfies the criteria or requires friction testing. The
miscellaneous reports perform several useful functions such as calculating
core average exposure (CAVEX), listing control rod drives that will exceed
a 30,000 MWd/T exposure during the next operating cycle, listing fuel bundle movements betwen cycles, etc. Additionally, these miscellaneous reports
provide the user with a guide to writing additional reports to query the
database. Each report is provided with a time and date stamp in order to
reduce confusion about the report results.

When a report is executed from within the system, the user is presented with an introductory data entry screen. Most of the data entry
screens do not actually request input information from the user, but rather
inform the user of the purpose of the report, indicate typical running
times, and allow for the opportunity to exit without executing the report.
When data is requested, it is usually in the form of a channel serial number, cycle number, or a date that corresponds to the most current isotopic
data.

The history reports results are grouped by channel location or by
channel serial number depending on the report. The reports are designed
to summarize the entire history of a given location, e.g., all the channels
that have passed through the new fuel vault; to provide the life history of
a given channel as it moved from receipt to permanent disposal; or to chron-
icle the life history of all channels received to date.

The channel management reports are designed to implement the vendor
and the FSAR channel shuffling recommendations previously mentioned. The
proposed channel locations in a reload core map, the <u>Core Analysis Map</u>
form, are compared to all other previous core locations for a given channel
to insure that the channel shuffling rules are not violated. An important
feature of the system is that it will accommodate the reuse of channels.
Specifically, a channel may have more than one fuel partner, does not have
to reside in the core for successive cycles, or may be stored in the spent
fuel pool, partnered or unpartnered with a fuel assembly, between irradiation periods in the core.

A typical report specification is shown below for the second recommendation mentioned above. The data entry screen informs the user that the
proposed reload core map will be sorted into control cells and that those
cells which contain two or more fuel channels that resided in the outer
row of the core for their first cycle of operation will be listed as requiring friction testing. The user is also asked to enter the cycle number
for the proposed reload. The actual report query is shown below.

```
Define "Hold" Number.
Define "Rule" Text 6.
Define "NumBundles" Number.
Assign Temp Hold :=0.
```

```
            Assign Temp Rule :=BLANK.

            For Core Analysis Map
                  With any Analysis<-Core Data named "FIND"
                        first loading = yes and
                  any FIND peripheral Location = yes and
                  any FIND cycle number < data entry cycle number;
            list records
                  Control Rod Index in groups.

            assign temp NumBundles := Hold + 1.

            list records
                  any Analysis<-Core Data number "BUNDLES"
                        Bundle Serial Number in order;
                  Core Location xx;
                  Core location yy;
                  any BUNDLES Channel Serial Number;
                  any BUNDLES Total Channel Exposure.

            If Temp NumBundles > =2 and
                  Control Rod Index  > 0 Then
                  Assign Temp Rule := "Yes".
            End

            list records

                  Temp Rule;
                  current date;
                  current time;
                  current page number.

            Assign Temp Hold := Temp NumBundles.
            Assign Temp Rule := BLANK.
```

The essence of this query is that only records (channels) in the
Core Analysis Map form (which describes the proposed reload core map) that
match records (channels) in the Core Channel Data form through the channel
serial number and spent their first cycle of operation in a peripheral location will be selected and listed in groups corresponding to the control
rod cell index. If the number of such channels in any one control cell is
greater than or equal to two, then a message indicating a rule violation
will be printed in the report. If no violation of the rule occurs, the
message area in the report is left blank. Additional information printed
in the report includes the control rod index number, bundle serial number,
proposed reload core location for the channel, channel serial number, and
total channel exposure in MWd/MTU. There is a header printed at the top
of each page of the report indicating the report title, current time and
date, and a footer at the bottom of the page indicating page number. The
report specification and format are shown below.

.header

```
                  Nuclear Fuels Channel Tracking System
                  GE Channel Shuffling Recommendations
                              Rule 1
                      __1____          ____2___

===============================================================
Control       Bundle    Reload      Channel     Total     Rule
  Rod         Serial     Core        Serial    Channel     One
 Blade        Number   Location      Number    Exposure  Violation
```

```
Number                        xx-yy              MWd/MTU
------------------------------------------------------------------
.footer
------------------------------------------------------------------
                              < _3>
.group header
.items
   _4_        _5_        6_-_7       _8___     __9__      __10___
.group trailer
- - - - - - - - - - - - - - - - - - - - - - - - - - - - - - - - -
.end
==================================================================
```

Field Descriptions

No.	Name	Type
1	Current time	Time
2	Current date	Date
3	Current page number Number Type: Integer	Number
4	Control Rod Index Number Type: Integer	Number
5	any Analysis<-Core data Bundle serial number	Text
6	Core Location XX Number Type: Integer	Number
7	Core Location YY Number Type: Integer	Number
8	any Bundles channel serial number	Text

9	any bundles total
	  channel exposure
	    Number Type: Integer
		Number

10	Temp Rule	Text

The other channel management reports process the data in a manner
similar to that described above. The other reports are somewhat more
complicated because channel reinsertion with the same or new fuel partners
must be accounted for in the analyses. These reports are not listed here
because of space limitations.

The effectiveness of the system is characterized by its ease of use and
time saved. All functions of the system are accessed through a menu-driven
user interface specifically designed for the Channel Tracking System so
that the actual rules and data manipulations are largely transparent to the
user. The time saved, compared to that required by previous hand methods,
can be substantial. A large BWR may have a core loading of 800 fuel bundles,
which means that all 800 channels must be validated against all of the channel
shuffling rules to insure that friction testing or deflection measurements
are not required before the next cycle start-up. Depending on the loading
pattern this validation can consume several man-days of effort. The
Channel Tracking System can perform the required analysis in less than an
hour with a current database. This capability, combined with the production of several other meaningful reports and a database query facility,
makes the Channel Tracking System a useful BWR utility tool.

References

1. DataEase, A Relational Database Management System, version 2.5 release
    2, DataEase International, Inc., 12 Cambridge Drive, Trumbull, CT
    06611.

2. General Electric, Nuclear Services Information Letter, Number 320,
    and Number 320 Supplement 1, Recommendations for the Mitigation of
    the Effects of Fuel Channel Bowing, December 1979 and February 1984.
3. Grand Gulf Nuclear Station, Updated Final Safety Analysis Report,
    Section 4.2.3.3.10, Channel Bowing.

FUEL INSERT SHUFFLER: A CASE STUDY OF EXPERT SYSTEM DEVELOPMENT

Joseph Naser[1], Robert Colley[1], John Gaiser[2],
Thomas Brookmire[3], and Steven Engle[4]

[1]Electric Power Research Institute
[2]Intellicorp
[3]Virginia Power Company
[4]Expert-Ease Systems, Inc.

Introduction

The potential for the use of expert systems in the nuclear power industry
is widely recognized. The benefits of such systems include consistency of
reasoning during off-normal situations when humans are under great stress,
the reduction of time required to perform certain functions and the
retention of human expertise in performing specialized functions. As the
potential benefits are more and more demonstrated and realized, the development of expert systems will become a necessary part of the nuclear
power industry. The development of the fuel insert shuffle expert system
will be used as a case study. In fact, it shows that the potential benefits are realizable. Currently, the development of the insert shuffle
plan requires three to four man-weeks of effort. Further modifications to
this plan are sometimes required due to either changes in the desired core
load pattern or damaged fuel assemblies or inserts. These changes generally require two to four man-days of effort and could be stressful if they
are critical path items on the outage schedule. Lastly, the personnel
that perform the plan development are often reassigned to other respon-
sibilities, taking with them their experience and developed intuitive
skills. It is anticipated that once the expert system is integrated into
the current process, the entire process will take no more than a day. The
expertise already captured will continue to be available and can be improved as more strategies are tried and the experience is encoded.

Knowledge Acquisition and Representation

A difficult part of developing an expert system is the knowledge acquisition and representation. In this case study, there were really two
experts, the utility expert who understood the processes involved with

insert shuffle planning, and an expert system tool expert who knew how to
represent the knowledge in the specific expert system shell. Both of
these experts were necessary, the first to adequately describe the problem
so that the expert system would be capable of solving the real problem and
the second to ensure that full advantage was made of the capabilities of
the expert system environment in solving the problem.

In the course of developing an expert system, the analysts themselves, of
necessity, learned a great deal about the problem. This learning occurred
naturally as the process under consideration was explicitly examined.

The first part of this learning process for the analysts was to understand
the domain of the process and the definition of the terms and languages
used in that domain. Each specialty develops a dialect which must be
learned in order to effectively communicate with the expert. In this case
study, the domain included learning everywhere fuel assemblies and inserts
could be placed, the various types of inserts, how the inserts were moved,
what tools were used, the physical constraints, the human factor considerations, and the parameters of the process, including those to be optimized. This also included studying some past insert shuffles to understand some of the techniques used.

Once this basic understanding was achieved, an interview with the expert
was performed. This interview helped to refine the information and data
that needed to be collected. It was also an opportunity to learn how the
process was currently being performed. As the current process was stepped
through; assumptions, constraints and heuristics were noted.

Description of the Physical Problem

A pressurized water reactor (PWR) nuclear power plant is fueled with a
large number of individual fuel assemblies. The plant being used as the
demonstration of this expert system development has 157 fuel assemblies.
Each of these fuel assemblies consists of a large number of fuel rods and
a fuel assembly insert. These inserts consist of either a control rod, a
burnable poison, a thimble plug or a secondary source. In addition, the
fuel assembly could have an open space instead of an insert. Typically,
these power plants operate for twelve to eighteen months without requiring
new fuel. At the end of this operation, called a cycle, approximately
one-third of the fuel assemblies are replaced by new fuel assemblies. At
this time most of the fuel inserts are moved from one assembly to another
by an overhead crane. This insert movement procedure is known as the
insert shuffle. The expert system described in this paper produces an
efficient crane movement pattern for this insert shuffle process.

The insert shuffle advisor expert system is applicable to a PWR where the
insert shuffle is performed in the spent fuel pool rather than inside the
reactor core. The reactor core is where the fuel being used to produce
power resides when the power plant is operating. The spent fuel pool is a
large storage area for fuel assemblies which have been previously used.
The plant being used for this expert system demonstration has 1044 fuel
assembly locations in the spent fuel pool. Other plants have considerably
more storage locations. A third fuel storage pool which plays a part in
the insert shuffle is the new fuel pool. This is the area where fuel
assemblies which have never been used in the reactor are stored. Some of
these new fuel assemblies already have the proper insert for the next fuel
cycle and won't participate in the insert shuffle process.

The insert shuffle process consists of four basic steps. The first is the
spent fuel pool preparation. The purpose of this step is to clear a

working space area in the spent fuel pool for the insert shuffle. This
step includes movement of the required new fuel and new inserts from the
new fuel pool. New fuel assemblies which already have the correct inserts
in them, and therefore do not participate in the insert shuffle, are
placed in the spent fuel pool near the fuel tunnel between the new and
spent fuel pools for convenience. The second step is the core off load.
This step takes all of the fuel assemblies and inserts currently in the
reactor core and puts them into the working space of the spent fuel
pool. The third step is the insert shuffle. In this step the overhead
crane takes the inserts from their current fuel assemblies and puts them
into the proper fuel assemblies for the next fuel cycle core configuration. The last step is the core on-load. This step takes the fuel
assemblies containing the correct inserts for the new core configuration
and puts them back into the reactor core for the next cycle of power plant
operation.

The core configuration for the next fuel cycle is designed by the electric
utility reactor physicists. The location of the fuel assemblies and the
proper insert for each assembly is determined to make efficient use of the
fuel during its life time in the power plant. This next cycle core design
is given to the reactor engineer who develops the insert shuffle plan.
This includes the planning of the preparation of the spent fuel pool and
the insert shuffle. This plan is developed by drawing heavily on past
experience with previous insert shuffles and typically requires three to
four man-weeks to develop. The plan is reviewed by other individuals and
verified by manually moving markers representing the fuel assemblies and
inserts.

Frequently the reactor physicist modifies the next cycle core configuration due to changed requirements. If the changes are large, the whole
insert shuffle plan has to be redeveloped. If they're small, then the
insert shuffle plan needs only to be modified which typically requires two
to four man-days. Sometimes the insert shuffle plan has to be modified
after the plant is shut down due to mechanically damaged fuel assemblies
or inserts. Again, this typically requires two to four man-days.

A full-scale insert shuffle advisor expert system prototype has been
developed to produce insert shuffle plans. This system has been developed
for a number of reasons. The first is to reduce the manpower requirements
needed to develop the plan. The second is to preserve the utility's
expertise as people move to new positions in the organization. The third
is to develop an automated way to verify the insert shuffle plan. The
fourth is to develop more efficient insert shuffle plans which can perform
the insert shuffle faster. The fifth is to allow required plan modifications to be made much more rapidly when the plant is shut down. The
latter two can have considerable financial consequences. A less efficient
plan or a longer time to modify the plan can, depending on other activities during the shut down, cause the power plant to be shut down for a
longer period of time. The cost for each additional day a power plant is
shut down is typically at least $500,000. This expert system will improve
personnel and power production productivity as well as preserving
expertise.

The Expert System Requirements, Constraints and Current Procedures and Practices

The expert system requirements are defined by the input information
available and the information required as output for the crane operator
and necessary recordkeeping. The system needs to be able to interface to
the existing engineering office and power plant procedures[1]. The

current interface to the function of developing an insert shuffle plan is
a paper system and the function is currently performed manually as stated
above.

The information input into the system consists of the current and final
core maps, the spent fuel pool map and new fuel pool map. These maps
consist of an array of vertical and horizontal cell locations, each of
which indicate the cell's content. The cell may be empty, or contain a
fuel assembly without an insert, or a fuel assembly with an insert. The
assembly and/or assembly with insert are each separately identified by an
alpha-numeric label. In the case of the final core map, the thimble plugs
and control rods are not uniquely defined. One objective of this expert
system is the unique identification of thimble plugs and control rods for
the final core.

The output from the system is the shuffle strategy consisting of the off-
load and after insert shuffle spent fuel pool maps and the sequence of
insert moves to achieve the after insert shuffle spent fuel pool map. The
sequence of insert moves and assembly moves are placed on forms called
fuel handling data sheets for use by plant personnel. The fuel handling
data sheets are then generated for the spent fuel pool preparation, the
core off-load, the insert shuffle, and the core on-load. Also produced
are the maps for the new and spent fuel pools and the core as they appear
prior to pool preparation, after pool preparation, after core off-load,
after the insert shuffle and after core on-load.

There are several constraints which must be considered when developing the
sequence of the insert shuffle. There are also physical constraints on
the placement of fuel assemblies and inserts. Some of these constraints
for this plant are as follows:

1. Due to the structure of the crane used to lift the inserts, it is
    quicker to move inserts in the trolley movement direction using a
    track hoist, than to move the entire crane bridge in the crane move-
    ment direction. There is also a human factor cost associated with the
    movement of the crane bridge. Aligning the bridge over the spent fuel
    pool is a tedious and mentally fatiguing task.

2. There is a large time penalty associated with the change-out of the
    tools used to grab the different inserts. This implies that to opti-
    mize the fuel insert shuffle time, it is necessary to minimize the
    number of insert tool change-outs.

3. There are three different tools used for the various insert types.
    There are separate tools for the thimble plugs and burnable poison
    inserts. The secondary sources and reactor control rod inserts use
    the same tool.

4. It is undesirable to move inserts outside of an assembly using the
    crane bridge in the crane movement direction as there is a possibility
    of damaging the insert.

5. The burnable poison tool cannot be used for inserts in cells next to
    the walls in the trolley movement direction.

6. There is a desire to minimize the residence time in the maneuvering
    reactor control rod positions called "D" bank positions.

    Finally, the new fuel is loaded into a certain area due to its access
    to both the reactor refueling channel and the new fuel pool to mini-
    mize movements between the new fuel pool and the reactor core.

Description of the Expert System

The insert shuffle planner expert system prototype has been developed
using the commercially available expert system development environment KEE
3.0[2]. A Xerox 1186 was the development machine used for this project.
Frame-based knowledge representation has been used extensively to represent objects in this system such as the old core, new core, spent and new
fuel pools, fuel assemblies, fuel assembly locations, inserts and insert
tools. Each object has a number of slot values for its attributes. Some
of these attributes are derived using active values. The object oriented
programming capabilities of KEE were used extensively. A combination of
rules and LISP functions are used to develop the insert shuffle plan.
Both have strengths and weaknesses, so an attempt was made to use the
capability which best fits the required need. Similarly when using the
rule system, both forward and backward chaining was used as required. The
KEE alternate worlds capability was also useful in certain portions of the
insert shuffle plan development. The graphical man-machine interfaces for
this expert system planner were created using both LISP functions and the
KEE graphics capabilities. The use of an expert system development environment allowed this prototype to be developed much more rapidly than
would have otherwise been possible.

The relationships between the various objects is represented as a graph of
the insert shuffle knowledge base as shown in Figure 1. On the figure,
there are objects such as the old core, new core, spent fuel pool, new
fuel pool, etc. In the diagram the relationship represented by the solid
lines are called parent-child links which indicate a superclass to subclass relationship. The broken lines are called member links and indicate
members of a class or set of objects. The attributes that are inherited
by the subclass from the superclass are called member slots, those that
characterize the individual object are called own slots.

As an example, consider the object cells. A cell is one of the locations
in a fuel pool which can hold one assembly and one insert, an assembly
alone or be empty. Cells have the attributes or slots: a cell identifier, the pool identifier, the x-coordinate and y-coordinate locations,
the boolean fact as to whether or not it is a "D" bank control rod cell,
the assembly identifier, the insert identifier and the class of the insert. These slots are inherited by the subclass of cell.

Beside representing physical objects, the graph in Figure 1 also shows
hierarchical decomposition and refinement of the rules. This can be seen
by following the rule superclass insert.shuffle.rules down to the specific
rule find.largest.segment.class rule. Rules like objects inherit from the
rule class. For example, the external form of a rule is broken into premises, assertions and actions. The organization of the rule classes into
hierarchies can be used to limit the size of the search space for a given
part of the problem that is being used in reasoning.

Insert Shuffle Planner Development

The overall objective of the insert shuffle planner is to place the
assemblies and inserts in the spent fuel pool so that the overall time to
perform the insert shuffle is minimized. The amount of time required is
directly related to the number of tool changes required and the number and
direction of the crane movements.

The placement of one assembly/insert pair next to another assembly/insert
pair is determined by comparing the new core assembly and insert configuration to the old core, spent fuel pool and new fuel pool assembly and
insert configurations. Each assembly in the new core has an insert that

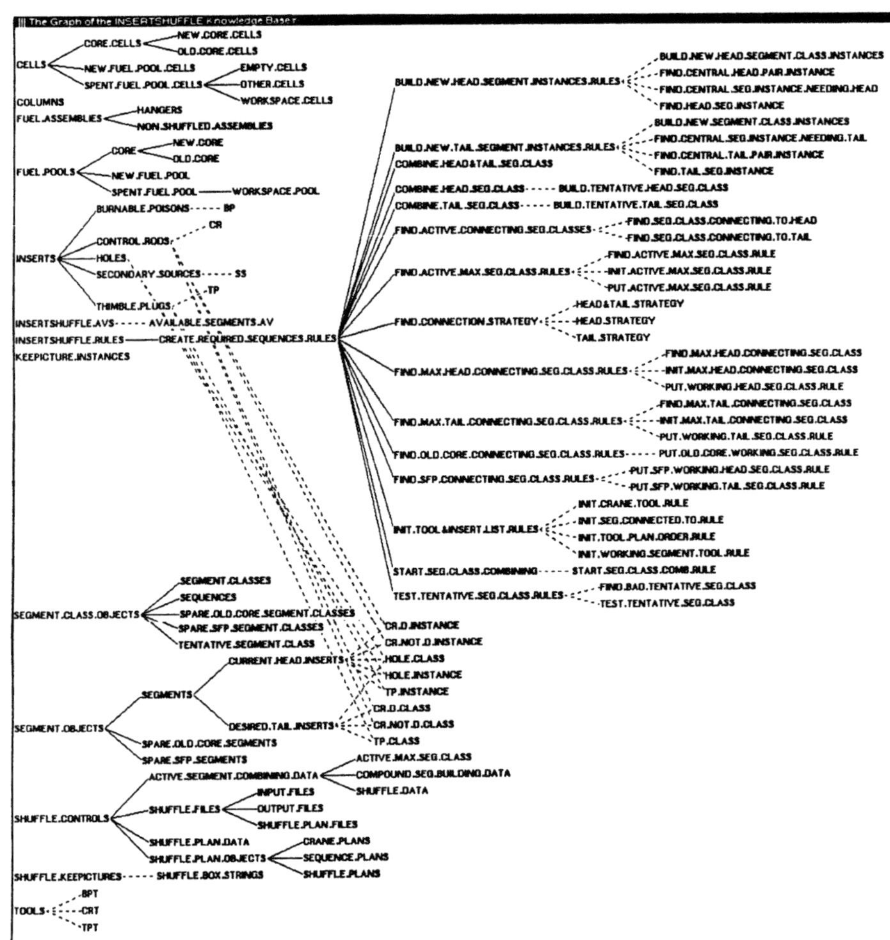

Figure 1. Graphical Display of the Insert
Shuffle Advisor Knowledge Base.

must come from an assembly in the old core or from the spent fuel or new fuel pools. For each of these new core assemblies it is first determined what insert it "has" and what insert it "needs".

The next step is to develop an ordered list matching the "has/needs" for all the assemblies needed to make up the new core. Due to the interchangeability of certain of the inserts, i.e., the thimble plugs and control rods, this is not one long continuous list but rather short segments with an interchangeable component on the end. These segments are then combined to form sequences which terminate with an insert which is not going into the new core. A large part of this segment building process is developing segments which all can use the same minimized number of tools in the same order. There are many possible heuristics for building sequences. There is currently one set of heuristics in the expert system. There are plans for several more sets of heuristics (or strategies) to be included in the expert system.

The sequences are then fit into the user defined work space available in the spent fuel pool. As follow-on work, it is desired to have the expert system specify the required shape and size of the work space to achieve a minimal fuel insert shuffle time.

From the placement of the assembly/insert pairs in the spent fuel pool and the determined number and order of tools used, the insert shuffle move plan is then developed. This plan consists of a step-by-step procedure to be used by the crane operators. This is an ordered list which specifies the location and identifiers of each assembly and insert to be moved with the appropriate tool and the location and identifier of the assembly into which it is to be placed.

This plan can then be checked visually by the engineering staff and inspected visually by the human operator, to verify that there are no physical or logical errors. These include attempting to place an insert into an assembly which already has one or a set of crane movements which are nonsensical.

A number of man-machine interface capabilities are included in this insert shuffle planner to make it more easily used and understood. A graphical display of the current core, final core, new and spent fuel pool configurations has been developed. Because these are too large to be conveniently displayed in their entirety on the screen, a scrolling capability is used. This display is shown in Figure 2. In addition, certain areas can be "zoomed" in for more convenient viewing. Each fuel assembly location in these displays gives the identification number of the fuel assembly and insert in that location. A spent fuel pool display shows the step-by-step crane movements during the insert shuffle procedure to allow a convenient visual method for verifying the insert shuffle plan. A set of screen "push buttons" are used to allow the user to initiate each step of the insert shuffle plan determination. Additional displays are used to show intermediate results of the plan development process.

The input and output has been designed to fit in with the currently used data bases and output forms. The configurations of the current core, new fuel pool, spent fuel pool and final core are downloaded from the mainframe to a PC. This information is transmitted over an RS232 port to the LISP machine and converted into the format required by the expert system. The final configurations and insert shuffle plan are sent back to the mainframe by the reverse of this process. Direct links between the mainframe and LISP machine would eliminate the need for the intermediate steps that were used in the testing of the insert shuffle planner. The actual crane movements developed by the insert shuffle planner are also

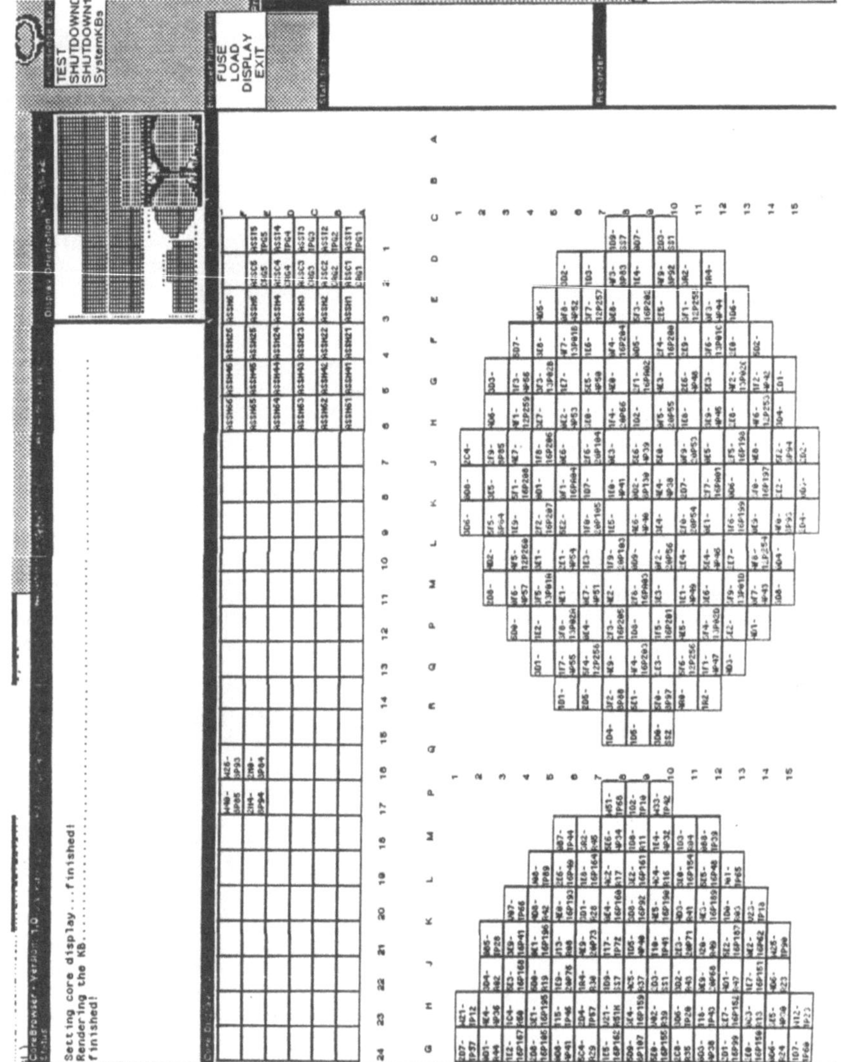

Figure 2. Scrollable Graphical Display of the Current Core, Final Core, Spent Fuel Pool and New Fuel Pool Configurations. (Icon of Total Display in Upper Right Hand Corner.)

printed out in a format which is the same as the crane operator has been
using in the past. This makes the change in the plan development methodology transparent to the crane operator.

<u>References</u>

1.   <u>Nuclear Fuel Operation Subsection Task Instruction No. 95 Development
     of Fuel Handling Data Sheets</u>, Revision 0, December 1985, Virginia
     Electric Power Company, Richmond, VA.

2.   <u>Intellicorp KEE Software Development System User's Manual</u>, KEE Version
     3.0, July 1986, Intellicorp, Mountain View, CA.

AN EXPERT SYSTEM FOR OPERATIONAL CONTROL OF PWR CORES

Orpet J. M. Peixoto

Stanford University
Stanford, California

Bjorn Frogner and Joseph M. Holzer

Expert-EASE Systems, Inc.
Belmont, California

Introduction

Plant operation is typically based upon heuristic rules[1] that have been developed through experience and that are easily understood by operators and engineers on the site. The use of optimal control methods for load-follow and power distribution control in the presence of xenon have been investigated by many researchers. Their work has helped gain substantial insight into the problem; however, it has not been employed in on-site operation. The main purpose of this paper is to present an expert system approach that combines the solutions of optimal control techniques and heuristic rules that might ultimately be used to assist the operators on-site. The details behind this paper are described by Peixoto[2].

Structure of the Reactor Control Expert System

An expert system has been developed whose objective is to determine the control parameters so as to satisfy the required power and the Constant Axial Offset Control (CAOC) limits during plant operation. This expert system employs both optimal and heuristic solution techniques. The general flow chart of the system is presented in Figure 1. Note that the selection of the solution technique is under user control. The system accommodates either data file or interactive data input.

Dynamic Model

A highly simplified variation of the one-group diffusion theory is used to represent the reactor core dynamics, and a two-point reactor model (upper and lower core) is used to represent the spatial effects[4]. Sensitivity coefficients derived from a much more accurate model was used to account for reactivity effects. This dynamic model has been implemented in the expert system to obtain both heuristic and optimal solutions as well as in the simulation code to test the approach. Full-length control rods and boron concentration are the control variables. The moderator temperature can optionally be used as a third control variable.

Optimal Control Solutions

The optimal control solution can be summarized as follows. Given the dynamic equations of a system,

$$dx/dt = f[x(t),u(t)] \quad ,$$

where x(t) is a vector with n components representing the state of the system at time t, u(t) is a vector of m components representing the control input to the system at time t and f is a vector function of the state x(t) and control u(t). We wish to find the optimal control that minimizes the following performance index[4]:

$$J(x,u) \; = \; G[x(T)] \; + \int_{t=0}^{t=T} L[x(t),u(t)] \cdot dt \quad ,$$

where T is the final time, and G and L are cost functions.

The control variables are subject to the following constraints: limited rate of boron dilution, limited rate of boration, and limited insertion and withdrawal of rod bank speed.

The solution to the iteration process, when running the optimal code, is the conjugate gradient method applied with the clipping-off technique[6].

Expert System

The knowledge representation is based on production rules. A backward chaining scheme is used to evaluate these rules. The objective of the expert system is to determine the values of the boron concentration, rod insertion and coolant temperature, so that the plant will meet the required power during the next time step without violating the CAOC (Constant Axial Offset Control) operating conditions.

Heuristics to accommodate the following strategies have been implemented: center axial offset control, minimum boron variation, and spinning reserve strategies. The basis for the heuristic planning strategies are presented by Storrick[3].

The expert system is designed to perform the multi-goal inferencing internally, while the algorithmic computations are performed externally. This is exclusively determined by the performance characteristic of the system. Since the expert system that is used was not designed to solve problems with extensive numerical calculations (such as the optimal control problem), it is necessary to leave the expert system environment while maintaining the current status of the solution. This enables us to calculate the optimal solution in a traditional programming language and then return to the expert system environment with the solution (see Figure 1). To execute this, a top level executive procedure is needed to coordinate the different applications and environments in the computer.

Results

For the load-follow condition, normal daily load-follow scenarios (a cycle consisting of 12 hours full-power, 3 hours going down, 6 hours half-power and 3 hours going up) for the three life cycle conditions and for various operation planning strategies were simulated. All the cases commence from full power, equilibrium conditions and an initial rod insertion of 1 node[7]. The initial boron concentration is dependent upon the cycle life (800 ppm at the beginning of cycle, 500 ppm at the middle, 300 ppm at the end of cycle). All the simulations were run with data for two large US PWRs. All the results discussed in this paper were run on a Macintosh Apple computer.

Full-length control rods with a maximum insertion for a rod bank being 10 nodes are used. Some rod positions indicated in the results have values beyond 10 nodes, which should be interpreted as if the first rod bank (called bank D) has been inserted to 10 nodes and the second (called bank C) has been inserted the indicated number of nodes minus 10.

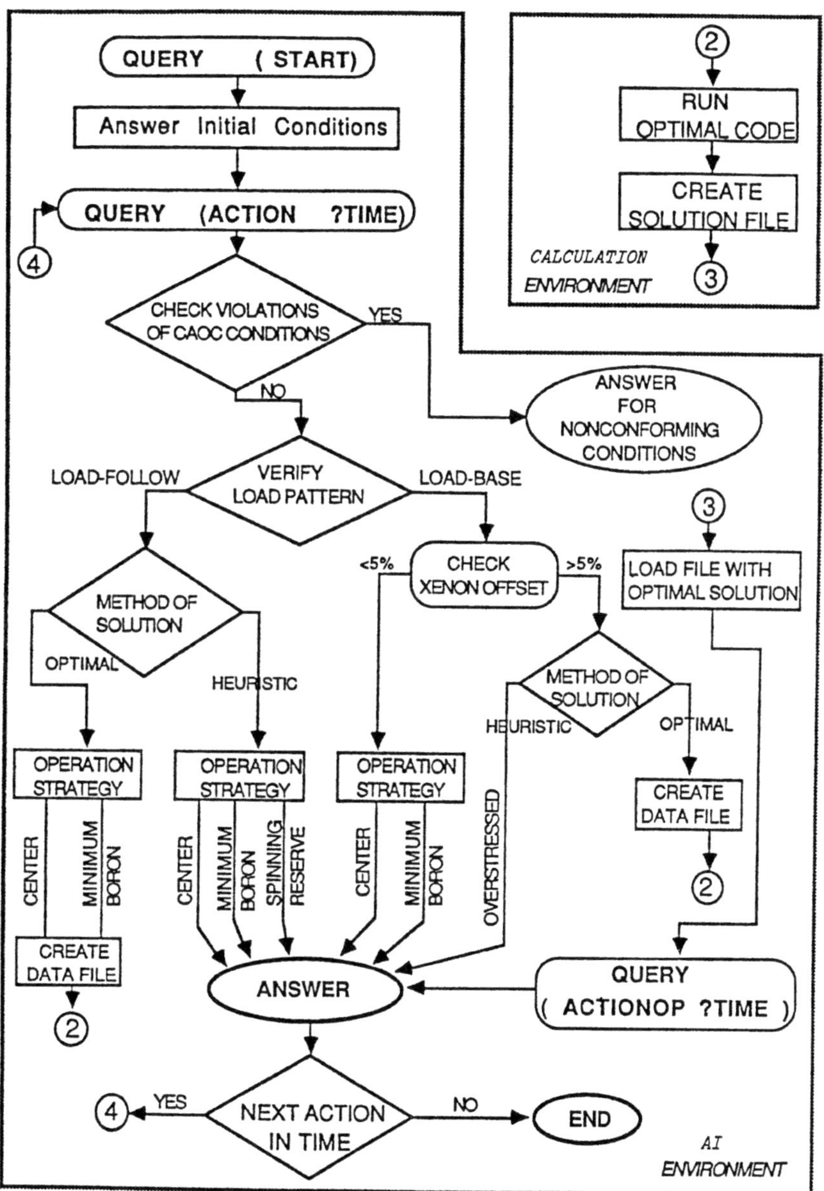

Figure 1. Flow Chart for the Reactor Control Expert System

Load-Follow Center Strategy

The principal objective in this case is to satisfy the dispatcher's power requirements while maintaining the axial offset as close as possible to its target value.

Figures 2 a through d provide plots of power, axial offset, boron and rods for a reactor at the end of core life. The dilution problem presents the most difficult case, and the anticipatory action of the optimal control is employed when the power is reduced or increased to minimize the performance index over the entire time sequence. The heuristic solution does not have anticipatory capability and one can observe that the resulting power deviates almost twice as much from the desired level of 1400 Mw. The anticipatory action in the optimal solution reduces this large deviation.

A similar behavior occurs with the heuristic solution for AO. There is no anticipatory action and the system maintains the exact target value until the appearance of the dilution problem; at this point the AO migrates to the edge of the permissible band. The heuristic solution does not permit the CAOC conditions to be violated, thereby resulting in a power reduction.

Load-follow Minimum Boron Strategy

The objective here is to satisfy the dispatcher's power requirements while avoiding large variations in boron concentration by permitting the axial offset to vary within the target band. The optimal solutions use power, axial offset and boron concentration in the performance index. The same initial conditions used for the optimal program have been applied in the heuristic solution.

Figures 3 a through d show plots of power, axial offset, boron and rods, for the reactor at the end of core life. To achieve approximately the same savings in boron, the optimal solution has to sacrifice its power level. Nevertheless, the axial offset is closer to the target value than in the heuristic solution. The heuristic solution requires greater rod bank movement than the optimal solution. The maximum deviation in the power for the optimal solution is 8%.

Considering the minimum boron planning strategy, one can see that the boron variation for the whole cycle is approximately 45 ppm (heuristic solution for end of core life) compared with 75 ppm obtained from the center heuristic solution (Figure 2c). This savings is at the expense of more maneuvering of the rod banks, and one can see in Figure 3d that it was necessary to use two control rod banks. The axial offset varies around the target value to large values, but inside the allowed band.

Load-Follow Spinning Reserve Strategy

The objective of this strategy is to satisfy the dispatcher's power requirements and to be prepared to quickly respond to unanticipated system demands. Boron, control rods and, at times, coolant temperature are used to achieve this goal. The same initial conditions used for the previous example have been used in the heuristic solution; no optimal solution was performed.

Figures 4 a through to d show plots of power, axial offset, boron and rods for the reactor at end of core life for three different heuristic strategies: center, spinning reserve without temperature reduction and spinning reserve with temperature reduction.

For this experiment, a fast return to full power is simulated, which means a return to full power at a rate of approximately 5% per minute. The required power is raised from the half-power condition at 09:00 hours to full-power at 09:30 hours.

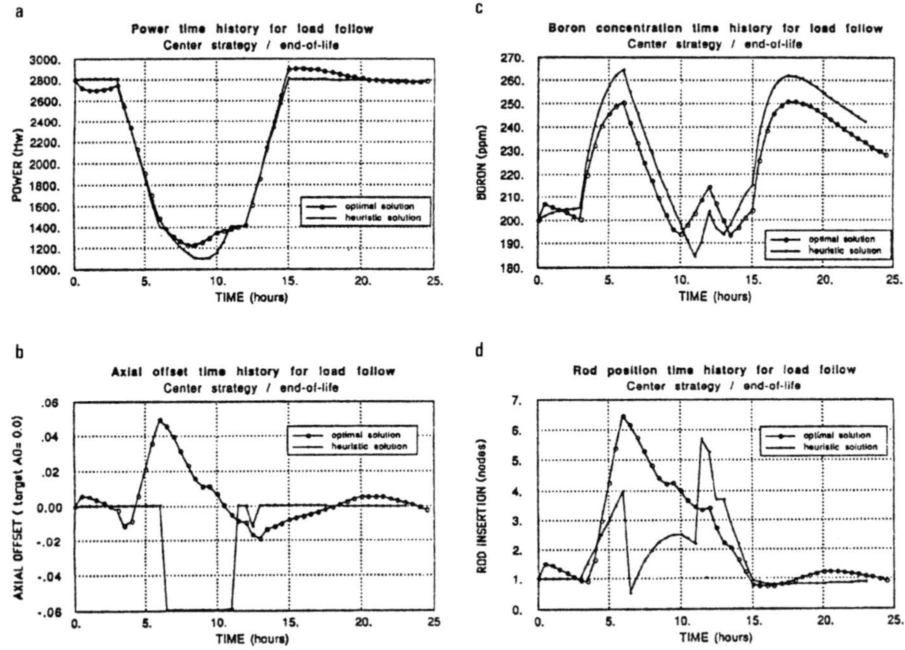

Figure 2. Results for Load-Follow Center Strategy

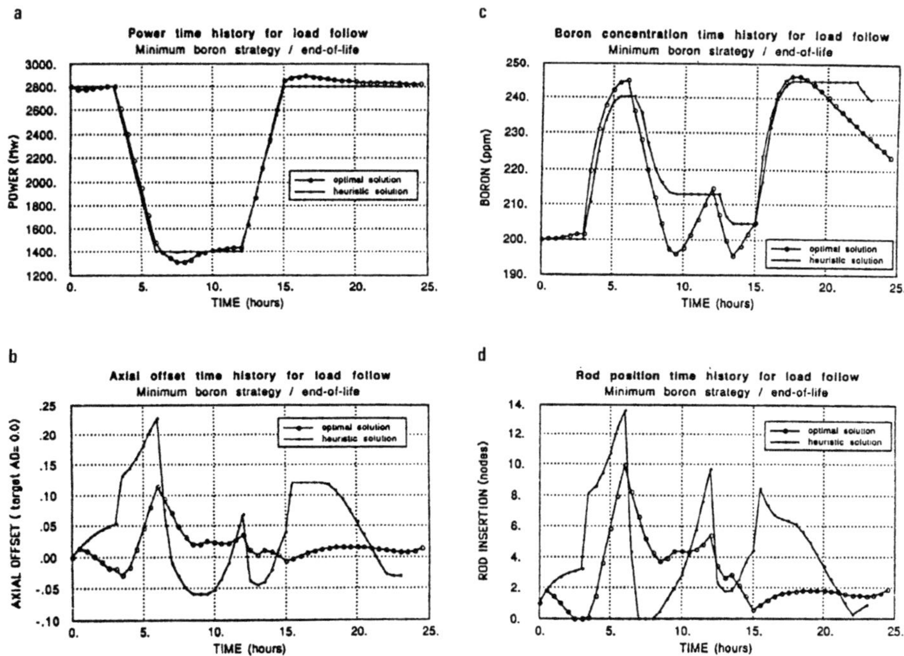

Figure 3. Results for Load-Follow Minimum Boron Strategy

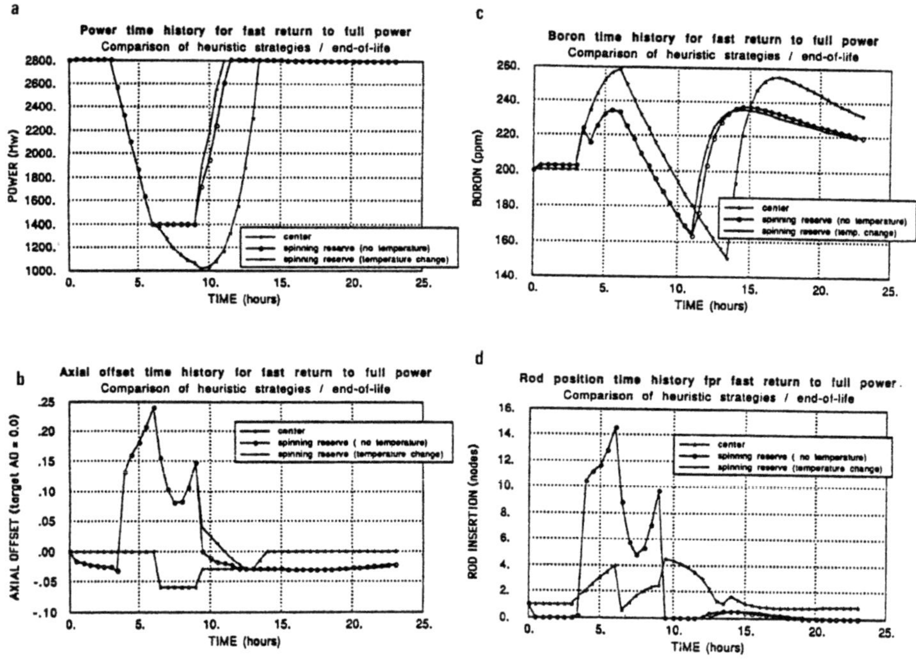

Figure 4. Results for Load-Follow Spinning Reserve Strategy

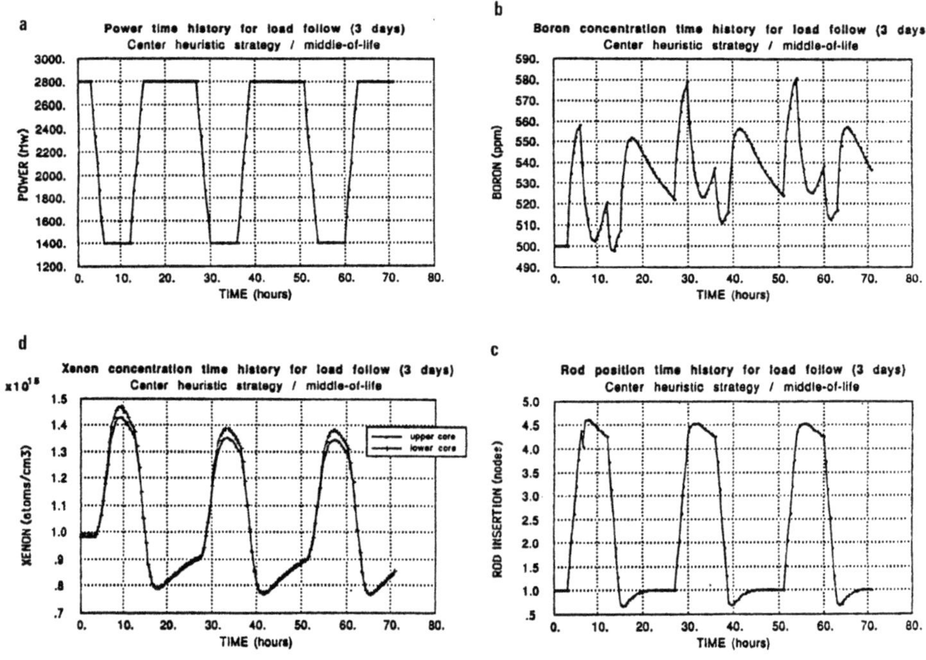

Figure 5. Results for Repeated Daily Load-Follow

Upon examining the plots, one can observe the benefits of the spinning reserve
strategy. For the center strategy, the full power return takes four hours; this
slow response is aggravated by the fact that the power drops up to 1000 Mw before
the reactor is able to recover from the effects of xenon build-up. The spinning
reserve strategy without temperature reduction takes two hours for a full return.
If the temperature is changed, it will reduce the time required to recover full
power after half an hour. The temperature reduction represents a gain of
approximately 10% in power for each step in time. It is interesting to note that
this case is one of the worst cases for full power return, due to low boron
concentration, extreme power ramp rates, and the xenon characteristics.

Repeated Daily Load-Follow

Figures 5 a through d show plots of power, boron, rods and xenon concentration for
the reactor at the middle of core life using the center heuristic strategy. The
expert system exhibits excellent performance for this case.

Damping of Xenon Oscillations

In this case, the initial conditions were taken from the load-follow response at
the time of return to full power. The problem at that point is to stabilize the
reactor. Figures 6 a to f show plots for both the optimal and heuristic solutions.
The total power (2800Mw) is a constraint in this case.

By comparing the optimal and heuristic solutions, one can observe that they have
the same pattern; however, the heuristic solution uses more rigid actions at the
beginning than the optimal solution.

Analyses of the results

The principal objective of the optimal solution is to optimize an aggregate goal
introduced by the designer in terms of a performance index. The various goals are
given individual weights to reflect their perceived relative importance.

The heuristic solution technique is also goal driven. The heuristic solution
focuses on achieving its primary goal. The secondary goals are introduced into the
problem as constraints. If no goal-constraints are violated, the heuristic method
will obtain the solution that achieves its main goal.

An advantage of the optimal solution is its ability to produce anticipatory action
that avoids future adverse situations (see Figure 2a). In this case, the coupling
between the optimal and the heuristic solutions may be performed by using the
response of the optimal solution as a goal or a goal-constraint in the heuristic
solution. For example, in the case of Figure 2a, the power versus time for the
optimal solution indicates that an anticipatory power reduction will minimize
subsequent dilution problems. When the power is reduced, the maximum value of the
boron concentration in the optimal solution may be used as an updated
goal-constraint in the heuristic solution. This action would force the heuristic
solution to make greater use of the control rods, thus avoiding severe dilution
problems. This coupling strategy would be implemented as a high level decision rule
in the expert system.

Examining the load-follow cases, one observes that the results are satisfactory for
all three categories. The following observations can be drawn:

- o For the center strategy, the optimal solution provides better results,
     especially at the end of core life, since it both avoids large drops in
     power and maintains good AO values;

- o In the minimum boron strategy, the savings in boron and power output
     indicate that the heuristic solution may be more appropriate;

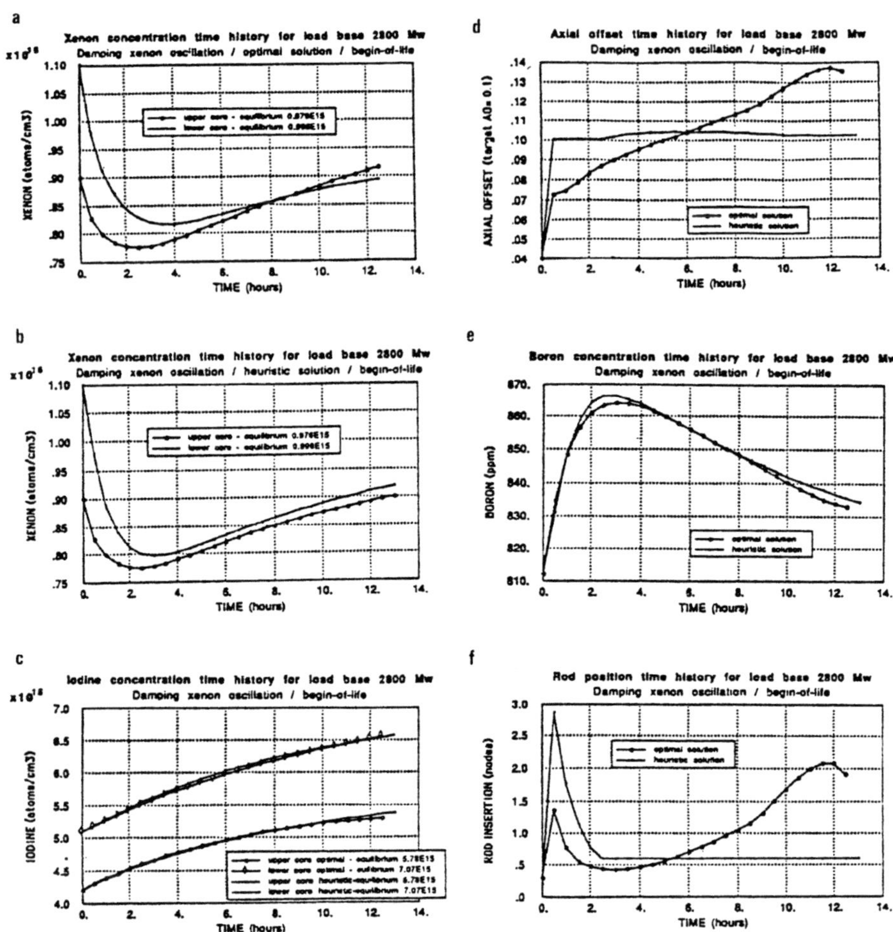

Figure 6. Results for Damping of Xenon Oscillations

o The spinning reserve strategy proves to satisfy fully the requirements of unanticipated return to nominal power.

The heuristic solutions provide particularly practical results because the control variables have periods in their time history when their values or their rates of change are constant.

Conclusions

The following conclusions and observations have been made:

1. An expert system can be used to merge the many heuristic rules currently employed in operation and to provide inter-comparisons between solution methods and among operation strategies.

2 The use of a simple two-point reactor core dynamic model enables implementation of a dynamic model inside a rule base expert system.

3 When the user desires to maintain a balance among different goals, the optimal solution appears to be the more suitable approach. For cases in which the dilution problem is more severe, i.e., at the end of core life, anticipatory action is required, hence the optimal solution is preferred.

4 A coupling between optimal and heuristic solutions may be obtained by observing both the time history of the goals in the performance index and the time history of the optimal control variables. These can then be used to formulate new constraints in the heuristic solution.

References

1. P.J. Sipush et al., "Load-Follow Demonstrations Employing Constant Axial Offset Power-Distribution Control Procedures," Nuclear Technology, vol. 31, pp 12-31, Oct. 1976.

2. O.J.M. Peixoto, "Operational Control of Pressure Water Reactors in Load-Follow and Load-Base," Degree of Engineer Thesis, Stanford University, Dec. 1986.

3. G.D. Storrick, "Generic Load Follow Strategy Report," S.O.300, Westinghouse Report, 1981.

4. "Functional Specifications for The Fast Predictor Module in the PWR-PSMS," Systems Control Inc., Palo Alto, California, 1979.

5. A.E. Bryson and Y.C. Ho, "Applied Optimal Control," Hemisphere Publishing Corporation, Washington, 1975.

6. V.H. Quintana and E.J. Davison, "Clipping-off Gradient Algorithms to Compute Optimal Controls with Constrained Magnitude," International Journal of Control, vol. 20, No 2, 1974.

7. N. Tsouri, J. Rootenberg, L.J. Lidofsky, "Optimal Control of a Large Reactor in Presence of Xenon," IEEE Trans. on Nuclear Science, vol. NS-22, Feb., 1975.

AN AUTOMATED SEARCH PROCEDURE FOR FUEL SHUFFLING

IN PWRs INCLUDING ROTATION EFFECTS

A. H. Robinson, J. O. Heaberlin, and G. L. Wang

Department of Nuclear Engineering
Oregon State University
Corvallis, OR 97331

ABSTRACT

An automated search procedure for determining a fuel loading pattern
with a minimum power peak, called NSHUFLE, has been developed which includes
both the rotation and exchange of PWR fuel assemblies. The procedure is
an extension of a previously developed approach based on heuristic rules
without assembly rotation effects.

The search for an improved core loading pattern is decomposed into
three subproblems: (1) distribution of high reactivity fuel throughout
the core (2) exchange and rotation of assemblies within the neighborhood
of the radial power peaking assembly with lower power elements throughout
the core, and (3) rotation of assemblies within the neighborhood of the
power peaking assembly.

The procedure is based on a depth-first search with one-level backtracking. Moves within each subproblem are described by a series of production rules. A move is accepted if the resulting power peaking factor is
decreased. The goal state is reached when no legal move within the subproblems results in a lower power peaking factor. Assembly exchange rules are
based on previous research. The assembly rotation rules are based on research described in this paper.

INTRODUCTION

The objective of a fuel management computer program is to provide an
automatic procedure for refueling reactors which minimizes the energy costs
and satisfies all of the safety and operating constraints. Any optimization
program that fulfills this objective must shuffle and replace individual
fuel assemblies to obtain the final reload pattern. Many optimization
methods have been applied to in-core fuel management, including dynamic
programming, linear and nonlinear programming, variational techniques, per-
turbation theory, and rule-based search. Each has contributed to advances
in fuel management strategies.

A modular approach to in-core fuel management can be achieved by interpreting the in-core fuel management problem as a state space search, using
knowledge-based problem solving techniques to determine the optimum strategy

for a given portion of the problem. As a step towards this approach, the authors of this paper have recast a previously developed scheme,[1] using heuristic rules, into a depth-first search with limited backtracking. The deterministic procedures previously applied, exchanging core positions of the fuel assemblies, have been reformulated as production rules. These rules are described in the section on "Assembly Exchange Rules."

In addition, new rules have been developed for positioning fuel assemblies with nonuniform exposure distributions or enrichments.[2] Substantial variations in exposure can occur within an assembly if it is located in positions having large power distribution gradients such as along the core periphery. The new assembly rotation rules are described in "Assembly Rotation Rules."

An automated search procedure combining both the fuel assembly exchange and rotation rules, called NSHUFLE,[3] has been developed to allow both exchange and rotation of fuel assemblies with nonuniform enrichments to be incorporated into the search for an improved fuel loading pattern. This procedure has been applied to the core of the TROJAN reactor for Cycle 8. The results of this application are reported in the section entitled "Application of the Search Procedure."

ASSEMBLY EXCHANGE RULES

The assembly exchange rules are based on previous work, SHUFLE, completed by Stout and Robinson.[1] In their work, the reactor core was modeled with one mesh point for each assembly. The power distribution was computed using coarse-mesh two-group diffusion theory. The initial Out-In fuel loading pattern is based on the following rules:

a. Locate new fuel on the core periphery.

b. Place low K-infinity fuel around high K-infinity fuel (checkerboard pattern).

Given the initial fuel loading pattern, the procedure would automatically search for a loading pattern with the lowest radial power peak. The rules listed below represent the general rules used in SHUFLE to perform binary exchanges of the fuel for an Out-In loading pattern. More details on the rules can be found in Reference 1.

a. Move fuel assemblies with high relative power toward assembly of lowest power.

b. Move new fuel assemblies only one position at a time.

c. Exchange high power, high K-infinity fuel with lower power, low K-infinity fuel that is at least several assembly positions distant. (This does not apply to new fuel.)

ASSEMBLY ROTATION RULES

TROJAN Cycle 8 was chosen as a reference core in order to (1) investigate how rotation of exposed fuel assemblies affects the radial power peak, and (2) develop production rules to address enrichment nonuniformity. The exposure distribution at the beginning of the cycle is illustrated in Figure 1. The relative power distribution of the loading pattern was computed for four nodes within each assembly; a radial power peak, of 1.478 occurs in node 2 of assembly F-5.

646

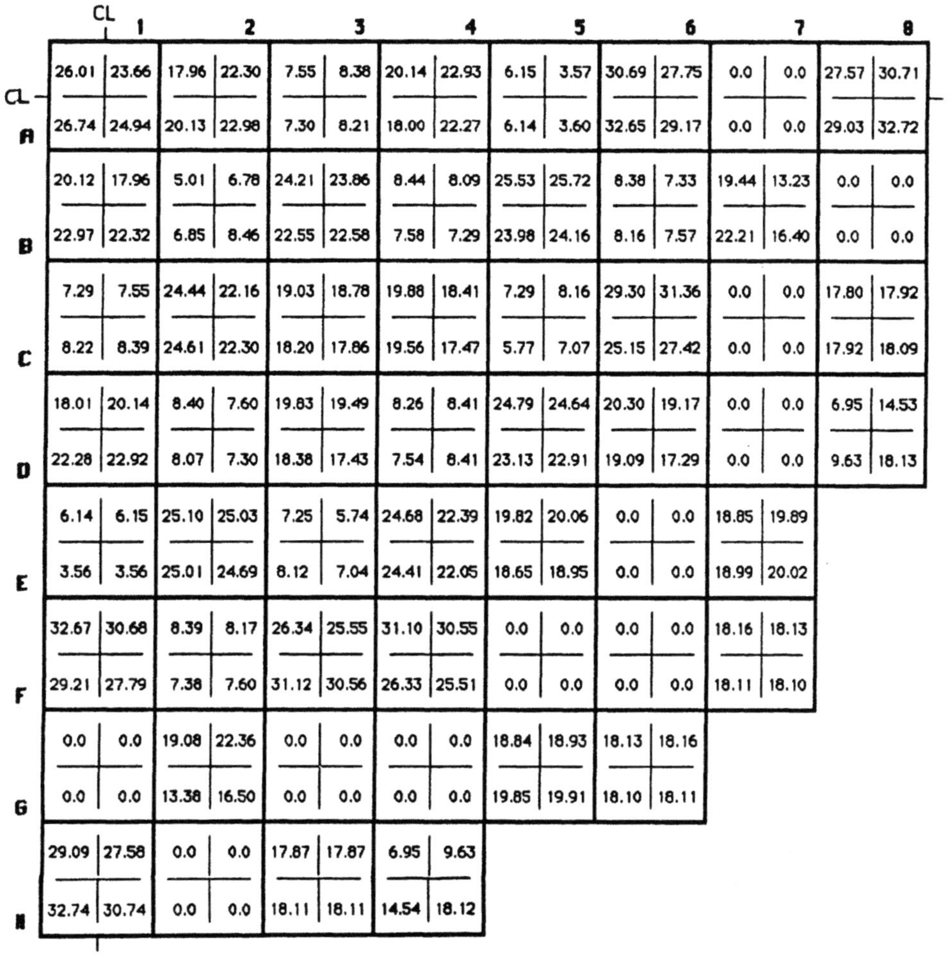

Fig. 1.   Exposure distribution at beginning of Cycle 8 for TROJAN.

Because assembly F-5 is a new fuel assembly and has no variation in exposure/enrichment across the assembly, rotating this element would not effect the radial power peak. Rotation of assemblies within a 5 by 5 array of the radial power peak, however, does change the radial power peak as shown in Table 1.

The following effects are observed from this exercise:

   a.   Rotating fuel assemblies around the radial power peak, in general, does not change the position of the radial power peak.

   b.   Rotating fuel assemblies around the radial power peak does, however, influence the value of the radial power peak. The value

Table 1. Effect of Assembly Rotation on Radial Peak Power

CORE POSITION	NODE EXPOSURE				RADIAL POWER PEAK		
	BEFORE ROTATION		AFTER ROTATION		BEFORE ROTATION	AFTER ROTATION	DIFFERENCE
D-3	19.83 19.49	18.38 17.43	17.43 18.38	19.49 19.83	1.4782	1.4759 *	-0.0023
D-4	8.26 8.41	7.54 8.41	8.41 7.54	8.41 8.26	1.4759	1.4760	+0.0001
D-5	24.79 24.64	23.13 22.91	22.91 23.13	24.64 24.79	1.4759	1.4701 *	-0.0058
D-6	20.30 19.17	19.09 17.29	17.29 19.09	19.17 20.30	1.4701	1.4652 *	-0.0049
E-3	7.25 5.74	8.12 7.04	7.04 8.12	5.74 7.25	1.4652	1.4637 *	-0.0015
E-4	24.68 22.39	24.41 22.05	22.05 24.41	22.39 24.68	1.4637	1.4619 *	-0.0018
E-5	19.82 20.06	18.65 18.95	18.95 18.65	20.06 19.82	1.4619	1.4319 *	-0.0300
E-7	18.85 19.89	18.99 20.02	20.02 18.99	19.89 18.85	1.4319	1.4285 *	-0.0034
F-3	26.34 25.55	31.12 30.56	30.56 31.12	25.55 26.34	1.4285	1.4403	+0.0018
F-4	31.10 30.55	26.33 25.51	25.51 26.33	30.55 31.10	1.4285	1.4193 *	-0.0092
F-7	18.16 18.13	18.11 18.10	18.10 18.11	18.13 18.16	1.4193	1.4193	0.0
G-5	18.84 18.93	19.85 19.91	19.91 19.85	18.93 18.84	1.4193	1.4259	+0.0066
G-6	18.13 18.16	18.10 18.11	18.11 18.10	18.16 18.13	1.4193	1.4193	0.0
H-3	17.87 17.87	18.11 18.11	18.11 18.11	17.87 17.87	1.4193	1.4309	+0.0116
H-4	6.95 9.63	14.54 18.12	18.12 14.54	9.63 6.95	1.4193	1.4198	+0.0005

* ROTATION MOVE ACCEPTED

of the peak power is reduced from 1.478 to 1.419 after assemblies D-3, D-5, D-6, E-3, E-4, E-7 and F-4 are rotated 180 degrees.

    c.   A lower radial power peak results when the lowest reactivity node in the 5 by 5 array is positioned towards the radial power peak assembly. The value of the peak power is reduced from 1.462 to 1.432 after rotating assembly E-5.

    d.   Smaller effects result if an assembly with less exposure nonuniformity and/or farther from the peak is rotated.

According to these results, the following new shuffling rules were developed:

    a.   Rotate fuel surrounding the radial power peak assembly so that the highest burned nodes are towards the peak power assembly.

648

b.   Rotate fuel assemblies only within the 5 by 5 array of the peak
      power assembly, starting with those assemblies closest to the
      peak power assembly.

APPLICATION OF THE SEARCH PROCEDURE

The search for an improved core loading pattern is decomposed into
three subproblems which are addressed by separate routines within NSHUFLE:
(1) distribution of high reactivity fuel, (2) exchange and rotation of high
power assemblies within a 5 by 5 array of the peak power assembly with other
lower power assemblies within the core (subroutine SHUFLB), and (3) rotation
of assemblies within the 5 by 5 array of the peak power assembly (subroutine
SHUFLC).

CL	1		2		3		4		5		6		7		8	
A (top)	26.01	23.66	17.96	22.30	19.03	18.78	20.14	22.93	6.15	3.57	30.69	27.75	0.0	0.0	27.57	30.71
A (bottom)	26.74	24.94	20.13	22.98	18.20	17.86	18.00	22.27	6.14	3.60	32.65	29.17	0.0	0.0	29.03	32.72
B (top)	20.12	17.96	5.01	6.78	24.21	23.86	8.44	8.09	25.53	25.72	8.38	7.33	19.44	13.23	0.0	0.0
B (bottom)	22.97	22.32	6.85	8.46	22.55	22.58	7.58	7.29	23.98	24.16	8.16	7.57	22.21	16.40	0.0	0.0
C (top)	7.29	7.55	24.44	22.16	7.07	5.77	17.47	19.56	7.55	8.38	27.42	25.15	0.0	0.0	17.80	17.92
C (bottom)	8.22	8.39	24.61	22.30	8.16	7.29	18.41	19.88	7.30	8.21	31.36	29.30	0.0	0.0	17.92	18.09
D (top)	18.01	20.14	8.40	7.60	19.83	19.49	8.26	8.41	22.05	24.41	18.95	18.65	0.0	0.0	18.13	9.63
D (bottom)	22.28	22.92	8.07	7.30	18.38	17.43	7.54	8.41	22.39	24.68	20.06	19.82	0.0	0.0	14.53	6.95
E (top)	6.14	6.15	25.10	25.03	7.25	5.74	22.91	23.13	25.51	26.33	0.0	0.0	18.85	19.89		
E (bottom)	3.56	3.56	25.01	24.69	8.12	7.04	24.64	24.79	30.55	31.10	0.0	0.0	18.99	20.02		
F (top)	32.67	30.68	8.39	8.17	26.34	25.55	17.29	19.09	0.0	0.0	0.0	0.0	18.16	18.13		
F (bottom)	29.21	27.79	7.38	7.60	31.12	30.56	19.17	20.30	0.0	0.0	0.0	0.0	18.11	18.10		
G (top)	0.0	0.0	19.08	22.36	0.0	0.0	0.0	0.0	18.84	18.93	18.13	18.16				
G (bottom)	0.0	0.0	13.38	16.50	0.0	0.0	0.0	0.0	19.85	19.91	18.10	18.11				
H (top)	29.09	27.58	0.0	0.0	17.87	17.87	6.95	9.63								
H (bottom)	32.74	30.74	0.0	0.0	18.11	18.11	14.54	18.12								

Assemblies moved from original location

x.xx	x.xx
x.xx	x.xx

Fig. 2. Final loading pattern determined by NSHUFLE.

NSHUFLE was applied to the loading pattern shown in Figure 1. This
loading pattern has a radial peak power of 1.669. The distribution of new
fuel without rotation further reduces the radial peak power to 1.523. The
final loading pattern recommended by NSHUFLE, shown in Figure 2, has a radial
power peak of 1.406. Table 2 identifies the resulting value of the radial
power peak for each move.

CONCLUSIONS AND PLANS FOR FURTHER RESEARCH

With a depth first search, it is impossible to prove the loading pattern recommended by NSHUFLE is a global optimum without doing an exhaustive
search. However, from a practical standpoint, it is not necessary that
the loading pattern be the best, only that it be acceptable and as near
the optimum as practically possible.

The automated search procedures in NSHUFLE have been successful in
recommending an improved fuel loading pattern by including assembly rotation
rules. This allows the procedure to solve the problem of how to load exposed
fuel assemblies with steep burnup gradients.

The modular approach used in recasting the shuffling logic in terms
of production rules and the interpretation of the in-core fuel management
problem as a state space search will now allow additional rules to be developed and tested within the given framework of NSHUFLE. Additional strategies for fuel shuffling can also now be added to NSHUFLE with the intent
of allowing the procedure itself to determine what strategy works best for
a given goal.

Table 2. Effect of Assembly Exchange and Rotation on Radial Power Peak

SUBROUTINE SHUFLB			
POSITION	RADIAL POWER PEAK		MOVE ACCEPTED?
	AFTER EXCHANGE	AFTER ROTATION	
E-5	1.523	1.496	YES
F-4		1.497	NO
F-4	1.499	1.495	YES
D-6		1.489	YES
C-5	1.489	1.490	NO
C-3		1.462	YES
A-3	1.436	NONE *	--
C-5		1.436	NO
E-4	1.436	1.434	YES
D-5		1.429	YES

SUBROUTINE SHUFLC			
POSITION	RADIAL POWER PEAK		MOVE ACCEPTED?
	BEFORE ROTATION	AFTER ROTATION	
C-4	1.429	1.412	YES
C-6	1.412	1.407	YES
D-8	1.407	1.406	YES

* NO ROTATION ALONG LINE OF SYMMETRY

REFERENCES

1. R. B. Stout and A. H. Robinson, Determination of Optimum Fuel Loading in PWR Using Dynamic Programming, Nucl. Tech. 20:86 (1973).
2. G. L. Wang and A. H. Robinson, "An Automated Procedure for Optimizing the Refueling of Pressurized Water Reactors which Includes Assembly Rotation Effects," OSU-NE-8701, Oregon State University (1987).
3. A. H. Robinson, J. O. Heaberlin, and G. L. Wang, An Automated Search Procedure for Selection of Pressurized Water Reactor Fuel Loading Patterns, to be submitted to Nucl. Tech.

PLANT STATUS MONITOR SYSTEM DESCRIPTION

William A. Prather and Gary A. Sly

EI Services, Inc.
Kent, Washington

In today's regulatory and financial environment, improving
plant efficiency and safety are necessary elements of plant
operations. Public utility commissions are making rate rul-
ings based, in part, on plant availability performance; and
the NRC is putting more emphasis on plant operational aspects.
This comes at a time when operating, maintaining, and managing a plant are becoming increasingly complex; moreover, the
desired number of experienced plant personnel are becoming
more difficult to find.

This situation can be partially resolved by using computer
software tools to assist plant operations, maintenance, engineering, and management personnel. These software tools provide information and interpretations based on plant and equipment status. They support improved plant availability, technical specification compliance, and administrative functions.
A key element of computerization is the ability to operate on
integrated information. If this ability exists, various plant
organizations can base their decisions on a consistent view of
the plant's status. In addition, computerization can reduce
the administrative burden by generating some of the required
reporting and tracking information.

The Plant Status Monitor (PSM) system has been developed by
EI Services for the Electric Power Research Institute (EPRI)
to address these issues. PSM tracks and maintains the operability status of all equipment and systems covered by the
plant technical specifications and other operational constraints that could affect the safe startup, operation, and
shutdown of the plant. Based upon the operational status of
the plant systems and components, PSM also determines compliance with the Limiting Conditions of Operation. In addition, the PSM can operate with tagging systems. The tagging
system used to demonstrate this capability is EI's proprietary
tagging system,TRIS (Tagging Request and Inquiry System).

PSM is a powerful and flexible software package which provides
the following capabilities:

- Operability status of the systems, trains, and components of the plant;

- Compliance with Technical Specification Limiting Conditions for Operation;

- Tracks timing requirements of Technical Specification action statements;

- A measure of the health of the plant.

- Operates with a tagging system.

Additionally, the PSM software tool provides plant operational information and support functions such as:

- Applications of probabilistic assessment of nuclear plant safety, reliability and availability.

- Tracking work orders, scheduling maintenance activities, spare parts, and commitment action.

- Tagging (equipment operating status, positioning, and alignment), optimizing configuration or restoration of equipment for various modes or power levels.

- Operational diagnostics and plant safety/risk management.

- Determines real-time plant parameter status, on-line availability, and scram reduction methods.

Proposed changes in component status are input by plant personnel and automatically incorporated into PSM's logic models and resident data base. PSM determines the impact of component status changes relative to plant technical specifications, operating modes, system status, plant availability, and other defined parameters. PSM provides current, accurate data on which to base important decisions.

PSM conveys information through a variety of displays which result in a system which is easy to use. These displays consist of a variety of menus, prompts, and results displays. The menus and prompts are carefully constructed to guide the user quickly and easily to the intended operation. The results displays show the results in easy to read, well-organized formats which readily convey the information requested. They include results for both current status and proposed status. Proposed status allows the user to ask "what-if" questions regarding the impact of potential changes in equipment status or plant operating mode without disturbing the current status. With PSM, users know "where they are", "where they are going", and "where they have been."

PSM is based on EPRI's GO Methodology, which uses logic models to faithfully represent the actual configuration of a specific nuclear power plant. Logic models are developed for each system which impacts the safety or power generation capability of

the plant. However, the detail of the models varies with the
complexity and significance of the system. Safety systems, for
example, are modeled in sufficient detail to completely represent
the status of the trains and subsystems, and the LCO's and
Action Statements. Other systems with a lesser influence on
status and LCO's are modeled as a single component. This permits
modeling to be sufficiently flexible to represent the plant to
the detail necessary, but still minimizes execution time by reducing model size where feasible.

Successful plant operations software services are based on knowledge of the basic technology, combined with experience in plant
operations and software development. EI Services currently is
engaged in plant operations software activities throughout the
utility industry. Over the past ten years, EI Services has become a leader in the development of computer systems and software codes for plant system analysis and operation, probabilistic risk assessment (PRA) studies, and plant operating support.
Cumulative software development and computer-related experience
within the company now exceeds 1,600 man-years. EI Services'
consulting and other pertinent technical support services include complex engineering analysis and support to assist clients
in improving plant operations, reliability, safety and availability.

CHAPTER 11

PLANT OPERATIONS AND SUPPORT

AN EXPERT SYSTEM FOR SPARE PARTS INVENTORY CONTROL

Kil Yoo Kim, Patrick Y.C. Chen[†] and David Okrent

Mechanical, Aerospace and Nuclear Engineering Department
University of California
Los Angeles, California 90024-1597

This paper describes an expert system which can handle spare part
requirements not only in corrective maintenance (CM) or preventive maintenance (PM), but also when failure rates[*] of components or parts are
updated by new data or by predictive maintenance (PDM), and which can also
decide optimum stocking level of each spare part. This expert system provides a maintenance (or inventory) manager with an improved basis for decision making in the maintenance related to spare parts. The definitions of
PM and PDM from NUREG-1212 [USNRC 1986] are used herein. This expert
system uses Intelligence/Compiler [Intelligence Ware, 1986] as a language/
tool in the IBM-PC.

FUNCTION OF THE PROPOSED EXPERT SYSTEM

The proposed expert system can decide which part to stock and how many
of each part to stock by a cost-effectiveness (CE) calculation and/or by
the user's experience and judgment and/or by vendor recommendation. That
is, at first the expert system determines optimum stocking level by the
CE calculation, and then decides the optimum stocking level by combining
the result of the CE calculation with the user's experience and judgment,
and/or the vendor recommendation in an inexact reasoning process by the
inference engine.

The expert system decides the optimum stocking level by using generic
data at the prestart-up phase. During operating periods, failure rate data
are updated with Bayesian method by checking reliability data, e.g., NPRDS
and plant specific data, and the optimum stocking level is decided with
these new data.

The expert system can suggest adequate actions when the user is
involved in CM, PM or PDM activity. When specific parts are required by CM
or PM activity, the maintenance manager may ask or use this expert system.

[†]Currently employed by Pickard, Lowe and Garrick, Inc.
[*]In calculation of spare parts inventory control, failure rates mean
 replacement rates with spare parts, because a clear-cut division has
 already been made between repair stock items and spare parts.

The expert system can then think like a maintenance manager and can show
how many and where the required spare parts are located. The expert system
can consider an inter-utility cooperative parts loan system such as the Pool
Inventory Management (PIM), can select suitable suppliers, and can show the
lead time of the spare part. If it is known that an item is entering a
wear-period in PDM, the expert system may suggest that the item be replaced
immediately, or that it be untouched until the next planned outage, etc.

COST-EFFECTIVENESS CALCULATION

The basic factor in decision making concerning which parts to stock and
how many of each part to stock is CE calculation, which is explained as
follows;

If a spare part or component were obtained, then the following cost
(= Total Inventory Cost = TIC) would be incurred at present value.

TIC = price + storage cost + maintenance cost

However, if there were a failure of the item in the future, there would
be the following loss of money (= LOSS) at present value, if the part were
not in inventory.

$$\text{LOSS} = (\text{price} + \text{risk cost}) \times \int_{P}^{r\ell+P} \left[\frac{1+j}{1+i} \right]^{t} f(t)\ dt$$

$$+\ (\text{rec} \times \text{loss fraction}) \times \int_{P}^{r\ell+P} \text{lead-time}(t) \left[\frac{1+j}{1+i} \right]^{t} f(t)\ dt$$

where $f(t) = \lambda(e)^{-\lambda t}$, λ = 'failure rate', i = 'interest rate', j = 'escalation rate', rl = MIN (remaining item lifetime, remaining plant lifetime),
p = 'period during which the item has been used', rec = 'replacement energy
cost rate', price = 'current purchasing price of the item' and loss fraction = 'fractional reduction of power due to failure of item'.

If LOSS is larger than TIC, obtaining the spare part is more cost
effective. In this case, we should continue to check whether acquiring
another spare is also cost effective. By this procedure, the optimum
stocking level can be calculated. It is not difficult to apply this model
even if many identical items are used in the power plant.

The previous model is applied only when the failure rate or hazard
function is constant. This is an acceptable assumption if the item is in
its useful period and it has not entered the wear-out period, or if the
situation is that the remaining lifetime of the item is longer than that
of the plant. Therefore, this model can be used in determining the optimum
stocking level for CM and PM items.

However, many PDM items can enter the wear-out period and require a
different evaluation. The hazard functions are no longer constant with
time in the wear-out period. The expert system assumes that the hazard
function of wear-out period increases linearly with time, i.e., h(t) = kt.
Then,

$$f(t) = k(t - t_{0})e^{-\frac{K}{2}(t - t_{0})^{2}} \qquad (= \text{Rayleigh distribution})$$

By using the Rayleigh distribution in the wear-out period, the CE can
be calculated for PDM items. This model also can be applied to the PM item
whose remaining lifetime is less than twice of its lead time, for conser-
vatism.

In the CE calculation, spare parts supplier's stability and the
increase in risk due to the lack of spare parts are also handled. Since
the technical specifications of a nuclear power plant may accept continuing
operation (e.g., 72 hours), in many cases, although a component fails in a
system which has redundant components, the lack of spare parts can increase
risk. The user can find an increased core melt frequency (ΔCMF) due to the
failed component by using SARA [Sattison, et al, 1987]. Then, the risk cost
can be calculated as follows;

$$\text{risk cost} = \Delta\text{CMF} \times (\text{reduced period})/365 \times \$5,000,000,000$$

where 'reduced period' means that if the spare part is available in inventory, then the time that the plant continues to run with a failed component
can be shortened by the amount of the 'reduced period', which reduces the
risk of core melt frequency.

The CE calculation used for determining the optimum stocking level
should be valid for rarely used items; the major portion of total dollar
value of spare parts inventory at a power plant may be for rarely used
items [Moncrief and Schroder, 1987].

STRUCTURE OF THE PROPOSED EXPERT SYSTEM

The expert system represents knowledge with multiple paradigms such as
rules, frames and logic. Since there are too many spare parts to handle in
the power plant, b-tree, which has some database functions, is used in
knowledge representation. Forward, backward, and inexact reasoning rules
are used in inference engine.

CONCLUSIONS

This expert system can provide a maintenance manager with an improved
basis in decision making for CM, PM, and PDM relating to spare parts inventory. Also, this expert system is very useful because it can handle PDM
effectively, and because recently predictive maintenance has tended to
become more important, especially for nuclear power plant life extension.
Finally, the conventional algorithm which is involved in the CE calculation,
is well coupled with 'inference engine'.

ACKNOWLEDGMENT

This work was supported in part by DOE Contract #DE-FG03-865F16610.

REFERENCES

Intelligence Ware, Inc., 1986, "Intelligence/Compiler User's Manual."
Moncrief, E.C., and Schroder, R.M., 1987, Spares Management for Improved
     Maintenance, ANS Trans, 54:281.
NUREG-1212, 1986, "Status of Maintenance in the U.S. Nuclear Power
     Industry 1985," U.S. NRC.
Sattison, M.B., Tullock, W.W., and Van Siclen, V.S., 1987, "SARA User's
     Manual", EGG-SPAG-7532.

INSPECTR: A POWER PLANT PERFORMANCE DIAGNOSTIC EXPERT SYSTEM USING EVIDENTIAL AND MODEL-BASED REASONING

Timothy Kessler, Gerald Liu, Bjorn Frogner, David Gloski,
Peter Wallace, and Dong-Ru Shieh

Expert-EASE Systems, Inc.
1301 Shoreway Road
Belmont, California 94002

INTRODUCTION

Chronic problems that adversely affect plant heat rate typically take a
secondary role compared to acute problems requiring minimal diagnostic expertise
simply because the acute problems are easier to find and treat without
allocating expensive expert resources. However, the cumulative cost of these
chronic problems often far exceeds that of the acute problems. The ability to
diagnose and correct chronic performance problems at the site without waiting
for expert assistance to be made available therefore represents a significant
benefit to any power plant.

The INSPECTR Interactive Performance Diagnostic Assistant incorporates the
latest advances in Artificial Intelligence technology (e. g. model-based
reasoning and evidential reasoning with uncertainty) to address the problem of
on-site diagnosis of performance problems. As a PC-based performance diagnostic
expert system, INSPECTR provides a practical means of combining the analytical
expertise of an experienced corporate office engineering staff with the hands-on
expertise of a site-based performance engineer. Working together, INSPECTR and
the performance engineer can identify, diagnose and correct performance problems
that would otherwise require a significant allocation of corporate staff and/or
outside consultant resources.

OVERVIEW OF THE INSPECTR PERFORMANCE DIAGNOSTIC ASSISTANT

INSPECTR is a hybrid rule-based and model-based expert system designed to
interpret information regarding the current operating state of a power plant in
order to identify and diagnose problems affecting plant heat rate. INSPECTR is
being developed by Expert-EASE Systems, Inc., with assistance from Yankee Atomic
Electric Company, under contract to the United States Department of Energy. It
will be available for use as a stand-alone system or as an integral component of
an integrated Nuclear Power Plant Performance Engineering Workstation
(THERMAC-N) being developed jointly by Expert-EASE Systems and Babcock and
Wilcox Corporation.

<u>INSPECTR Design Objective</u>

The overall design objective of THERMAC-N was to provide an integrated
workstation to assist performance engineers in monitoring and improving plant
thermal performance. This workstation includes an analytical model to perform
detailed cycle heat balance calculations and a simplified data acquisition and
analysis package to perform routine monitoring of key plant performance
parameters. However, in order to maximize the usefulness of THERMAC-N in
maintaining optimum plant performance, an expert system was proposed as a means
of supplementing the capability of the average performance engineer to interpret
the information provided by the system. Specific features of this expert system
that were considered to be of paramount importance in meeting the overall design
objective were:

o   The expert system should be developed using a commercially-available
shell program designed for the IBM PC/AT in order to maintain hardware
compatibility with the remaining workstation components. Shell
programs that require special purpose computers (e. g. LISP computers)
should not be used.

o   The expert system initially should provide the capability to identify
and diagnose a representative number (25-35) of categories of
performance problems that are of generic concern to most power plants,
with expansion capability to address additional less-common or
plant-specific problems as required. This "generic" knowledge base
should, to the extent possible, be independent of the plant-specific
physical configuration.

o   In order to account for plant-specific variations in physical
configuration, availability of measurements, etc., the expert system
should include a direct information link with a detailed,
plant-specific analytical model. The analytical model is used to
supply needed information about the plant configuration (e. g. how
many feedwater heaters are there, with which turbine stages are they
associated, etc.) and to generate information about the current plant
state that is necessary for problem diagnosis but that may not be
available from direct measurements.

o   The expert system should be capable of addressing uncertainty both in
the accuracy of available measurements and in the likelihood of a
particular performance problem producing a particular set of
observable symptoms. Multiple levels of uncertainty should be
identified and combined using state-of-the-art techniques.

o   The expert system should be sensitive to varying levels of domain
expertise on the part of the user by providing for multiple levels of
external problem focusing. The user should, at his option, be able to
depend entirely upon the expert system to diagnose a problem, work
with the system interactively to co-diagnose the problem, or he should
be able to use the expert system solely as a means of confirming his
own suspicions.

o   The expert system should maximize operational efficiency by sharing
data with all other components of THERMAC-N. It should also include
the capability to acquire key performance data on-line.

o   The expert system should provide a functional and friendly user
interface that includes English language information query and
explanation facilities.

The following section provides a detailed description of how these

features have been implemented in the INSPECTR expert system component
to meet the overall design objective of THERMAC-N.

DESCRIPTION OF THE INSPECTR PERFORMANCE DIAGNOSTIC ASSISTANT

<u>Expert System Shell</u>

In order to maintain system hardware costs at an acceptable level and
to retain hardware compatibility with THERMAC-N, a fundamental
requirement of the INSPECTR expert system shell program was that it be
compatible with an IBM PC/AT micro-computer. Accordingly, a number of
commercially available PC-based shell programs were evaluated for this
purpose. Although none of these programs were found to include all of
the features desired for INSPECTR, Neuron Data Inc.'s NEXPERT[1] shell
program was judged to be the best available tool for this application.
The following NEXPERT features were considered to be essential:

- o   Object-oriented structure, which allows the user to structure
      his knowledge base according to the hierarchical structure
      common to most power plants;

- o   Ability to specify a context structure for rules, which
      allows the user to control problem focusing in a manner
      similar to the hierarchical object structure described above;

- o   Ability to access external routines or perform other
      user-specified functions based upon the successful "firing"
      of individual rules;

- o   Ability to combine forward (symptom-oriented) and backward
      (goal-oriented) rule chaining in the same consultation
      session;

- o   Ability to volunteer data to the system prior to the start of
      the session or to access data files directly during the
      course of the consultation;

- o   Ability to focus the diagnosis externally by suggesting
      likely conclusions prior to the start of the session.

<u>Expert System Knowledge Base</u>

The INSPECTR knowledge base uses "production rules" to relate common
performance problems to a specific set of observable or calculable
symptoms. These rules were developed in conjunction with the
performance engineering staff at Yankee Atomic Electric Company, whose
expertise in the area of power plant operating efficiency and heat rate
reduction is recognized throughout the utility industry. Input from
Yankee Atomic was combined with information available from several
published reports to develop a "generic" performance diagnostic
knowledge base that addresses approximately thirty common categories of
performance problems experienced by most nuclear power plants.

This generic knowledge base contains the following special design
features that make it uniquely suited to diagnosing performance
problems at power plants:

- o   The INSPECTR knowledge base is hierarchical in structure,
      consisting of two types of rules. "Focusing" rules deal with
      deviations in boundary conditions of the major steam cycle

subsystems (i.e., boiler, turbines, condenser, feedwater train) to provide problem focusing at the subsystem level. "Problem Identification" rules deal with symptoms observed at the component level within a subsystem to identify a specific performance problem. This "two-tiered" approach increases operational efficiency by facilitating a breadth-followed-by-depth solution strategy with the breadthwise search performed at a high level.

o   In contrast to the more conventional rule structure of combining several conditions in conjunctive fashion to yield a single conclusion; i.e.,

if A and B and C then D

INSPECTR rules treat only one symptom at a time; i.e.,

if A then D
if B then D
if C then D.

The conventional rule structure requires that all symptoms be observed in order for a problem to be considered. However, using the INSPECTR approach, observation of any symptom (A, B, or C) is treated as evidence of the possibility of the problem (D), with the likelihood of the problem increasing as each successive symptom is observed. The INSPECTR rule structure therefore allows for the possibility of one or more symptoms of a particular problem being missed without removing that problem from further consideration.

o   INSPECTR generic rules deal with symptoms that are observed or calculated in the immediate vicinity of the affected component. This allows the generic knowledge base to be virtually independent of the plant-specific physical configuration. Symptoms of a problem in a particular component that might be observed in other plant components are treated separately by a model-based (or deep) reasoning process that is described in the next subsection.

<u>Model-Based Reasoning</u>

The generic knowledge base described in the previous subsection is sufficiently detailed to identify a number of performance problems which produce a truly distinct set of locally observable symptoms. However, many common performance problems cannot be diagnosed by using only locally observable symptoms, either because their symptoms are similar to those caused by other problems or because local measurements are unavailable. Such problems are typically identified through observation of the behavior of other plant components or through performance of a series of "what-if" calculations using a detailed analytical model.

In order to address these more complex problem diagnoses, INSPECTR incorporates a direct information and control link with the detailed analytical model contained in THERMAC-N. The analytical model is used to complement the INSPECTR generic knowledge base in the following ways:

o   Measurements that may be missing or unreliable in a particular plant can be simulated by the model, thereby

enabling the knowledge base to operate on a consistent set of
plant state information under all conditions for all plants.

o   Calculated parameters (e.g., heat transfer coefficients,
    stage efficiencies, etc.) that provide important clues to a
    possible problem can be generated by the model and used by
    the knowledge base to supplement the information available
    from direct measurements and observations.

o   The model can be used to adjust the "reference" plant state
    (i.e., the expected plant state in the absence of any
    performance problems) to reflect current operating conditions
    that differ from design conditions but are beyond the control
    of the plant staff. For example, the current circulating
    water temperature can be automatically factored into a
    revised reference calculation to provide a more appropriate
    base case against which to compare the current plant state.

o   After the generic knowledge base has narrowed the list of
    candidate problems based upon an evaluation of locally
    observable or calculable symptoms, the model can be used to
    simulate each of the likely candidates in a series of
    "what-if" calculations. These calculations provide
    information about the response of all cycle components to
    each of the candidate problems, thereby providing additional
    symptoms for evaluation and comparison.

The information and control link between the THERMAC-N analytical model
and the INSPECTR expert system is completely automatic, requiring no
user input to perform any of the functions listed above. However, the
user has the option of bypassing any or all of these "deep reasoning"
functions if a "shallow" diagnosis is sufficient for the particular
situation at hand.

Uncertainty Handling

A common and serious limitation of many rule-based expert systems is
that the rules can only be processed in a purely deterministic
manner. For example, the rule:

if A then B

is interpreted as:

if I know that "A" is true, then I know that "B" is true.

However, in power plant (and most other "real world") applications one
is never really certain about either the actual value of "A", or the
relationship between "A" and "B". In these situations, the above rule
should actually be interpreted as:

if I observe that "A" is true, then "B" might also be true.

Although a small number of expert system shell programs incorporate a
provision for treating uncertainty, none (including NEXPERT) treat
uncertainty in a mathematically rigorous manner that is consistent with
the requirements of a power plant diagnostic application. Required
features of an uncertainty model for power plant performance diagnosis
include:

o   The model must be capable of treating measurement uncertainty

(i.e., if I observe that "A" is true, how certain am I
that "A" is actually true) and relational uncertainty
(i.e., if I know for certain that "A" is true, how
certain am I that "B" is true) as separate components of
an overall rule uncertainty. This separation of
uncertainty components is necessary because measurement
uncertainty may vary significantly from instrument to
instrument and plant to plant while relational
uncertainty remains relatively constant.

o   The model must be capable of treating uncertainty in a form
that is conveniently supplied by the domain expert. For
example, experience has shown that performance engineering
experts find it difficult to quantify the relational
uncertainty of the rule expressed above (i.e., if I observe
symptom "A", what is the likelihood that it is caused by
malfunction "B") because symptom "A" may be a condition
common to several malfunctions. However, experts feel much
more comfortable in quantifying the uncertainty of the
converse relationship (i.e., given that malfunction "B" is
true, how certain am I that I should observe symptom "A").

o   The model must be able treat a situation of partial ignorance
about a particular measurement or relationship. For example,
given the following two rules:

if A then B

if not A then C

If "A" is observed to be true with 80% certainty, one should
not automatically assume that "A" is false with 20% certainty
because this assumption "creates" evidence in favor of
conclusion "C" that may not really exist. Unless there exists
some "reason to believe" that "A" is actually false, this
remaining 20% certainty should be treated as ignorance about
the value of "A".

INSPECTR addresses all of the above requirements by evaluating rule
uncertainty using the Dempster-Shafer Theory of Uncertain Evidence[2].
Dempster-Shafer Theory is ideally suited to power plant diagnostic
applications because:

o   It was developed specifically to support an "evidential
reasoning" process in which a conclusion is reached based
upon the accumulation of supporting evidence rather than an
"all-or-nothing" deterministic approach. Dempster-Shafer
Theory is therefore completely consistent with the structure
of the INSPECTR knowledge base.

o   It explicitly treats the concept of partial ignorance through
use of a dual-value measure of certainty (i.e., certainty
about the actual state of a particular parameter is expressed
as two values; the first representing the degree of certainty
that the observed state is true and the second representing
the degree of certainty that the observed state is false).
Since the two certainty values are not required to sum to
unity, any remaining "unassigned" certainty is attributed to
ignorance.

o   It provides an expression for combining uncertainties

(Dempster's Rule) that is a natural extension of Bayesian
Probability Theory and has been demonstrated to be
mathematically rigorous[3]. Dempster's Rule is also
sufficiently straightforward to allow it to be manipulated to
suit the needs of a particular application.

o   It can be implemented in the NEXPERT expert system shell
     program through external routines that are executed after
     successful firing of individual production rules. These
     external routines are included in the EASE+NEXPERT graphical
     data base and interface package described in the following
     subsection.

Dempster-Shafer Theory represents the current state-of-the-art in
uncertainty analysis. Its use in INSPECTR represents a significant
improvement over deterministic or simple Bayesian approaches.

<u>Information Transfer and User Interface</u>

The previous subsections have identified a number of different
information links between the INSPECTR knowledge base and the "outside
world" in the form of the remainder of THERMAC-N and the performance
engineer. These information links are established using Expert-EASE
Systems' EASE+NEXPERT application generation toolkit, portions of which
were also used to develop the user interface and control system for
THERMAC-N. EASE+NEXPERT is a graphics-based data base manager which
provides a friendly interface between the user and a complex
application program. In INSPECTR, EASE+NEXPERT serves as the overall
operating and control environment as well as performing the following
specific functions:

o   Knowledge Base Editing: EASE+NEXPERT provides the capability
     to revise the generic INSPECTR knowledge base to include new
     or plant-specific performance problems as required.
     EASE+NEXPERT also provides security protection to limit the
     availablilty of this feature to the appropriate users.

o   Initiation of the Diagnosis: EASE+NEXPERT triggers execution
     of the NEXPERT shell program upon command from the user. This
     command is provided through user selection of an appropriate
     option from a menu. When INSPECTR is used in conjunction with
     THERMAC-N, this menu selection is offered immediately upon
     entering THERMAC-N environment and immediately upon exiting
     any of the other THERMAC-N modules.

o   Focusing of the Diagnosis: EASE+NEXPERT provides an interface
     between the INSPECTR knowledge base and a graphic
     representation of the plant-specific steam cycle. This
     graphic representation consists of a series of interconnected
     icons representing individual cycle components in the manner
     of a typical heat balance diagram. The user focuses the
     diagnosis on a particular subsystem or component by simply
     placing the cursor on the appropriate icon. When the initial
     focus has been established, EASE+NEXPERT also allows the user
     to specify whether to restrict the diagnosis to the area of
     focus or to broaden the focus if no problem is found there.

o   Control of the Diagnosis: EASE+NEXPERT automatically prepares
     a table of NEXPERT commands to control the diagnosis. User
     interaction with NEXPERT is limited to responding to queries
     for information needed for the diagnosis that is not
     available in the data base.

o   Data Transfer during the Diagnosis: EASE+NEXPERT provides a
    direct link between the INSPECTR knowledge base and the
    detailed analytical model of THERMAC-N. This link greatly
    reduces the need for INSPECTR to query the user for
    information needed for the diagnosis since much of this
    information is readily available in the model. During a
    typical diagnostic session, querying is restricted to
    information that is only available through visual observation
    (e.g., position of an isolation valve) or measurements
    unrelated to a heat balance calculation (e.g., bearing
    temperature of a pump). EASE+NEXPERT also transmits
    information in the reverse direction (knowledge base to
    model) when confirming calculations are required to support a
    particular diagnosis.

o   Results of the Diagnosis: EASE+NEXPERT informs the user of
    the results of the diagnosis by providing a list of likely
    candidate problems and a qualitative or quantitative
    assessment of the likelihood of each candidate. If further
    explanation is required, EASE+NEXPERT provides a complete
    list of symptoms for each candidate together with the
    qualitative or quantitative probability that each symptom was
    observed.

SUMMARY

The INSPECTR Interactive Performance Diagnostic Assistant incorporates
the latest advances in Artificial Intelligence technology (e.g.,
model-based reasoning and evidential reasoning with uncertainty) to
address a problem of generic importance to nuclear power plants.
INSPECTR is being developed specifically to treat the problem of
diagnosing chronic performance problems or precursors to acute problems
based upon an incomplete and uncertain set of information about the
plant state. It is designed to work with the site-based performance
engineer to identify, diagnose and correct performance problems that
would otherwise require a significant allocation of corporate staff
and/or outside consultant resources. When integrated with THERMAC-N,
INSPECTR can automatically accumulate the best and most complete
information available to help the performance engineer perform complex
problem diagnoses.

INSPECTR currently recognizes approximately thirty categories of common
performance problems experienced at typical fossil and nuclear power
plants. As the knowledge base gains maturity through further
development and site testing, the number of diagnosible problems is
expected to increase significantly.

ACKNOWLEDGEMENTS

The authors wish to acknowledge the contributions of Mr. Ron Lucier of
Yankee Atomic Electric Company in developing the INSPECTR generic
knowledge base and those of Ms. Shiao Huang of Expert-EASE Systems in
implementing this knowledge base in the NEXPERT shell program. The
authors also wish to acknowledge the contribution of the technical
staff of Neuron Data, Inc. for their assistance in developing the
EASE+NEXPERT expert systems environment, and that of the United States
Department of Energy for providing the funding for this development
effort through DOE Contract No. DE-AC03-85ER80247.

REFERENCES

1. <u>NEXPERT OBJECT Fundamentals.</u> Palo Alto, CA: Neuron Data Inc., 1986.

2. G. Shafer. "Mathematical Theory of Evidence." Princeton, NJ: Princeton University Press, 1976.

3. G. Liu. "Causal and Plausible Reasoning in Expert Systems." Berkeley, CA: University of California at Berkekey, 1987.

4. <u>EASE+ TOOLS User Application Manual.</u> Belmont, CA: Expert-EASE Systems Inc., December, 1986.

PRODUCTION-RULE ANALYSIS SYSTEM FOR NUCLEAR PLANT
TECHNICAL SPECIFICATIONS TRACKING

M. Ragheb, M. Abdelhai and J. Cook

Department of Nuclear Engineering
University of Illinois at Urbana-Champaign
103 S. Goodwin Ave., Urbana, Illinois 61801

Illinois Power Company, Clinton Power Station
P. O. Box 678, Clinton, Illinois 61727

ABSTRACT

A Model-Based Production-Rule Analysis System is developed for the tracking of the Limiting Conditions for Operation (LCOs) and the Surveillance Requirements in power plants. These LCOs and Surveillance Requirements are delineated in Power Station's Technical Specifications. The system is constructed as a Production-Rule Analysis System having the same structure as the E-Mycin system with a Natural Language Interface. The Goal-Tree constituting the Knowledge-Base is generated and is searched by the Inference Engine using Backward-Chaining with a Deduction-Oriented Antecedent-Consequent Logic. The performance of the system is demonstrated in the identification and tracking of the LCOs and Surveillance Requirements and the associated applicable actions for maintaining the operability status of the components in the plant reactivity control systems.

INTRODUCTION

Model-Based Production-Rule Analysis System[1,2], using the Rule-Based Paradigm from the field of Knowledge Engineering[3] is constructed for the tracking and scheduling of the limiting conditions for operation and the surveillance requirements in power plants. Model-Based systems are distinguished from Expert Systems by their modelling of engineeering devices based on knowledge about their structure and behavior, rather than on human expertize alone. Technical Specifications of power plants contain a large volume of interrelated and sometimes stringent specifications that are often modified and are presented in a relatively complex manner in a large manual. This creates situations where utilities' plants unintentionally fall into a noncompliance status. The need to determine the plant status, assure compliance and provide increased operating flexibility, justifies the adoption of the presented approach. The methodology also allows the operators to account for plant modifications through simple modifications to the Knowledge-Base. This is a distinct advantage of symbolic programming methodologies[4], in contrast to procedural programming methods which possess a rigid structure. The latter structure does not readily lend itself to changes once a program is written and debugged. In addition, the symbolic programming approach offers the capability of explaining to a user how the inferences are reached and why a certain inquiry is generated by providing to the user direct access to its Knowledge-Base.

The Technical Specifications vary from plant to plant, but in general they comprise the following items: 1. Definitions, 2. Safety Limits and Limiting Safety System Settings, 3.

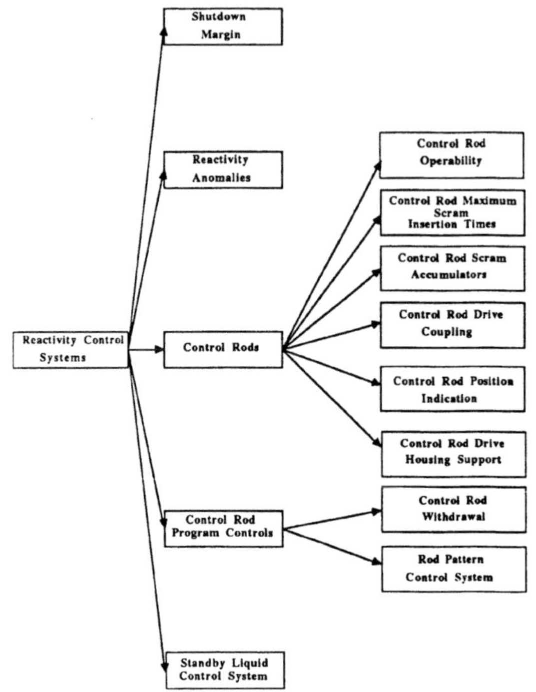

Fig. 1   Relationships between plant functions for the Reactivity
Control Systems

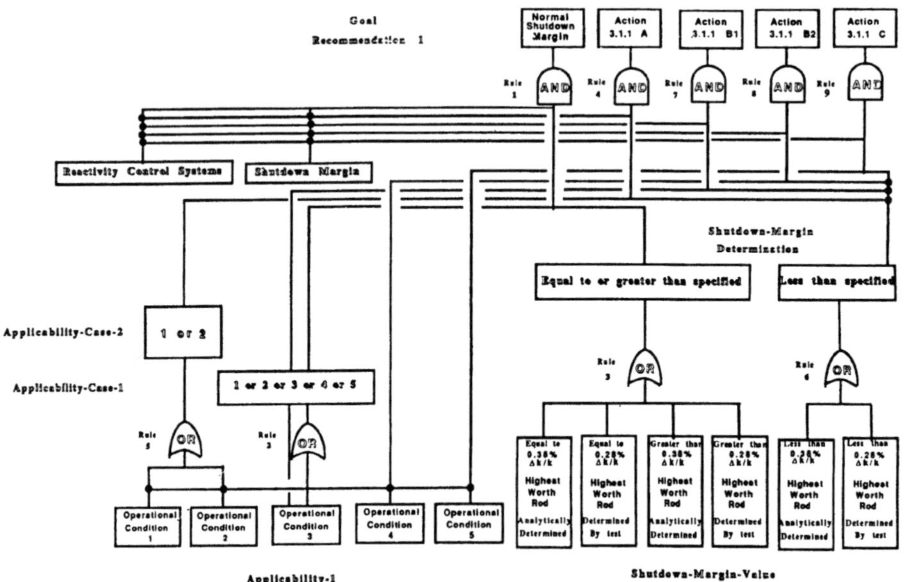

Fig. 2   Goal-Tree Representation of the Plant's Shutdown
Margin Technical Specifications

Limiting Conditions for Operation (LCO) and Surveillance Requirements, 4. Design Features, 5. Administrative Controls.

Each of these item has further subsections which allows a description of the system in the form of a tree structure. For instance, the LCOs and Surveillance Requirements include the following subsections: 1. Reactivity Controls, 2. Power Distribution Limits, 3. Instrumentation, 4. Reactor Coolant System, 5. Emergency Core Cooling Systems, 6. Containment Systems, 7. Plant Systems, 8. Electrical Power Systems , 9. Refuelling Operations , 10. Special Test Exceptions.

In turn, each item contains further detail. As an example, the Plant Systems comprise: 1. Service Water Systems, 2. Control Room Ventilation System, 3. Reactor Core Isolation Cooling System, 4. Snubbers, 5. Sealed Source Contamination, 6. Fire Suppression Systems, 7. Fire Rated Assemblies, 8. Main Turbine Bypass System.

Each one of these in turn would be further detailed as to: 1. Limiting Conditions for Operation, 2. Applicability, 3. Action Items , 4. Surveillance Requirements.

To demonstrate the proposed methodology, and to avoid complexity, a single subsystem will be considered in the following section, followed by corresponding sample case results.

PROBLEM FORMULATION

We consider for the present analysis the limiting conditions for operations and the surveillance requirements part of the Clinton's Plant Technical Specifications. We further concentrate on the Reactivity Control Systems and their subsystems as shown in Fig. 1. These include the Shutdown Margin, the Reactivity Anomalies, the Control Rods, the Control Rod Program controls, and the Standby Liquid Control System.

The Control Rods system is further expanded as to Control Rod Operability, Control Rod Maximum Scram Insertion Times, Control Rod Scram Accumulartors, Control Rod Drive Coupling, Control Rod Position Indication and Control Rod Drive Housing Support. Similarly, the control Rod Program Controls system is expanded into the Control Rod withdrawal and the Rod Pattern Control subsystems.

A Goal-Tree is constructed with the top events as the Action Levels corresponding to the considered subsystem. In the case of Shutdown Margin subsystem these are identified in Table 1, and shown at the top of the Goal-Tree of Fig. 2 as Goals or Recommendations as to the appropriate course of action. Notice that a normal shutdown margin state is defined for the purpose of this work, even though not normally included in the Technical Specifications.The reason for this addition is that the monitoring of a system requires in general the identification of both its normal and abnormal states.

The Inference Engine adopted is Personal Consultant[3]. It possesses the same structure as the E-Mycin system[6], with an added Natural Language interface. It operates in the Backward-Chaining mode approach. It hypothesizes consecutively that each one of the goals is true, and moves down the Goal-Tree to test whether its hypothesis is true or false. Once the bottom of the tree is reached, facts are needed with respect to applicability conditions and the shutdown margin values. These facts can be obtained by the system through polling sensors data, or querying the user for their values. The facts constituting the operational conditions 1 to 5 are defined in Table 2.

The Goal-Tree which constitutes a model of the functional behavior of the system is next translated into a knowledge-base by expressing each of the logical gates in the tree in the form of an if/then rule. Figure 3 shows the form that the AND gate in Fig. 2 designated as Rule 4 takes when translated into and if/then rule. It is expressed in both a LISP language version[4] and an English language version available through the Natural Language interface.

Table 1 . Actions required for the system shutdown in accordance with
the given surveillance requirements

Action Name	Action Description
1. Normal Shutdown Margin	No Action is needed
2. Action 3.1.1 A	Reestablish the required SHUTDOWN MARGIN within 6 hours or be in at least HOT SHUTDOWN within the next 12 hours
3. Action 3.1.1 B1	Immediately verify all insertable control rods to be inserted and suspend all activities that could reduce the SHUTDOWN MARGIN
4. Action 3.1.1 B2	Establish SECONDARY CONTAINMENT INTEGRITY within 8 hours
5. Action 3.1.1 C	Suspend CORE ALTERATIONS(Except movement of IRMs, SRMs or special movable detectors) that could reduce the SHUTDOWN MARGIN and insert all insertable control rods within 1 hour. Establish SECONDARY CONTAINMENT INTEGRITY within 8 hours

Table 2 Definitions of Operational Conditions used in applicability
items

Condition	Mode Switch Position	Average Reactor Coolant Temperature
1. Power Operation	Run	Any Temperature
2. Startup	Startup/Hot Standby	Any Temperature
3. Hot Shutdown	Shutdown	>200°F
4. Cold Shutdown	Shutdown	≤200°F
5. Refuelling	Shutdown or Refuel	≤140°F

```
RULE004 [LIMITING-CONDITIONS-FOR-OPERATION-AND-SURVEILLANCE-REQUIREMENTSRULES]
-------
If 1) LIMITING-CONDITION-SECTION is
        REACTIVITY-CONTROL-SYSTEMS, and
    2) REACTIVITY-CONTROL-SYSTEM is SHUTDOWN-MARGIN, and
    3) APPLICABILITY-CASE-2 is ONE-OR-TWO, and
    4) SHUTDOWN-MARGIN-DETERMINATION is LESS-THAN-SPECIFIED.
Then 1) it is definite (100%) that the recommended action is
        ACTION 3.1.1 A, and
     2) Inform the user of this decision.


PREMISE:  ($AND
            (SAME CNTXT LIMITING-CONDITION-SECTION
            REACTIVITY-CONTROL-SYSTEMS)
            (SAME CNTXT REACTIVITY-CONTROL-SYSTEM
            SHUTDOWN-MARGIN)
            (SAME CNTXT APPLICABILITY-CASE-2 ONE-OR-TWO)
            (SAME CNTXT SHUTDOWN-MARGIN-DETERMINATION
            LESS-THAN-SPECIFIED))
ACTION:   (DO-ALL
            (CONCLUDE CNTXT RECOMMENDATION "ACTION 3.1.1 A"
            TALLY 1000)
            (SPRINTT "Since the SHUTDOWN MARGIN is determined
            to be less than specified and the APPLICABILITY
            is considered as the OPERATIONAL CONDITION 1 OR
            2, then the needed action is ACTION 3.1.1 A (Please
            use the HELP KEY for mor information about this
            action)."))
```

Fig. 3 The Lisp and the English Language versions of Rule 4

CASE RESULTS

When an operator starts a consultation session with the Production-Rule system, the first screen that will appear is the one shown in Fig. 4. This screen introduces the user to the system and together with the subsequent screens, guide the user through an interactive process.

Figure 5 shows the query to the operator as to the system that he wants to check under the title of Reactivity Control Systems. The user chooses one of these systems for further analysis upon moving a cursor up or down on the screen.

Figure 6 shows the query by the system to the user as to the facts about operational conditions previously defined in Table 2, and regarding the numerical values of the shutdown margins.

Having reached an inference about a recommended action, the system displays the screens shown in Fig.7 where the inference in one case is that no action is needed and in the other case action 3.1.1 C as defined in Table 1 is recommended. In this case the recommended action is to Suspend core alterations that could reducee the shutdown margin and insert all insertable control rods within an hour, and to establish secondary containment integrity within 8 hours.

A major characteristic of the system is its transparency and its ability to provide explanation to a user about its inferencing process. This is exemplified by its ability to answer questions by the user as to how it has reached certain inferences. Figure 8 displays the screen that offers the user such an opportunity, as well as the screen that explains to him how the inference about the recommendation was reached through the application of Rule 1.

Another useful characteristic is the capability of the system to explain why certain facts are being sought by the system from the user. Such a feature helps the user in providing an adequate answer to the inquiry. Figure 9 displays an answer as to "why" the question about the Reactivity control system is needed. In this case the system is checking the premises of Rule 1, and needs such informatioon from the user to conduct its inductive logic process.

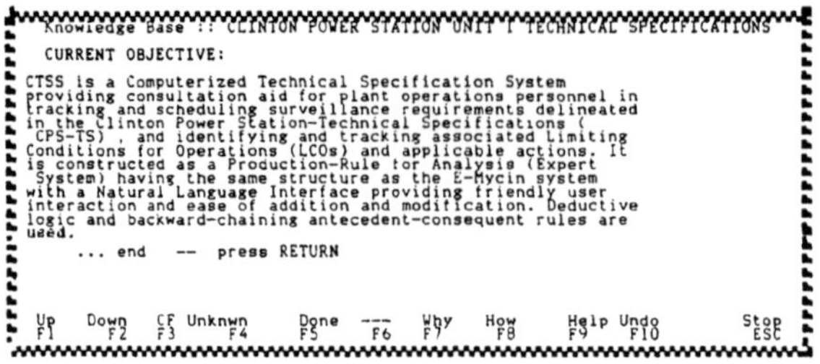

```
Knowledge Base :: CLINTON POWER STATION UNIT I TECHNICAL SPECIFICATIONS

 CURRENT OBJECTIVE:

CTSS is a Computerized Technical Specification System
providing consultation aid for plant operations personnel in
tracking and scheduling surveillance requirements delineated
in the Clinton Power Station-Technical Specifications (
 CPS-TS) , and identifying and tracking associated Limiting
Conditions for Operations (LCOs) and applicable actions. It
is constructed as a Production-Rule for Analysis (Expert
 System) having the same structure as the E-Mycin system
with a Natural Language Interface providing friendly user
interaction and ease of addition and modification. Deductive
logic and backward-chaining antecedent-consequent rules are
used.

    ... end   --   press RETURN


 Up    Down   CF Unknwn     Done   ---   Why   How    Help Undo       Stop
 F1     F2    F3      F4     F5     F6    F7    F8     F9   F10         ESC
```

Fig. 4 The screen display corresponding to the Promptever statement

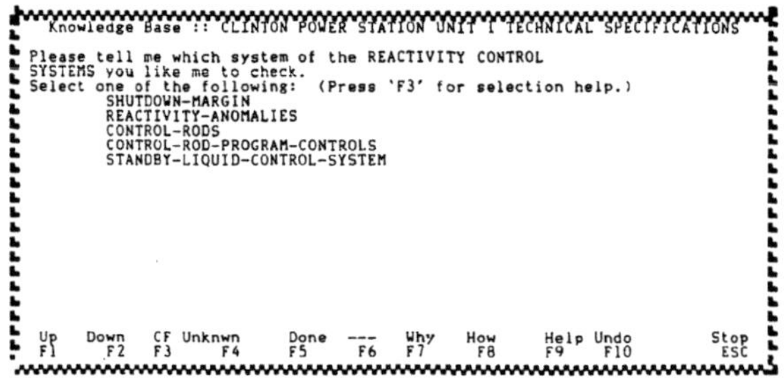

```
 Knowledge Base :: CLINTON POWER STATION UNIT I TECHNICAL SPECIFICATIONS

Please tell me which system of the REACTIVITY CONTROL
SYSTEMS you like me to check.
Select one of the following:  (Press 'F3' for selection help.)
        SHUTDOWN-MARGIN
        REACTIVITY-ANOMALIES
        CONTROL-RODS
        CONTROL-ROD-PROGRAM-CONTROLS
        STANDBY-LIQUID-CONTROL-SYSTEM


 Up    Down   CF Unknwn     Done   ---   Why   How    Help Undo       Stop
 F1     F2    F3      F4     F5     F6    F7    F8     F9   F10         ESC
```

Fig. 5 Choice of systems under Reactivity Control Systems for
further analysis

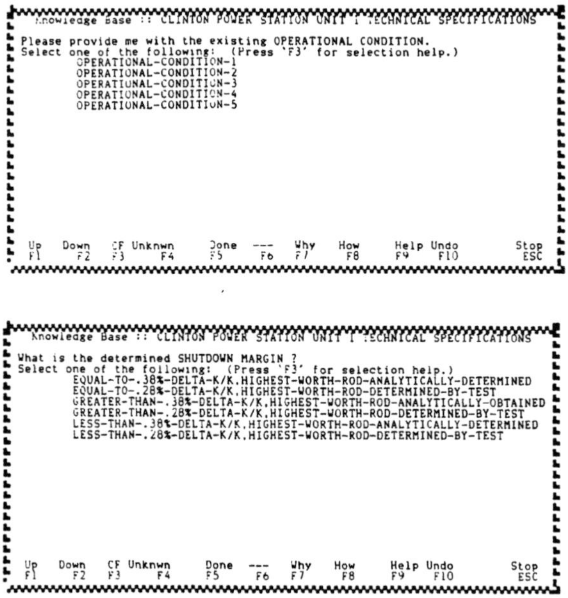

```
 Knowledge Base :: CLINTON POWER STATION UNIT I TECHNICAL SPECIFICATIONS

Please provide me with the existing OPERATIONAL CONDITION.
Select one of the following:  (Press 'F3' for selection help.)
        OPERATIONAL-CONDITION-1
        OPERATIONAL-CONDITION-2
        OPERATIONAL-CONDITION-3
        OPERATIONAL-CONDITION-4
        OPERATIONAL-CONDITION-5


 Up    Down   CF Unknwn     Done   ---   Why   How    Help Undo       Stop
 F1     F2    F3      F4     F5     F6    F7    F8     F9   F10         ESC
```

```
 Knowledge Base :: CLINTON POWER STATION UNIT I TECHNICAL SPECIFICATIONS

What is the determined SHUTDOWN MARGIN ?
Select one of the following:  (Press 'F3' for selection help.)
        EQUAL-TO-.38%-DELTA-K/K,HIGHEST-WORTH-ROD-ANALYTICALLY-DETERMINED
        EQUAL-TO-.28%-DELTA-K/K,HIGHEST-WORTH-ROD-DETERMINED-BY-TEST
        GREATER-THAN-.38%-DELTA-K/K,HIGHEST-WORTH-ROD-ANALYTICALLY-OBTAINED
        GREATER-THAN-.28%-DELTA-K/K,HIGHEST-WORTH-ROD-DETERMINED-BY-TEST
        LESS-THAN-.38%-DELTA-K/K,HIGHEST-WORTH-ROD-ANALYTICALLY-DETERMINED
        LESS-THAN-.28%-DELTA-K/K,HIGHEST-WORTH-ROD-DETERMINED-BY-TEST


 Up    Down   CF Unknwn     Done   ---   Why   How    Help Undo       Stop
 F1     F2    F3      F4     F5     F6    F7    F8     F9   F10         ESC
```

Fig. 6 System queries regarding the operational condition and
the shutdown margin

```
Knowledge Base :: CLINTON POWER STATION UNIT I TECHNICAL SPECIFICATIONS

Since the SHUTDOWN MARGIN is determined to be less than specified
and the APPLICABILITY is considered as the OPERATIONAL CONDITION
5, then the needed action is ACTION 3.1.1 C (Please use the HELP
KEY for more information about this action).
    ... end   --   press RETURN




Up    Down   CF Unknwn      Done  ---   Why   How    Help Undo      Stop
F1      F2   F3     F4      F5     F6   F7    F8     F9   F10         ESC
```

Fig. 7 Displays of system recommendation as to the appropriate actions

```
Knowledge Base :: CLINTON POWER STATION UNIT I TECHNICAL SPECIFICATIONS

 HOW did the system determine the value of ...
 Select one of the following:
        RECOMMENDATION of LIMITING-CONDITIONS-FOR-OPERATION-AND-SURVEILLANCE
        SHUTDOWN-MARGIN-DETERMINATION of LIMITING-CONDITIONS-FOR-OPERATION-A
        SHUTDOWN-MARGIN-VALUE of LIMITING-CONDITIONS-FOR-OPERATION-AND-SURVE
        APPLICABILITY-CASE-1 of LIMITING-CONDITIONS-FOR-OPERATION-AND-SURVEI
        APPLICABILITY-1 of LIMITING-CONDITIONS-FOR-OPERATION-AND-SURVEILLANC
        REACTIVITY-CONTROL-SYSTEM of LIMITING-CONDITIONS-FOR-OPERATION-AND-S
        LIMITING-CONDITION-SECTION of LIMITING-CONDITIONS-FOR-OPERATION-AND-
SURVEILLANCE-REQUIREMENTS-1




Up    Down   CF Unknwn      Done  ---   Why   How    Help Undo      Stop
F1      F2   F3     F4      F5     F6   F7    F8     F9   F10         ESC
```

```
Knowledge Base :: CLINTON POWER STATION UNIT I TECHNICAL SPECIFICATIONS


RECOMMENDATION was determined to be:   NORMAL SHUTDOWN-MARGIN
with certainty 100% by using RULE001

If 1) LIMITING-CONDITION-SECTION is
      REACTIVITY-CONTROL-SYSTEMS, and
   2) REACTIVITY-CONTROL-SYSTEM is SHUTDOWN-MARGIN, and
   3) APPLICABILITY-CASE-1 is
      ONE-OR-TWO-OR-THREE-OR-FOUR-OR-FIVE, and
   4) SHUTDOWN-MARGIN-DETERMINATION is
      EQUAL-TO-OR-GREATER-THAN-SPECIFIED.
Then 1) it is definite (100%) that the recommended action is
        NORMAL SHUTDOWN-MARGIN, and
     2) Inform the user of this decision.

     ... end   --   press RETURN

Up    Down   CF Unknwn      Done  ---   Why   How    Help Undo      Stop
F1      F2   F3     F4      F5     F6   F7    F8     F9   F10         ESC
```

Fig. 8 Choices offered to the user on "How" inferences were reached,
        and ensuing explanation for an inference

```
Knowledge Base :: CLINTON POWER STATION UNIT I TECHNICAL SPECIFICATIONS

Please tell me which system of the REACTIVITY CONTROL
SYSTEMS you like me to check.
 Why this question is needed:

REACTIVITY-CONTROL-SYSTEM is needed to determine the
recommended action

RULE001
If 1) LIMITING-CONDITION-SECTION is
      REACTIVITY-CONTROL-SYSTEMS, and
   2) REACTIVITY-CONTROL-SYSTEM is SHUTDOWN-MARGIN, and
   3) APPLICABILITY-CASE-1 is
      ONE-OR-TWO-OR-THREE-OR-FOUR-OR-FIVE, and
   4) SHUTDOWN-MARGIN-DETERMINATION is
      EQUAL-TO-OR-GREATER-THAN-SPECIFIED,
Then 1) it is definite (100%) that the recommended action is
        NORMAL SHUTDOWN-MARGIN, and
     2) Inform the user of this decision.
     more ...   --   press RETURN
Up    Down   CF Unknwn      Done  ---   Why   How    Help Undo      Stop
F1      F2   F3     F4      F5     F6   F7    F8     F9   F10         ESC
```

Fig. 9 Example of system response to a "Why" inquiry by the user

SUMMARY AND CONCLUSIONS

A Model-Based Production-Rule Analysis system is developed for the tracking and scheduling of the Limiting Conditions for Operation and the Surveillance Requirements as part of the Technical Specifications of power plants. As presently implemented, the system functions as a consultation aid for the plant operations personnel. The Knowledge-Base models the functions of the plant Shutdown Margin, the Reactivity Anomalies, the Control Rod Programs Controls, and the Standby Liquid Control Systems. The Control Rods subsystem considers the control rods operability, their maximum scram insertion times, their scram accumulators, the Control Rod Drive coupling, the position indication, and the drive housing support. The control rod Program controls in turn include the rod withdrawal and the rod pattern control system The Production-Rule analysis system is capable of tracking and maintaining the operability status of the components affecting the plant reactivity control systems. An advantage of the system for the intended application can be identified as providing quick access to information that is normally accessed by looking through thick manuals in search of specific information . As such it provides the prospect of the availability of efficient anf speedy consultation aids to plant operators under both normal and accident operational situations. Another advantage is that it lends itself to easy modification of its Knowledge-Base by plant personnel as plant modifications are implemented. An equivalent procedural programming system would possess a rigid structure and would not provide the plant personnel with this capability. Another useful feature is the ability to provide explanation and tutoring through answering questions about why a query is generated and how an inference is reached. These advantages are countered by a need for high processing speeds and large storage memories for the handling of such heuristic problems . With the introduction of such capabilities in newer generation computers it will be possible for these applications to be beneficially used.

ACKNOWLEDGEMENTS

This work was partially supported by the Illinois Power Company, Texas Instruments, and the Illinois Alumni Foundation. The help and support by J. Cook, R. Morgenstern and D. Hall from the Illinois Power Company is gratefully acknowledged.

REFERENCES

1. Ragheb, M. and Gvillo, D.,"Development of Model-Based Fault Identification Systems on Microcomputers," Applications of Artififial Intelligence III, SPIE Vol. 635, p.268 (1986).

2. Ragheb, M. and Gvillo, D. "Symbolic Simulation of Engineering Systems on a Supercomputer," Applications of Artificial Intelligence III, SPIE Vol. 635, p.368 (1986).

3. Sims. H. A., "Knowledge Engineering with the TI's Personal Consultant," TI professional Computing, Dec. (1985).

4. Winston, P. H. and Horn, B. K. H., "Lisp," second edition, Addison-Wesley Publishing Co.(1984).

5. Ragheb, M. and Youel, D. "Model-Based Analysis Systems for Plant Instrumentation," Proc. Frontiers of Power Conference, Oct. 27-28, Stillwater, Oklahoma (1986).

6. Buchanan B. G. and Shortliffe E. H., "Rule-Based Expert Systems, The Mycin Experience of the Stanford Heuristic Programming Project," Addison-Wesley Publishing Co., NY (1984).

THE USE OF PROLOG FOR COMPUTERIZED TECHNICAL SPECIFICATIONS

L.M. Lidsky, D.D. Lanning, A.B. Dobrzeniecki, K. Meyers,
and T.G. Reese

Department of Nuclear Engineering
The Massachusetts Institute of Technology
Cambridge, MA 02139

ABSTRACT

The task of deciding whether a particular nuclear plant condition is
in compliance with its Technical Specifications is, in principle,
algorithmic. However, the implicit algorithm is so complex and the
required level of confidence in code validity is so high that traditional
methods of computerized analysis have proven unsatisfactory.
PROLOG provides a close match to the explicit semantic structure of the
written specifications, a good tool for analysis of the implicit subsystem
operability trees and, by virtue of its declarative style and imbedded
logical structure, greatly simplifies the problem of code validation.
LCOM, a PROLOG representation of a subset of BWR Technical Specifications,
was developed to explore the applicability of PROLOG in this problem
domain and to provide benchmark data for full-scale implementations. The
declarative style and built-in unification mechanism of PROLOG does
substantially simplify the process of program construction and validation.
Extrapolations indicate processing times for a complete Tech Spec
implementation of $\lesssim$ 1 minute in systems capable of $\gtrsim$ 20 KLIPS.

INTRODUCTION

As part of the licensing procedure for commercial nuclear power
plants, the Nuclear Regulatory Commission (NRC) requires a complete safety
analysis of the design and installation. This requirement is detailed in
the Code of Federal Regulations, Title 10, Part 50 (10CFR 50). Based on
the results of the approved safety analysis, "Technical Specifications",
as defined by 10CFR 50.36, must be prepared and approved by the NRC. The
purpose of the Technical Specifications is to define the plant status and
operating conditions that must be maintained in order to ensure that the
plant remains within the design envelope that has been specifically
analyzed in the safety analysis reports.

The Technical Specifications for nuclear power reactors are a
collection of explicitly-defined, interrelated rules which must be
interpreted in light of plant conditions, plant history, and subsidary
sets of plant-specific rules. It is necessary to know, at all times,
whether specific plant configurations or postulated reconfigurations are
in compliance with the Technical Specifications. This is not always an
easy task, even for simple plant maneuvers and, as the number of

simultaneous deviations from the "nominal" operating state increases, the application of the rules becomes increasingly complex. As a result, utilities occasionally find themselves in totally inadvertent violation of their Technical Specifications. Even when plant safety is only minimally compromised, such violations can subject the utility to substantial financial penalty. Another financial penalty is incurred when maintainence scheduling is suboptimal because Tech Spec related actions are complex and difficult to manage.

STRUCTURE

Although Technical Specifications vary from plant to plant, there has evolved a uniform format, exemplified by the Standard Technical Specifications for General Electric Boiling Water Reactors (NUREG-0123, rev 3). NUREG-0123 was intended to serve as a framework for individual license applications, needing only slight revision to suit particular design solutions. These changes would appear as an appendix to the pre-approved generic BWR/5 specifications provided by NUREG-0123. These "standard Tech Specs" were chosen as the test bed for our system because of their wide availablility and broad applicability.

The standard Tech Spec consists of six numbered sections, covering the areas of:

1. Definitions
2. Safety Limits and Limiting Safety Design System Settings
3. Limiting Conditions for Operation
4. Surveillance Requirements
5. Design Features
6. Administrative Controls

Sections 3 and 4 are the core of the Tech Specs, detailing the allowable operating regimes and surveillance requirements, and specifying the particular actions to be followed when the required conditions are violated.

Sections 3 and 4 of the standard Technical Specifications comprise 122 separate sections, each detailing the operational envelope, surveillance requirements, and required actions relevant to identified plant subsystems. These requirements occupy 275 pages of text.

Figure 1 reproduces the Limiting Conditions for Operation (LCO) portion of section 3.5.1 of the Standard BWR/5 Tech Specs. The full text also includes a listing of the actions that must be taken if the LCO's are violated. The words appearing in upper case in the text are all explicitly defined in Section 1 of the Tech Specs. The numbered operational conditions refer to such operating modes as power operations, startup, hot shutdown, etc.

The structure of the section illustrated is rather more complex than is apparent for two reasons, one having to do with the explicit definition of OPERABLE, the other depending upon the necessity to derive the intent of unspecified terms. OPERABLE is defined in Section 1 of the Tech Specs by "...A system, subsystem, train, component or device shall be OPERABLE or have OPERABILITY when it is capable of performing its specified function(s) and when all necessary attendant instrumentation, controls, a normal and an emergency electrical power source, cooling or seal water, lubrication or other auxiliary equipment that are required for the system, subsystem, train, component or device to perform its function(s) are also capable of performing their related support function(s)..."

3.5.1 ECCS divisions 1,2 and 3 and the automatic depressurization system (ADS) shall be OPERABLE with:

    a.   ECCS division 1 consisting of:

1. The OPERABLE low pressure core spray (LPCS) system with a flow path capable of taking suction from the suppression chamber and transferring the water through the spray sparger to the reactor vessel.

2. The OPERABLE low pressure coolant injection (LPCI) subsystem "A" of the RHR system with a flow path capable of taking suction from the suppression chamber and transferring the water to the reactor vessel.

3. (At least) (7) OPERABLE ADS valves.

    b.   ECCS division 2 consisting of:

1. The OPERABLE low pressure coolant injection (LPCI) subsystems "B" and "C" of the RHR system, each with a flow path capable of taking suction from the suppression chamber and transferring the water to the reactor vessel.

2. (At least) (7) OPERABLE ADS valves.

    c.   ECCS division 3 consisting of the OPERABLE high pressure core spray (HPCS) system with a flow path capable of taking suction from the suppression chamber and transferring the water through the spray sparger to the reactor vessel.

APPLICABILITY: OPERATIONAL   CONDITION   1, $2^{*\ \#}$ and 3^{*}.

---

[*]The ADS is not required to be OPERABLE when reactor steam dome pressure is less than or equal to (113) psig.

[#]See Special Test Exception 3.10.6.

FIGURE 1. LCO Portion of Section 3.5.1 of Standard Technical Specifications for General Electric BWR/5 Boiling Water Reactors.

The definition of OPERABLE thus extends the domain of interest from the explicitly named subsystems in the Tech Specs themselves to the much larger domain of auxiliary systems, back-up devices, individual components, and even to the alignment of multiple position components such as valves or switches. There are typically 10,000 - 30,000 individually tracked components in a nuclear reactor power plant data base that might, under some conditions, affect Tech Spec interpretation.

The second complicating issue has to do with the need to infer the intent of the NRC in interpretation of terms not explicitly defined. An example of this is the construct "upon being manually realigned" which appears repeatedly in the Tech Specs. It is not sufficient to show that the specific function named could be carried out if the requisite valves were brought into a specific configuration. It is a reasonable assumption that the NRC intended the licensee to show that the valves could be so re-aligned, under pertinent plant conditions. Conceivably, one would have to show that the valves were in an accessible location, that the radiation level in the valve area was sufficiently low, that an additional operator could be deployed to the valve location, etc.

There are two common causes for utility violation of reactor Tech Specs. The first lies in their obvious complexity; for example a minor change in one subsystem in a particular reactor operating mode may place additional constraints upon another subsystem which is applicable only in a different operating mode. Violations are encountered when the mode is changed. Although changes in operating condition are a common cause of Tech Spec violation, it is easily possible for a reactor operator, no matter how experienced, to overlook some of the more complex interactions taking place as various components belonging to one or another system are taken out of service for one reason or another during normal operation. The second common cause of violation is differing interpretation of undefined concepts by different operators at the same facility.

CHOICE OF PROLOG

The task of Technical Specification mechanization involves a large and complex set of logical rules, subordinate sets of rules, and a large database with complex arguments. It is anticipated that both top level rules and subordinate rules will be modified from time to time and it is desirable that such changes be easily accomodated by the system. (The database will, of course, be constantly changing). It is particularly important in the case of nuclear power plant Technical Specifications that the system logic and <u>programming</u> be easily verified.

The possible role of computers in assuring consistent compliance with reactor Technical Specifications is apparent, and several systems attempting to meet this goal have been produced. However, these systems, using traditional programming tools, have not been entirely satisfactory. Classical programming methods are not well suited to handle tasks which require resolution of logical complexity or involve substantial symbolic manipulation. The resulting software is difficult to develop and, of greater significance, difficult to verify and to validate and very hard to maintain in the event of change. The tightly interconnected logical systems tend to make imperative-style programs very brittle; i.e., small changes can have cascadingly large and sometimes unanticipated effects.

Attempts have also been made to computerize Technical Specifications using various commercially available "Expert System Shells". These have also proved disappointing because the logical relationships and data structures available within this format are not sufficiently complex to map the equivalent structures of the Technical Specifications. The logical relationships in the Technical Specifications are rigorously defined and the database is given. Therefore, reliance on heuristics is neither necessary nor desired.

The combination of complex logical structure, requisite clarity of representation, and complex multi-element database makes it apparent that PROLOG might be well matched to the problem domain.

LCOM SYSTEM DESIGN

LCOM is a Limiting Condition of Operation Monitor for a subset of the Tech Specs. The heart of LCOM is a representation of the logical structure of the Tech Specs is such form that compliance can be determined solely by program-driven query of the relevant database. In principle, the Technical Specification document itself is the explicit program specification. To the extent that explicitly defined terms ("named entities") were involved, the actual specifications were treated as precisely that and translated as directly as possible into line by line PROLOG equivalents. Every named entity in the Tech Spec documentation was treated as the root of a tree and required to have a truth value. The implementation at this level was dominated by the concern to preserve, as much as possible, the transparent nature of the logical structure of the Tech Specs themselves. Both the named entities and the undefined concepts are represented by operability tress which are, in essence, the main processing tool used in LCOM. The explicit Technical Specifications can be considered as the set of macroscopic goals that the underlying operability trees attempt to prove as true or false.

Unlike the macroscopic goals set by the Technical Specifications, the operability trees are not fixed with respect to structure or content; the trees will vary according to the specific plant being considered and will be periodically modified to reflect changes made in the plant. The operability trees used in LOOM can be divided into two broad categories: those that represent a strict application of failure modes and effects analyses to entire systems and subsystems, and those trees that are more loosely structured and attempt to capture knowledge of the plant through the use of carefully chosen rules. Both types of trees look about the same when actually coded, but those underlying the named entities draw more directly upon the explicit plant component database.

An operability tree is invoked by a call to its first clause, which generally is the name of the tree it represents. Thus, lpci_A_inop is the head clause of the operability tree that determines the status (inop or op) of train A of the low-pressure core injection system. Following the head clause, there are a number of additional clauses that are evaluated as part of the called tree. Each clause will, in general, involve a search of the plant database to determine if the conditions required by the clause are met or not. The conditions of the clause are usually related to component status or parameter values. Some clauses will perform comparisons involving data retrieved from the database. In this way, the initial call to a tree sets in motion a detailed analysis of plant conditions; the operability tree returns a single conclusion to the function that initially invoked the tree. The operability trees thus begin with a top level fault goal which is backward-chained until a terminal node is reached. These trees differ from the tree structures that are usually involved in backward-chaining inference procedures because, unlike conventional node and leaf structures, they do not return the first terminal node that is found, but rather the single unique value that is associated with the plant database. This is a key design feature because it insures that the processing done by the operability trees is controlled bythe procedural semantics of PROLOG, providing a testable, transparent and auditable solution path.

Another important feature of the standard operability trees is that the search mechanism through the operability trees is implicitly specified by the built-in PROLOG resolution mechanism. This permits a knowledgable programmer sufficient control to ensure rapid search.

The operability trees underlying the _undefined_ entities or concepts in the Technical Specifications cannot be based on a strict functional analysis, but rather must represent a heuristic development based on particular plant structure. These trees, when invoked, often require a higher level of interaction with the plant operator, because they often depend on none-database plant conditions. The advantage of including such heuristicly based trees and their explicit coding is that it ensures uniform interpretation of the specifications by all plant personnel and provides written documentation of the procedures for the NRC.

Other design decisions were driven by required functionality and by existing practices in nuclear power stations. The issues were primarily involved with the database and auxillary structures. The NRC requires an auditable (i.e. permanently stored historical) trail of all safety related changes in plant status including the change itself, the time of the change and the authorization for the change. To the extent that every item in the database affects the interpretation of the Tech Specs, every item in the database needs a traceable history. On the other hand, the operators prefer to interact with the database on a fairly high conceptual level (named entity or major subsystem) while the engineering staff needs to consider items or a single component for purpose of required repair, and for scheduling NRC mandated tests.

The solution to this problem is the provision of several databases
with the "actual" database heavily safeguarded, frequently archived, and
access-limited by coded hardware "keys" of the sort usually intended for
software protection. Each key is associated with a given password and
assigned to authorized operators. Changes to the auditable database
require both hardware key and proper password. Other "hypothetical"
databases are provided which allow on-;ine, rapid, high-level interaction
with the Tech Specs for tactical and strategic planning.

The other database related problem was simpler in origin but far more
difficult of solution. Nuclear plants do not have available an accurate
up-to-date on-line database of all the items needed to determine Tech Spec
compliance [This, of course, is a major reason for the observed frequency
of violation.] The task of creating, and maintaining, the requisite
database is of such magnitude that the incentive of much improved Tech
Spec compliance is not sufficient motivation. The solution to this
problem is the development of a database with broad enough utility that
its development and maintainance could be easily justified. The net
effects of the constraint were the requirement that the database not be
tailored to optimize of Tech Spec monitor at the expense of other uses.

LCOM IMPLEMENTATION

LCOM was implemented in PROLOG 2 and utilized in a specially modified
version of the ES/P advisor shell provided by Expert Systems International
for prototyping purposes. PROLOG 2 is an expanded Edinburgh syntax
PROLOG that operates under DOS, VAX VMS and UNIX. Figure 2a illustrates
the translation of the Fig. 1 Tech Specs directly into code at the upper
syntactical level. Figure 2b is a representative operability tree. Note
that a "system level" declaration of (in)operability obviates the need for
component-level database search.

LCOM as it currently exists is approximately 400K Bytes in size.
This figure, which excludes the commercial shell under which the monitor
runs, represents two sections of the BWR Generic Tech Specs (Section

```
/*------------Determining Plant Status Parameter/Sec 3.5.1------------*/
eccs_1_351 : 'The first division of section 3.5.1 is inoperable'
        fact
        rule
            true is (lpcs_inop or lpci_A_inop or
                    (num_adsvlvs_out > 1 and rx_p > 113)),
            false.
eccs_2_351 : 'The second division of section 3.5.1 is inoperable'
        fact
        rule
            true if (lpci_B_inop or lpci_C_ inop or
                    (num_adsvlvs_out >1 and rx_p > 113)),
            false.
eccs_3_351 : 'The third division of section 3.5.1 is inoperable'
        fact
        rule
            true if (hpcs_inop),
            false.
```

Figure 2. Translation of Sec. 3.5.1 (Fig. 1) in PROLOG-based
shell syntax

```
/*------------------------------LPCIA FAULT TREE------------------------*/
/*Examines the operability of equipment in the LPCI subsystem A in order
/*to determine the functionality of the entire system /*returns the values
'true' for the parameter 'lpci_A_inop' if the system /*is found to be
inoperable
/*-------------------------------------------------------------------*/
lpvi_A_tree(true):-entity("LPC*LPC-001A",_,[system,_],inop,_,_,_,_),!.
lpci_A_tree(false):-entity("LPC*LPC-001A",_,[system,_],op,_,_,_,_),!.
lpci_A_tree(true):-inop("LPC*PMP-001A");
                   inop("LPC*MOV-011A).
lpci_A_tree(false).
```

Fig. 3 PROLOG operability tree for "Named Entity"

3.5.1, ECCS-Operating; Section 3.5.2, ECCS-Shutdown). The 400KB is broken
down as follows :

o   The database consisting of various ECCS equipment comprises
    100KB (25% of LCOM); this includes 100 separate data entries,
    with component status information and other records stored along
    with each entry. The entries are duplicated in three distinct
    files, allowing hypothetical databases to be created and
    consulted without modifying the actual database.

o   The primary rules and supporting code that are the transparent
    embodiment of the actual written Tech Specs take up 25KB (6% of
    LCOM); these rules are a complete translation of Sections 3.5.1
    and 3.5.2 of the Generic Tech Specs, and aside from some minor
    questions regarding interpretations of requirements, these rules
    should not change over time.

o   The secondary rules and operability trees comprise 25KB (6% of
    LCOM); this support code is not directly related to the written
    Tech Specs, but rather represents the plant specific manner in
    which the primary Tech Spec rules are evaluated. These rules
    are highly dependent upon the size and scope of the plant
    database, and will be fairly fluid over time as plant
    operational and organizational procedures change.

o   The LCOM interface code is approximately 150KB in size (38% of
    LCOM); this code, which exists within numerous modules,
    provides the primary interface with the user, ranging from menu
    selections to help information.

o   The database support code consists of 100KB (25% of LCOM); most
    of the functions that this code performs are transparent to the
    user, but accomplish the tasks of updating, evaluating, and
    maintaining the database; such utilities as print routines,
    audit trails, and security checking are part of the database
    support code.

SCALING CONSIDERATIONS: SIZE AND SPEED

The central part of LCOM, that portion that will not change much across
different plants, consists of the primary rules and supporting code. As
we have found, the written Tech Specs can be translated in a nearly
automatic fashion into transparent PROLOG code. For the current prototype
version of LCOM, the embodied primary Tech Spec rules represent

approximately 10% of the entire Generic Tech Specs. Thus, it would be a
relatively easy extrapolation to include the top-level rules and
requirements for the entire Tech Specs of an operating plant.The difficult
parts of configuring LCOM for an actual plant's Tech Specs would be the
tasks of developing the coded operability trees and managing the large
plant database. Conservatively, the database that LCOM currently uses is
less than 1% the size of the database that would exist for an actual plant
(Larger artificial databases were constructed for scaling benchmarks).
This estimate is based upon the numbers and types of components that are
typically used in failure modes and effects analyses, and that are
considered in determining system operability from a base level. Thus,
intelligent management and processing of the database and operability
trees would be essential in order to avoid a system that was unacceptably
slow or cumbersome to use.

Various tests and simulations were performed to determine the speed and of
an LCOM-like system operating on an actual plant's Technical
Specifications. Databases were constructed that approached 10,000
entries, and primary Tech Spec rules were created to represent about 50%
of actual Tech Spec requirements. The tests showed that LCOM processing
time increased linearly both as a function of database size and number of
rules. Mean-linear scaling, given the complexities of the tasks
involved, can be understood by considering the procedural interpretation
of the LCOM processing code. That is, because the operability trees and
primary rules were constructed to provide a strict Boolean response that
could be readily verified upon investigation of the code,
non-deterministic and time-consuming features such as heuristic tree
searches with backtracking were avoided. Of course, a Tech Spec monitor
that provides a more in-depth analysis of plant conditions, with possible
operational suggestions, would be more difficult to optimize than the
current LCOM prototype - all probable outcomes and recommended responses
could not be defined in advance for such a monitor.

The LCOM processing speed tests showed that for a PC based system (running
PROLOG at approximately 1 KLIPS), a full-scale analysis of plant
conditions to determine Tech Spec compliance would require on the order of
6 to 10 minutes. It should be noted that this value is for *processing*
time only and does not include the time necessary to load database changes
and/or initiate the LCOM run. For a powerful desktop microcomputer or
workstation running PROLOG at 20 KLIPS or above, the single pass
processing time dips to below one minute.

CONCLUSIONS

The LCOM research effort demonstrated the feasibility of designing a
Tech Spec monitor using PROLOG. The declarative style and built-in
unification mechanism of PROLOG did, as conjectured, substantially
simplify the process of program construction and validation. Scaling
tests indicate that full-scale implementation for operating plants will
have adequate to good performance on advanced workstations running PROLOG
at approximately 20 KLIPS. The major issues in such full-scale
implementation would involve not the programming and maintainence of the
Technical Specification logic structure themselves but such ancillary
issues as database integrity and development of implicit Tech Spec
requirements.

ACKNOWLEDGMENT

This work was supported by the USDOE under contract
DE-AC02-85NE37944. Particular appreciation is owed to the contract
monitor, John Lewellen, for his support of exploratory studies and his
interest in efficient information transfer.

EMERGENCY PROCEDURES TRACKING SYSTEM FOR NUCLEAR POWER PLANTS

William Petrick[1], Karen Ng[1], Chuck Stuart[1],
David Shen[2], David Cain[3], Bill Sun[3]
and Carl Christensen[4]

[1]Nuclear Software Services, Los Gatos, CA
[2]Taiwan Power Company, Republic of China
[3]Electric Power Research Institute, Palo Alto, CA
[4]General Electric Company, San Jose, CA

ABSTRACT

This paper describes a computer-based emergency operating procedures
(EOP) tracking system for nuclear power plant applications that is based on
artificial intelligence technology. The project to develop and prove a decision support system for emergency procedure tracking was co-sponsored by
the Electric Power Research Institute and the Taiwan Power Company for application to the Kuosheng BWR nuclear power plant. The software was developed by Nuclear Software Services of Los Gatos, CA and the system is being
integrated into the overall Safety Parameter Display System (SPDS) developed by the General Electric Company of San Jose, CA.

Because the BWR emergency procedures are symptom based, it is possible
to be in many places in the procedures concurrently, a situation which may
require the operating crew to take a number of different actions to control
the plant. The EOP tracking system provides the capability to analyze and
evaluate realtime plant conditions relative to the written operating procedures used during emergencies. As a result of the online logic processing,
pointers or cues back to the written procedures are provided to the operator for guidance.

Key design requirements, such as integration with the SPDS system,
"realtime" operation, minimum computer resources, and verifiable and maintainable logic, were identified and evaluated. It was determined that these
requirements were best met by using artificial intelligence technologies
that are better suited than the conventional software technologies to solve
this type of problem. As a result, a production rule, expert system was
developed which includes a rule-based knowledge base and inference engine
written in the "C" language for use on a DEC VAX computer containing the
SPDS database. The system is to be evaluated in tests at the Kuosheng
simulator and, when accepted, will be installed at the Kuosheng nuclear
power plant in Taiwan, Republic of China.

PROJECT GOALS AND OBJECTIVES

The Electric Power Research Institute (EPRI) has pioneered research in computerized procedures tracking over the last four years [ref.1][ref.2][ref.3]. The Taiwan Power Company (TPC) of Taiwan, Republic of China, has recognized a need to provide a decision support system to assist nuclear power plant operating personnel during emergency situations. In 1985, the two organizations agreed to co-sponsor work to develop a decision support system that could be verified at the Kuosheng simulator and then be installled at the Kuosheng nuclear power plant.

The project goals of this effort were to improve the decision making process during emergencies by automating the logic processing that an operator is required to do, by computerizing the bookkeeping an operator is required to do, and by providing online tracking and explanations to assist in responding to emergency conditions.

Specific project objectives were established to: (1) develop the computer capability for emergency operating procedures tracking during emergencies and for evaluation to support the decision making process; (2) modify the logic and supporting algorithms as necessary to support the procedures and SPDS functional requirements developed for the Taipower BWR/6 Kuosheng plant; (3) evaluate the emergency operating procedures tracking function in the Kuosheng nuclear power plant SPDS and simulator; (4) verify and validate the system for inplant use at the Kuosheng plant; and (5) transfer the technology to U.S. utilities.

PROBLEM DESCRIPTION

The basis for operator decision making during nuclear power plant emergencies is the plant Emergency Operating Procedures (EOPs) which were derived from the generic Emergency Procedure Guidelines (EPGs) developed by the BWR utilities Owners Group [ref.4][ref.5]. The operating crew must interpret the plant state within the context of these procedures while responding to the upset conditions. Over the past seven years there has been a fundamental change in BWR plant emergency procedures from eventbased procedures to symptom-based procedure formulations.

Symptom based procedures focus on symptoms in an attempt to stabilize plant operations within acceptable limits to avoid or minimize hazards. No diagnosis is needed to enter the procedures, and no specific action is required to continue in the procedures. As conditions change, so do the possible operator responses.

Although the symptom based procedures are logically structured, they are inherently complex because each symptom results in an action, the failure of which produces a new symptom, and so on. The successive application of remedies results in a decision network that has many branches, loops, and logic at each branch point. In practice it is burdensome for operators to identify new symptoms, evaluate the various symptoms that apply and interpret the procedures to carry out the recommended actions.

The EOP tracking system is designed to support the operator's decision making process by continuously monitoring and comparing the current plant state against the symptoms in the emergency procedures and providing the recommended operator action. This bookeeping chore unburdens the operators from collecting plant information and interpreting procedures logic, allow-

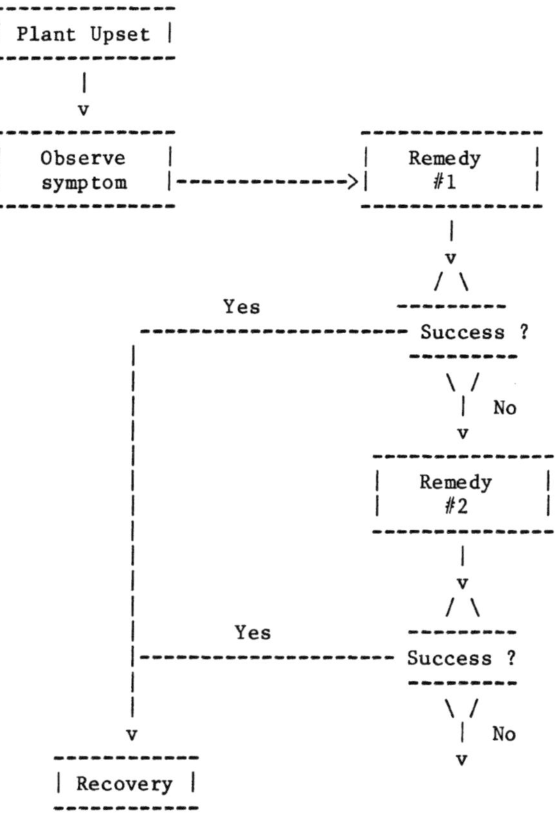

Figure 1. SYMPTOM-BASED PROCEDURES

ing them to focus their attention on taking corrective actions to achieve
the goals established for each symptom. A diagram of the symptom-based procedures is shown in figure 1.

DESIGN REQUIREMENTS

At the start of the project, there were no preconceived notions about
how to implement a computer-based EOP tracking system. Valuable experience
had been gained with previous FORTRAN implementations, so a list of design
requirements was generated as a top level design document [ref.6]. As the
design developed, a number of additional requirements were added to improve
the usefulness of the system to the operations crew. The requirements were
divided into three areas: (1) user support, (2) interface to plant data,
and (3) the application.

Under user support, the following requirements were identified. The
system must:

- provide an index into the written procedures rather than attempting to provide step-by-step advice

- be able to track multiple symptoms concurrently

   - NOT presume the operator agrees with or responds to any output

   - provide the capability to trace logic paths

   - provide supplementary information for any conclusions

   - allow multi-user capability

   - allow priority filtering of output messages

The interface to the plant data required the system to be compatible with the SPDS data and SPDS displays available to the operator in the control room. To do this, the system needs to use the same database as the SPDS displays, and it must operate in the SPDS environment. In addition, in order to test the system, it must also operate in both the real plant environment and the simulator environment.

The primary requirements set by the application were that the "software" must be verifiable and maintainable by plant personnel, recognizing that the EOP´s will undergo periodic revisions.

In any data acquisition system there will always be the possibility for signal failures, hardware problems, etc. that can lead to erroneous data. This "bad" data must be recognized and accounted for in the logic processing.

SYSTEM DESIGN

Based on the above design requirements, it was determined that traditional software techniques using procedural implementations would not meet the verifiable and maintainability requirements. Instead, a system design was developed which borrowed from artificial intelligence techniques used in production rule expert systems [ref.7][ref.8].

The design includes a knowledge base which is defined by a set of rules which, syntactically, look very much like the written emergency procedures. An inference engine, or logic processor, is always online as a background task on the SPDS computer monitoring for entry conditions. The user interface is a separate interactive screen manager task which can be invoked from any VT200-series terminal attached to the VAX computer.

The rule processing is data driven and uses a forward chaining algorithm to arrive at conclusions. The rules are written as IF-THEN statements which provide for clustering, much in the way the BWR emergency procedures are structured. In AI terminology, this clustering is known as context shifting. The rules are "compiled" for efficiency and speed, and the rules do not need to be ordered. By this separation of the application rules from the software that performs the logic processing, the requirements for verification and maintainability are readily satisfied.

Since the rules are designed to be maintained by plant personnel, they are stored as ASCII files on the VAX computer where they can be edited with any standard text editor. Additional software tools were written to provide syntax checking and debugging of the rules prior to their being used in the production system.

The user interface is a VT200-series terminal which displays messages associated with any conclusions reached by the logic processor. The contents of the messages are determined by the application and are maintained

692

as an ASCII file by the plant personnel. The standard operator display contains the list of current messages (conclusions). In addition, the user has the ability to: (1) query the system for a traceback of the logic path that led to the conclusion, (2) ask for supplementary information for any of the conclusions, (3) raise or lower a filter level for the messages, (4) filter out conclusions based on bad data, or (5) supply data to the system if required by a rule that contains an ASK_USER function.

One key feature of this system that is not readily available in commercial expert systems is the handling of data quality. Each piece of data or parameter used in the rules has an associated quality tag that is used in evaluating the premises. This quality information is also used to assess the validity of the conclusions to avoid the propogation of bad data throughout the system.

The plant data that drives the system is derived from the same database that is used in the Kuosheng SPDS system supplied by the General Electric Co. The EOP tracking system is integrated with this comprehensive plant database so that it complements the information available to the operators in the control room. Most of the parameters referenced in the rules of the EOP tracking system are composed points based on detailed logic embedded in the GE database. A flowchart showing the interface to the GE plant data system is given in figure 2.

RULE SYNTAX

A key element in the maintainability of the rules is the clarity of the rule syntax. The syntax was chosen to follow the written procedures so that verification of the rules can be performed by a knowledgeable engineer. The rule syntax is simple and straightforward:

```
RULE    parent-name   child-name
   IF
     (premise)
     (premise)
        .
   THEN
     (action)
     (action)
        .
```

The IF-portion of the rule contains any number of premises which evaluate to TRUE or FALSE. All of the premise evaluations are ANDed together, and if the result of all the premises is TRUE, then the action or actions specified in the THEN-portion of the rule are taken. A premise can contain database parameter names, numerical constants, logical expressions, or a function evaluation. The relational operators that are available include equal, greater than, less than, greater than or equal, less than or equal, and not equal. Functions that are available include NOT(parameter), COUNT(list), MAX(list), MIN(list), QUALITY(parameter), WAIT(parameter), and ASK_USER(parameter).

The possible actions include setting a parameter value or evaluating (firing) a rule cluster.

The following three rules demonstrate the rule syntax and its relation to the written procedures for: (1) an entry condition which, when satisfied, fires other rule clusters, (2) a rule that sets a state, and (3) a rule which allows for operator judgement.

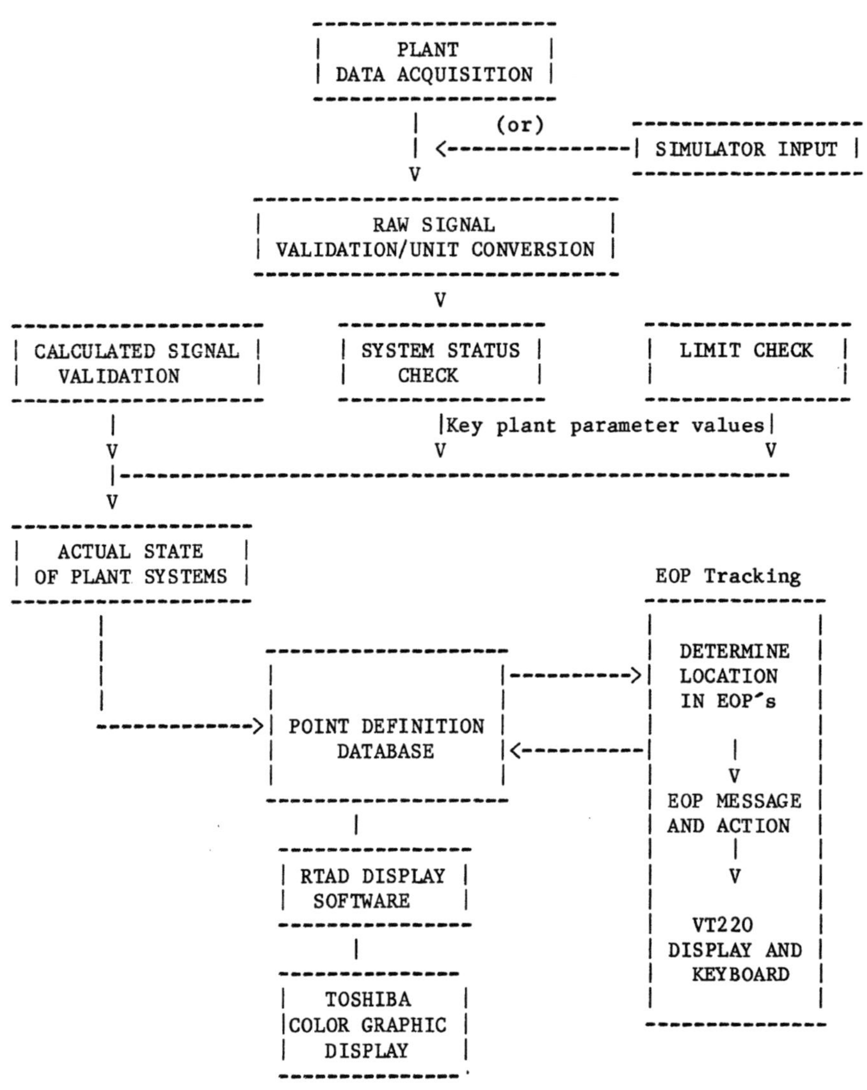

Figure 2. EOP TRACKING SYSTEM INTERFACE TO GE PLANT DATA SYSTEM

Example 1: SIMPLE ENTRY CONDITION

                        Procedures Text:

Drywell pressure above 0.122kg/cm2 (1.74 psig drywell high pressure
scram setpoint), OR ...

                        EOP Rule:

RULE RPV_ENTRY 3
  IF
    (DRYWELL PRESSURE IS ABOVE D/W HIGH PRESSURE SCRAM SETPOINT)
  THEN
    (SET RPV ENTRY)
    (ENTER RC/Q, RC/P, RC/L)

Example 2: RULE TO SET A STATE

                        Procedures Text:

After RPV pressure has been reduced to below the shutdown cooling
interlocks setpoint 9.34kg/cm2 (133 psig) AND the RHR shutdown cooling
mode has been established, proceed to normal cold shutdown.

                        EOP Rule:

RULE   RC/L 10
  IF
    (RPV PRESSURE < 9.34)
    (RHR SHUTDOWN COOLING IS ON)
  THEN
    (SET 3 III.B.3)

where the parameter named "3 III.B.3" is the cross reference to a
specific paragraph in the written procedures.

Example 3:   RULE REQUIRING OPERATOR JUDGEMENT

                        Procedures Text:

If suppression pool water level and RPV pressure cannot be maintained
below the suppression pool load limit BUT ONLY IF adequate core cooling
is assured, THEN terminate injection into the RPV from sources external
to the primary containment EXCEPT from boron injection systems and CRD.

                        EOP Rule:

RULE SP/L 3
  IF
    (S/POOL WATER LEVEL IS ABOVE THE S/POOL LOAD LIMIT)
    (RPV PRESSURE IS ABOVE THE S/POOL LOAD LIMIT)
    (WAIT(DELAY_TO_CLEAR_LOAD_LIMITS CONSTANT))
    (ASK_USER(CORE COOLING IS ASSURED))
  THEN
    (SET 4 III.C.3.e.2)

CONCLUSIONS

The inherent complexity of the written BWR emergency operating procedures suggest that a computer-assisted decision support system would be useful to plant operations during an emergency. Previous attempts at FORTRAN-based implementations proved inadequate, so a system design based on newer artificial intelligence technologies was developed.

The problem was divided into two parts which were addressed concurrently. The first part was the development of a production rule expert system shell which evaluates IF-THEN rules to arrive at conclusions. The second part was the translation of the BWR written procedures into IF-THEN rules which can properly guide the operator during emergency situations. Quantitative evaluation tests of the system are planned at the Kuosheng simulator near the end of 1987.

The transfer of this new technology to the U.S. BWR plants for emergency procedures tracking may involve NRC licensing considerations, and will require a rewriting of the rules to conform to the specific plant design, plant procedures, and available plant database.

A more general transfer of the technology can occur for applications at plants which have a computerized plant parameters database. The expert system shell developed in this project can be used for other operational problems which can be formulated as IF-THEN rules. Applications such as alarm filtering and post-scram analysis have been suggested.

REFERENCES

1. K.B.Ng et al, "The IMAGE Information Monitoring and Applied Graphics Software Environment", Electric Power Research Institute, report EPRI NP-4758-CCM, volumes 1-4, 1986.
2. D.G.Cain, "BWR Shutdown Analyzer Using Artificial Intelligence Techniques", Electric Power Research Institute, Special Report NP-4139-SR, July, 1985.
3. "Functional Specifications for AI Software Tools for Electric Power Applications", Electric Power Research Institute, Final Report NP-4141, August, 1986.
4. P.L.Chen, "Emergency Operating Procedures for Kuosheng Nuclear Power Plant", Taiwan Power Company, February 3, 1986.
5. BWR Owners Group, "Emergency Procedures Guidelines", Unpublished Draft, Rev 3L, BWR1-6, May 10, 1984.
6. K.B.Ng et al, "Emergency Operating Procedures Tracking System", Electric Power Research Institute, Interim Report, NP-5250M, June 1987.
7. V. Daniel Hunt, "Artificial Intelligence and Expert Systems Sourcebook", Chapman and Hall, New York, N.Y., 1986.
8. Charles Rich and Richard C. Waters, editors, "Readings in Artificial Intelligence and Software Engineering", Morgan Kaufmann Publishers, Inc., 1986.

A COMPUTERIZED SYSTEM FOR IMPROVED

MANAGEMENT AND EXECUTION OF PLANT PROCEDURES

M.H. Lipner, R.G. Orendi, A. J. Impink, and R.A. Mundy

Westinghouse Electric Corporation
P.O. Box 355
Pittsburgh, PA   15230

INTRODUCTION

Industry reports attribute a high percentage of the root causes
associated with significant events to human performance problems. Many
of these problems are procedure related. Plant operating procedures
define the methods, means and limits of plant operation and provide the
operator with the information network he needs to operate a nuclear
power plant safely and efficiently. These include normal operating,
abnormal operating, alarm response, emergency operating, and surveillance procedures. Since the total number of plant procedures can run
into the thousands, operators easily become confused and frustrated when
attempting to locate and then effectively follow a particular procedure,
especially in high stress situations. To address these problems, an
on-line computerized procedures system has been developed which guides
the operator through the procedures, provides necessary parallel information, selects the appropriate procedures, and alleviates the operator's memory burden by automatically keeping track of equipment and
system status.

This paper describes the Computerized Procedures System, discusses
common deficiencies related to the use of paper procedures and illustrates the resolution offered by the computerized procedures system.

COMPUTERIZED PROCEDURES SYSTEM DESIGN

Purpose

The Westinghouse Computerized Procedures (COMPRO) System has been
designed to assist plant operators in executing procedures more efficiently and cost effectively. It is an on-line tracking system that
allows the utility operating staff to access and follow procedures in an
easy and logical manner. The Computerized Procedures System can be used
for all types of procedures including systems operating, plant operating, surveillance, emergency operating, abnormal operating and alarm
response.

The purpose of the system is two-fold:

o   To guide the user step by step through the procedures by moni-
    toring the appropriate plant data, by processing the data, and
    by identifying the recommended course of action.

o   To provide the necessary parallel information which allows the
    operator to assess other plant conditions which may require
    attention.

Figure 1 is a description of the Computerized Procedures System CRT
screen contents.

<u>Computer Software/Hardware Configuration</u>

The Computerized Procedures System program is currently implemented
on the Westinghouse Gould/SEL 32/97 engineering simulator computer which
is a dual-processor (CPU/IPU) 32-bit mini-computer running the standard
SEL MPX real-time operating system. The Gould/SEL interfaces with a
display generator to provide the color graphics output of the procedures
program on a high-resolution color monitor. All of the procedures software is written in the high level Fortran-77 language and uses standard
operating system routines to execute.

PROBLEMS RELATED TO PROCEDURE EXECUTION

Described below are common documented procedural problems identified
in reports and audits as being major contributors to poor operator
performance. Interwoven throughout the discussion are the features of
the Computerized Procedures System which help to resolve each problem.

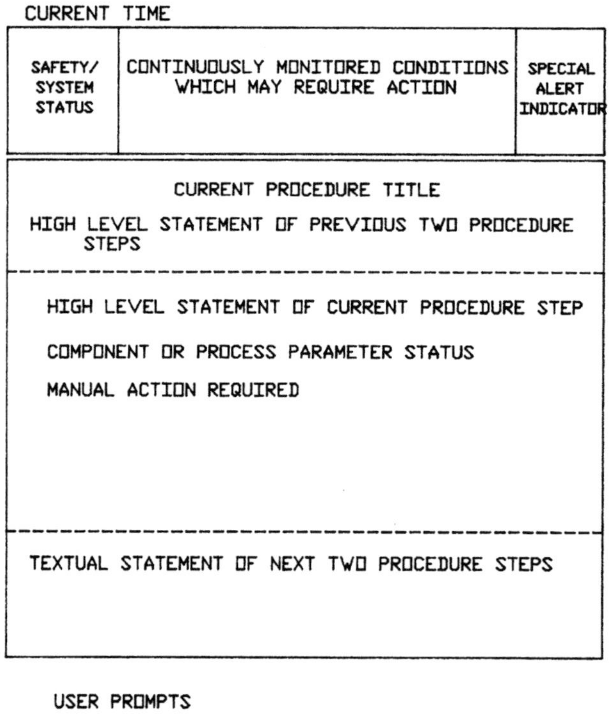

CURRENT TIME

SAFETY/ SYSTEM STATUS	CONTINUOUSLY MONITORED CONDITIONS WHICH MAY REQUIRE ACTION	SPECIAL ALERT INDICATOR

CURRENT PROCEDURE TITLE

HIGH LEVEL STATEMENT OF PREVIOUS TWO PROCEDURE
    STEPS

HIGH LEVEL STATEMENT OF CURRENT PROCEDURE STEP

COMPONENT OR PROCESS PARAMETER STATUS

MANUAL ACTION REQUIRED

TEXTUAL STATEMENT OF NEXT TWO PROCEDURE STEPS

USER PROMPTS

Fig. 1. Description of Computerized Procedures
         System CRT screen contents.

Procedure Unfamiliarity

Procedure unfamiliarity and procedure knowledge weakness are common
inadequacies leading to plant trips and transients. This is particularly true with surveillance procedures performed by instrument and
control technicians. Typically, they are not as well trained on equipment and systems as control room operations personnel. Nonetheless,
they are permitted to perform checks on reactor protection systems and
primary instrumentation. For example, during the performance of a reactor coolant system overtemperature and overpower protection channel
test at a particular plant, several problems were noted. The instrument
and control technicians had to install test equipment in an attempt to
establish the appropriate conditions for performing the test. The technicians spent approximately one and one-half hours making various test
equipment installations, removals and adjustments without using procedures. The procedure also required the technicians to initial each
step to verify completion of actions – this was not done. The technicians also did not maintain a log of events to document these
actions. With the aid of COMPRO, these inadequacies could have been
minimized. The actual plant conditions would have been sensed and
instructions for simulating appropriate conditions would have been provided. COMPRO requires verification of actions by the user prior to
proceeding to the next step and COMPRO automatically maintains a chronological log to record problems for future reference and resolution.

Missed Notes and Cautions

Procedures typically contain numerous notes and cautions which are
designed to alert and inform the operator of potential hazards to personnel or equipment and to present advisory or administrative information. As the notes and cautions are encountered in a procedure, the
operator is expected to remember this additional information and take
action accordingly. This information could be as simple as adding water
to a tank when a low level setpoint is reached or as complex as blocking
of safety injection signals every time an automatic unblock occurs.
Failing to add water to the tank when required could cause problems in
maintaining an adequate heat sink for the nuclear reaction. Blocking of
safety injection system signals during recovery actions is required to
prevent challenges to safety systems and to prevent severe pressure and
temperature perturbations in the reactor coolant system. The information contained in the notes and cautions taxes the operator's memory
and, if missed, could easily lead him into unnecessary or unplanned
conditions. In one plant, the operators jeopardized an emergency diesel
generator by restarting the unit without verifying key operating parameters contained in precautions and initial conditions. Further, the
unit was secured without first checking that the cylinder exhaust temperatures were within allowable limits – information also contained in a
caution. COMPRO alleviates the operator's memory burden by automatically keeping track of notes and cautions and other information required
to be continuously monitored. If one of these items needs attention,
the Computerized Procedures System informs the user by energizing the
"Special Alert Indicator" window (Figure 1) and by stating the problem
and the required action in the "Continuously Monitored Conditions"
window.

Missed Steps and Transitions

During high stress situations, it is not uncommon for an operator to
miss a procedure step or a transition to another step or procedure.
These conditions occur when the operator incorrectly interprets a step
or symptom wording or when he thinks the step has already been com-

pleted. The results of these errors of omission can be catastrophic
since the action required may be to mitigate a severe challenge to one
of the plant critical safety functions, which are designed to maintain
the three barriers to fission product release. Other consequences may
include damage to equipment (not stopping a reactor coolant pump when
required) or a delay in plant recovery (not starting a plant cooldown
when conditions permit). In a recent simulator exercise, the operators
were frequently confused and uncertain about which steps or procedures
to transition to due to ambiguous plant conditions. Even when proper
transition conditions were apparent, hesitation by the operator delayed
the recovery effort. In addition, some of the more experienced operators frequently anticipated procedure steps, performed steps out of
sequence, skipped steps and performed actions not required. The COMPRO
System can help with these problems by presenting the procedural information clearly and logically in one place.

The Computerized Procedures System displays the title of the currently active procedure, allowing utility personnel to quickly identify
which procedure is in use. In a situation where transitions or moves to
other procedures are required, the new procedure title is displayed
immediately upon selection by the user. The system allows the user to
quickly understand the status of the current procedure step by displaying on the CRT screen a high-level statement of the step. Figure 2
illustrates a typical COMPRO display. For example, the current procedure step is to verify Auxiliary Feedwater (AFW) flow greater than 470
gpm. Since the AFW flow is less than 470 gpm, the high level statement
reads:

    AFW FLOW   LESS THAN 470 GPM

The Computerized Procedures System allows the user to understand the
status of the parameters or components which led to the high-level
statement discussed above. For example, in the case of AFW flow, the
components of concern are the AFW pumps. Hence, if the AFW pumps are
not running, then

    MOTOR DRIVEN AFW PUMP A   NOT RUNNING
    MOTOR DRIVEN AFW PUMP B   NOT RUNNING
    TURBINE DRIVEN AFW PUMP   NOT RUNNING

appear on the screen. This is the supporting information which the user
of the system may require in order to determine why the AFW flow is less
than required.

COMPRO allows the user to understand what actions are required, if
any, in response to the current procedure step. For example, if the AFW
pumps are not running, then the action required is

    MANUALLY START PUMPS AND ALIGN VALVES AS NECESSARY

and this message appears on the screen.

Either a keyboard or a touch sensitive screen can be used for operator input to the system. Figure 2 illustrates a typical COMPRO display
with a touch sensitive screen. If an action is required, the system
prompts the user to touch the ACTION COMPLETED box when the action is
completed, or to touch the ACTION OVERRIDDEN box to override if he does
not wish to take the action. Note that the user remains in control at
all times. He only informs the system whether an action has been taken
or not, and, hence, he controls the subsequent course of the procedure.

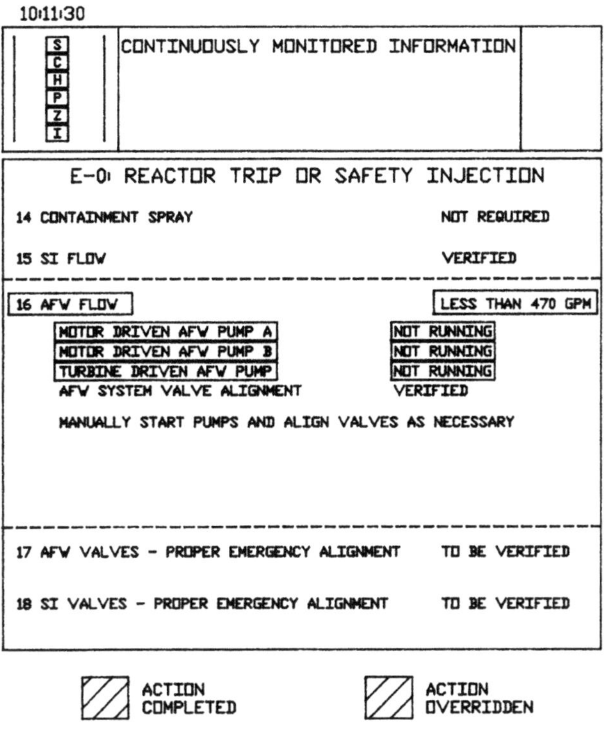

Fig. 2. Example of emergency procedures
status as shown by the COMPRO system.

Multiple Procedures

Complex operations often require more than one procedure to be in
effect at the same time. Operators tend to get lost when performing
procedures in parallel. When this situation occurs, procedure steps can
get repeated or missed. Place-keeping aids can help, but do not alleviate the need to have several procedures in effect at the same time.
During emergency transients, the operator has the additional burden of
keeping track of critical safety functions as well as parameters and
equipment that need to be continuously monitored. This information
alerts the operator to transition to other procedures when certain conditions are present. An operator or supervisor must be extremely well
coordinated and organized to make all the right moves at the appropriate
time and conditions. For example, in a recent simulator exercise, a
supervisor temporarily lost his place in the procedures, causing a delay
in the recovery actions. He then became confused at the end of a procedure when he had a choice of several procedures to enter.

It is this type of atmosphere in which the features of COMPRO are
used to their fullest extent. The guidance provided by the programmed
procedures logic will lead the operator on an acceptable path to plant
stabilization and recovery. Overall safety status is continuously
monitored and displayed and any deviation is immediately apparent.

Should the user need to choose among several potential new procedures, the Computerized Procedures System systematically asks the user, one at a time, which procedure he would like to implement.

High Level Plant Status

The tedious task of following procedures during transient events deters the operator from concentrating on applying his knowledge to comprehending the overall situation, detecting anomalous conditions, and determining appropriate courses of action. With the emphasis on verbatim procedural compliance, the operator becomes so intent on following procedures that changes in systems or component states can occur undetected. It is important for the operator to follow the procedures, but he should also be cognizant of the "big picture" and not lose control of the situation. Changing plant conditions during events requires the operator to be flexible and to be able to change recovery techniques as required. During one emergency, a step requiring safety injection initiation had mistakenly been skipped, and the senior reactor operator spent approximately five minutes assisting the reactor operators in recovering from the error. As a result, the senior reactor operator was not able to stand back and keep the broad picture of what was occurring, which was his designated job for the emergency.

Another situation which requires the operator to have a broad picture of plant and equipment status is during a reactor startup. Figure 3 represents a typical display of what an operator might see at the beginning of a reactor startup.

The reactor core startup rate (SUR), average moderator temperature (TAVG), steam generator water level (SG LVL), reactor coolant system pressure (PRESS), and pressurizer level (PZR LVL) are monitored continuously. Each safety parameter is color coded (red, orange, yellow, or green) to represent the relative severity of the condition. Any required action to respond to the abnormal condition will appear in the "Continously Monitored Information" window. An appropriately color coded "ALERT" will flash in the "Special Alert Indicator" window if anything other than a green condition results. All of the safety parameter conditions must be green before withdrawing rods.

Note that an "ALERT" is indicated in the "Special Alert Indicator" window. As the operator prepares to proceed to step 5 of the procedure, Tavg begins to decrease. The "ALERT" flashes, and a color-coded message informing the operator that temperature is decreasing appears in the "Continously Monitored Information" window. The action to be taken by the operator also appears in the window. The user prompts also change color to match the alert condition so that the operator knows he is responding to that particular situation. The Tavg bar has expanded and changed color (to yellow) to also indicate to the operator that something is amiss. After the operator has returned Tavg to its normal band, the alert disappears and he can proceed to step 5 of the startup procedure.

Post Procedure Reviews

Following transient events at nuclear power plants, it is very difficult to reconstruct the sequence of events, operator actions, and plant conditions at various times. This type of information is critical in determining root causes of events, filling out post-trip reports, and ensuring proper corrective measures are taken to prevent re-occurrence. Without a detailed record of operator actions and results of the

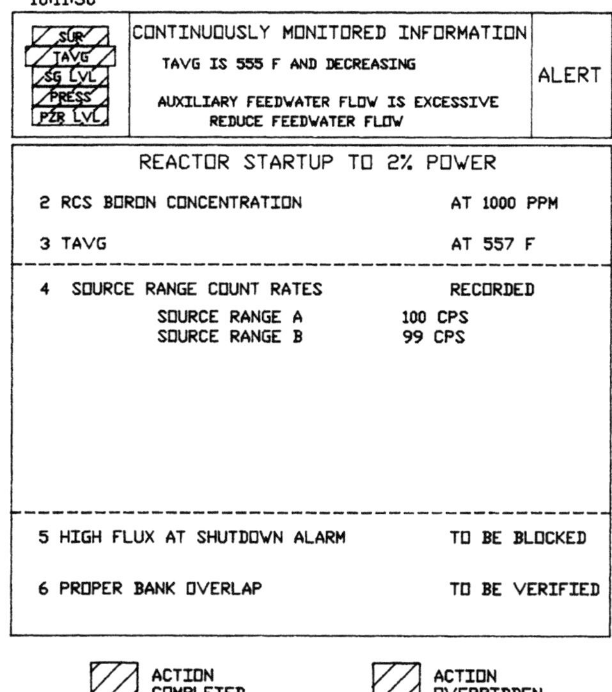

Fig. 3. Example of initial reactor startup
status as shown by the COMPRO system.

actions, it is difficult to analyze if the actions taken were proper and
if the operator performed the actions as specified. By reviewing and
analyzing the detailed event sequence, a better recovery strategy may be
developed, thus reducing the severity of the transient should it happen
again. Also, equipment sequencing and performance could be reviewed to
ascertain if it operated properly.

The COMPRO System performs complete and understandable conditions
recording. The system records the contents of the display exactly as
the user sees it. This includes the time, a bar display (or equivalent)
of the major system states, any caution text which appears in the caution text window, the procedure title, the current step information
including the high level statement and all relevant plant status information, and the operator's response to the system's prompt.

Verification of Actions

Actions that are assumed to have been taken but have not occurred
can lead to adverse conditions not expected or anticipated by the operator. When in high stress or time critical situations, it is extremely
important that the operator is advised that the required automatic and
manual actions have occurred. This is especially important when actions
occur outside the control room and communications are difficult or
direct feedback is not indicated in the control room. Procedure actions

are based on a pre-defined step sequence and confirmation of previous
actions is critical for subsequent actions. Positive confirmation of
action during transients gives the operator confidence in using and
understanding the procedures.

The COMPRO System provides for continuous execution of the current
procedure step logic and updating of the data displayed on the display
device. If, for instance, the current step requires the operator to
start a particular pump, the display will initially indicate that the
pump is not running and it will prompt the operator to start it. The
display will also instruct the operator to respond either that the
action is completed or that he has chosen to override the prompt. If
the response is to override the suggested action, the system moves on to
the next appropriate step. If the response is to indicate a completed
action, the system uses the plant data to determine whether the action
is indeed complete. If the action is complete, the system moves to the
next appropriate step. If it is not complete, the screen updating
resumes until the next acceptable response is entered. Note that the
operator can not leave a step without either completing an action and
having it verified in light of the plant data or overriding the action.

CONCLUSION

In summary, we have shown how the deficiencies reported in this
paper can be resolved by using the Computerized Procedures System.
Specifically, the Computerized Procedures System has the following capabilities:

o   The human operator is freed from the "non-cognitive" component
     of the response generating activity by transferring the
     rule-based functions to the Computerized Procedures System.

o   Notes and cautions, when encountered in a procedure, are stored
     in the computer's memory. If a note or caution requires atten-
     tion, the user is informed of this fact, along with a statement
     of the action necessary to solve the problem.

o   The Computerized Procedures System displays the current pro-
     cedure title as well as the current step. When multiple pro-
     cedures are in effect, the system can access any of the pro-
     cedures and the current step in each of the procedures is dis-
     played.

o   The Computerized Procedures System contains a conditions logger
     which accepts data from the plant data processor and procedure,
     step, plant, and system status and user information from the
     active procedure processor and generates a continuous record of
     conditions and actions during the operations that are being
     performed.

o   The computer provides an independent verification of the course
     of action chosen by the human operator or automatic actuations
     as required by safety systems. This positive feedback provides
     assurance to the operator that an action has occurred.

The Westinghouse Computerized Procedures System solves many of the
human performance problems associated with procedures. It has been
designed to assist plant operators in executing procedures more efficiently and cost effectively. The system brings all of the necessary
information to one location for easy and continuous assessment of plant
conditions and permits the operator to concentrate on the "big picture".

COMPUTER AIDED PROCEDURE EXECUTION

Fridtjov Øwre

Institutt for energiteknikk
OECD Halden Reactor Project

INTRODUCTION

Written procedures provide an important information source for operators in process plant control rooms. They have, however, been much criticized in recent years, especially from a human factors' standpoint (ergono-
mics: text vs. flowchart presentation, labeling, letter size, referencing).

It is hypothesized that a number of the weaknesses with today's use of
written procedures can be eliminated if they are transferred to a computer
in a proper manner. The OECD Halden Reactor Project in Halden, Norway, has
since 1985 been working with a research and development programme addressing
questions such as : 1) Is it feasible to computerise all types of procedures
currently in use in nuclear power plant control and operation?, 2) Is it
feasible to present and follow procedures effectively on CRTs and thereby
make the use of hardcopy manuals redundant? and 3) Will a computerised procedure system improve operability and, thereby, reduce risk?

This paper describes a computerised procedure system as implemented in
the Halden Projects full-scope PWR simulator facility. The system has been
implemented by means of the programming languages PROLOG and LISP. It depends heavily on the Flavor object-oriented programming technique as well as
the Windows graphic system as installed on SYMBOLICS 3640 computers.

SCOPE, MOTIVATION AND AIMS

The Halden Project has as one of its 1985-87 year programme items the
development and demonstration of a prototype computerised procedure system
as a part of an integrated surveillance and control system [1].

The scope of the project is limited to the development of a computerised procedure system that can handle todays' written procedures in the context of an advanced computerised and CRT based control room. It is not concerned with the actual construction/design process of procedures.

Many organisations have identified a number of problems concerning present day manual procedure systems. This includes for example, identification
of the correct procedure, time to locate procedures, time consuming data
gathering when following procedures, difficulties in observing whether the

initiated action has the desired effect, problems with multiple disturbances, problems when moving from one procedure to another etc.

Procedures have also been much criticized from human factors stand-
points e.g. form, style, content and structure [2], and, since procedual tasks
also are a natural part of the advanced control rooms, the motivation behind
the project is to investigate how procedures can be presented on CRTs and
executed integrated with other computerised operator support systems.

Our aim with the computerised procedure system is to assist operators
in central control rooms in identifying a relevant procedure as well as in
executing it once chosen. We also believe that providing quick access to
stored procedural information and relieving operators of trivial tasks in
connection with executing procedures such as collecting data, waiting for
responses and doing response checks, are important additional advantages.

OVERALL SYSTEM DESCRIPTION

The computerised procedure system has been designed to assist operators
in identifying and executing all types of procedures. It is an on-line computerised system that enables operators to retrieve procedures from a procedure database and follow one, or several procedures in parallell, through
an interactive session. An important feature of the system is the possibility to automatically read the actual status or values from the process. The
system proposes the next procedure step according to what is found. However,
it is always the operator who initiates the proceeding. This means that
there is a built-in control of the procedure execution, which can, however,
be overridden by the operator if necessary. The operator gets feedback when
conditions are fulfilled or targets reached.

Integration of this tool with other computerised support systems in use
was an important design objective. This is solved by providing the opportunity to access relevant control and surveillance display formats directly
from the procedure step display. Components and automatics can also be operated through the system if the procedure step requires such manipulation.

THE COPMA SYSTEM

A number of requirements must be fulfilled for a computerised procedure
system to function efficiently. A prerequisite for safe operation is correctness of procedures. This implies that procedures should be verified to
the extent possible. The procedures must be stored in such a way that they
allow for fast execution and easy access by the operators. Also, modifications and extensions of the procedures must be as simple and safe as possible, to ensure that the total quality and reliability of the software
system is as high as possible. Such requirements are arguments for the development of a specific language, tailor-made for computerisation of procedures. A specification language for procedures, called PROLA, has therefore
been developed in connection with the research program.

COPMA stands for computerised manuals. The system is divided into two
parts, called the off-line and on-line parts, see Figure 1. The off-line
part is used for various preparatory work while the on-line part is the
system ment to be in continous operation in the control room.

The purpose of the off-line system is to move the procedures from some
document in the real world into a very specific format in the computer. This
is done in two steps:

In the first step the procedures are written once again according to
the rules of the procedure language PROLA. The intention is to satisfy, in a
systematic and formal way, the needs which procedure designers have when
expressing the various tasks the computer and the operator should accomplish
together. This formalism is necessary since the computer in this case is
supposed to take over tasks that is presently carried out by humans.

All operations, actions and jumps to be carried out by the man and
machine are now written down according to the step structure of the procedure.

The second step involves running the edited procedure through a syntax
analyser. Here the procedure is checked according to the rules of the PROLA
language, just like a compiler checks a source code in any programming language. The syntax analyser is implemented in PROLOG. This programming language turned out to be very efficient for such an application.

The output of the syntax analyser is the procedure translated into a
set of lists. The reason for this format is that the on-line program is a
LISP program that needs lists as its datastructure to operate on. The stand-
ardised list format implies that the one LISP program is able to handle all
statements in all procedures. Figure 2 shows a very simple example of the
three different formats a procedure takes during the off-line process.

The second line in Figure 1 indicates what is called the Graph Editor
This is an interactive off-line module enabling procedure designers to
construct procedure graph trees to be used later during on-line execution.

A procedure graph tree is a graphic representation of the procedure
step structure. It gives an overview of how procedure steps are linked
together. When using the editor the user's task is to place the steps in a
resonable position relative to each other. The graph tree has several
functions in the on-line mode. This will be discussed later.

The second major part of COPMA is the on-line system, and the main module here is the central procedure following system - the COPMA kernel -
which executes the control of the system. It handles each instruction in the
procedure according to the rules, much like an operating system handles
monitor calls in the computer.

When a procedure is called for it is first read from the procedure
database. This copy resides in the system as long as the procedure is being
used and during procedure execution various parts of the copy are accessed.
Instructions like AUTOMATIC-CHECK entail automatic reading of the process
data from a process data base, while MANUAL-CHECK means that input must be
supplied by the operator. If it finds instructions like e.g MONITOR or
MONITOR-SEQUENCE it will generate certain elements called monitor elements.
The presence of these elements will involve checking whether conditions are
fulfilled in due time. The monitor elements are put into a monitor element
database by the COPMA kernel and the actual checking of their content is
performed by a special part of the program.

The procedure information is presented through a graphic screen in the
COPMA system and all the interactions are carried out by means of a mouse.
This interface is described in detail later on.

COPMA MAN-MACHINE INTERFACE

In order for an automated system like COPMA to have any chance of being a success when introduced in the control room, it must be accepted by

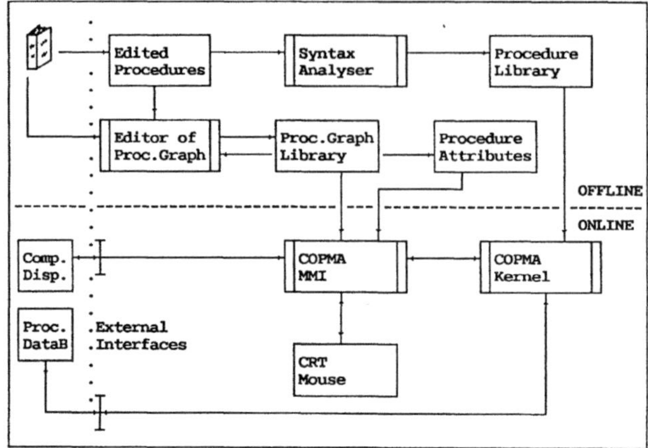

Figure 1. Block diagram of the COPMA system

Written Procedure format:

STEP   1:   PRE-CHECKS   BEFORE   TURBINE   RUN-UP

```
MAIN PROCEDURE              CURRENT STEP      1 OUT OF 4
Run up Turbine SA10         Pre-checks before run-up


 CURRENT ACTION   Check main plant parameters and turbine controllers
 1  out of 4
                  SA50 EP= 217mw The turbine running at 90%
 Format - SA10    SA10 EP=   0mw Shutdown

                  YA13 P = 124 Bars Normal primary pressure
                  RA00 P =  50 Bars Normal secondary pressure

                  SA10U001  Off      Group controls off   SHUTDOWN
                  RA10U001  Off
                  RB10U001  Off
```

PROLA Procedure language format:

```
STEP 1

( MESSAGE "Check Main Plant Parameters and Turbine Controllers" )

( AUTO-CHECK   NOT (SA50E001 EP >= 217      AND
                    SA10E001 EP =  0         AND
                    YA13P001 P >= 124        AND
                    RA00P001 P >=  50        AND
                    SA10U001= F              AND
                    RA10U001= F              AND
                    RB10U001= F)

      => ( INITIATE STEP 100 FINISH) )

( MESSAGE "Check Main Executive Control is Off" )
```

Syntax Analyser format:

```
STAT(MESSSAGE,x,(y INITIATE z)
STAT(AUTO-CHECH,x1,(y1 INITIATE z1)
    ......
    ......
```

Figure 2. Three different procedure formats

operators and found <u>useful.</u> The critical part is the man-machine interface involving:

- how the procedure is presented and
- how to work with the system

These requirements are met by means of a multi-pane window on a graphic screen. Each of the seven panes serves a particular function, some of them more than one.

All input is by means of mouse clicks. Commands or answers are issued by placing the mouse upon corresponding command/answer field items, which causes the command/answer to be highlighted when the mouse button is used.

Figure 3 gives an example of how the COPMA screen will look in the middle of execution of a procedure. This figure is used as a reference for the description of the purpose and use of the various panes.

Dialogue Pane

Its purpose is to display a few commands to the operator for instance enabling him to get started. The commands in question are: 'list-passive-procedures' and 'list-active-monitor-elements'. Executing the 'list-passive-procedure' command is comparable of searching in the main procedure index in the manual system.

Work Pane

All available (passive) procedures are listed here when requested from the dialogue pane. The operator can now <u>select</u> one of the procedures and either just look at its content or he can start executing it. If execution is started the activated procedure is deleted from this list and included in the list of active procedures displayed in the active procedures pane. Two mouse clicks are sufficient. In the manual system this is comparable of retrieving the correct volume and open it on the correct page.

Activated monitor elements will also be listed in the work pane. If new monitor elements are activated or made passive while displaying the monitor element list, the list is updated automatically. The description of the monitor element includes the condition being monitored and which procedure and step initiated the monitor element.

Overview Pane

The main purposes of the overview pane are to show how different steps in a procedure are linked together, how far the execution of a procedure has reached and which branches have been examined. This is visualised by the graph tree having nodes corresponding to procedure steps, and highlighting of the nodes and branches examined until now.

When the operator selects a procedure in the work pane, the corresponding graph tree is immediately displayed in the overview pane. To get on working he must select one of the steps. This will cause the step name to be displayed in the active procedures pane under the related procedure name. In Figure 3, the node corresponding to step S-1 is surronded by a frame because the content of this step is displayed in the step survey pane. S-1 is also displayed in reverse video because it is the step currently worked upon. This is defined as one <u>activity.</u> Many activities can be defined at the same time. They can belong to different procedures.

The instructions belonging to the selected step are then displayed in

the step survey pane. It takes at most three mouse clicks to get to this
point.

The graph tree might be moved around within the pane. This is necessary
because the graph usually requires more space than available within the
pane. Only a part of the graph is then visible at a time.

In written procedures the operators have the possibility to put their
own bookmarks in the text to help them find these places later on. This is
taken into account in the COPMA system by providing possibilities for bookmarks. A bookmark icon is placed in connection with a node in the graph and
a reference is placed in the active procedure pane. Later, when accessing
this reference, the snapshots of the overview pane and the step survey pane
will reappear in their respective panes.

The execution sequence, i.e., the path followed through the graph, is
indicated by thicker arrows. The operator has the option to reconsider this
sequence. He can step backwards in the sequence and start over from whatever
node he wants. Reconsiderations are done one step backwards by each click.

Procedure and Step Heading Panes

The name of the procedure currently being worked upon is displayed as a
headline on the screen. The next line of the screen presents a short explanation of the step displayed in the step survey pane at the moment. When
changing to another active procedure, these lines are updated accordingly.

Step Survey Pane

The step survey pane displays the instructions of the procedure-step
selected in the overview pane. The progress of procedure execution is controlled by the operator. He gives orders to execute or skip instructions. If
the result of carrying out an instruction is accepted, the operator gives
the command OK to make the system proceed to next instruction. He can override results of automatic checks by pointing the mouse symbol to YES and
click the mouse to force the check to become true, or to SKIP to force it to
become false. When a new instruction is to be performed, the operator must
take action within a predefined time interval. If not the system will give
an audible warning.

The figure shows a typical example of the step survey pane, displaying
the step S-1 of procedure RUN UP TURBINE SA10. The current instruction recommends that the format SA10 be displayed on a remote screen. Process formats recommended for display on remote screens are listed as mouse sensitive
items. The command to actually display them on a remote screen is given via
this pane. This is done by pointing the mousesymbol upon the item in question and pressing a mouse button.

Plant components might also be manipulated from this module. The components in question are listed as mouse-sensitive items together with the
available instructions. Feedback is given showing the status of the manipulated components.

Active Procedures Pane

Since the system can handle many procedures at the same time, it is
necessary to give the operator an overview of those he is executing right
now. The active procedure pane gives the operator just that- an overview of
all the active procedures and their associated activities.

The activities listed here may be separated into two groups. Activities

created by clicking the mouse on a step in the overviewgraph comprise one
group. The other group includes those activities created by a monitor element condition being fulfilled. An activity from the last group will be
transferred into the first group when the activity is worked upon the very
first time. These two different kinds of activities are displayed in a
different manner for convenience.

Figure 3 shows two active procedures. The first procedure, MINOR LEAK
IN PRIMARY CIRCUIT, has one activity started at step S-1. The second, RUN UP
OF TURBINE SA10, has two activities started at steps 1 and 5 respectively.
Step 5 is not visible in the graph in this example.

There exists a command to start working with a new activity, or change
from one activity to another. One can also start working with an activity
due to a monitor element. In the example an activity has been started from
S-1 and the information related to S-1 has been brought into the overview
and step survey panes.

When a bookmark is inserted into a procedure from the overview pane, an
associated text is displayed together with the activity in question. From
the active procedure pane it is then possible to switch back to the activity
by clicking the mouse on the associated bookmark text. This replaces the
content of the overview pane and step survey pane with the snapshots connected to the bookmark. In Figure 3 a bookmark is connected to the activity
started at S-1 and refers to step S-3 in the overview pane.

Future Extentions

The future work on the MMI will include some additional features. A
brief listing of these future functions is given in the following:

A hardcopy of any selected procedure step can be requested by the operators.

The operator may make his own notes in the margin of the step suvey
pane. Notes or messages of interest in connection with the procedure may be
written into the workpane and stored. This information may later be
redisplayed on operators request.

The work pane is intended to hold a calculator helping the operator
doing simple algebraic calculations as well as a trend facility to show
development of any process variable of the operators personal choice.

In addition to the monitor elements described above it will be possible
for the operator to insert his own "private" monitor elements to monitor any
variable in the process.

IMPLEMENTATION

An implementation based on object oriented techniques can be seen as a
set of discrete objects communicating in various ways in order to solve a
common task. An example from the implementation of the COPMA kernel will
illustrate this idea. The common task to be solved is the execution of one
procedure step. Each step can be regarded as a sequence of instructions. The
objects involved are as described in Figure 4.

Execution of a step is related to a particular activity. This activity
is represented as an object with pointers into a static structure describing
the step. The figure indicates that the next instruction to be performed is
instruction 3 within step i.

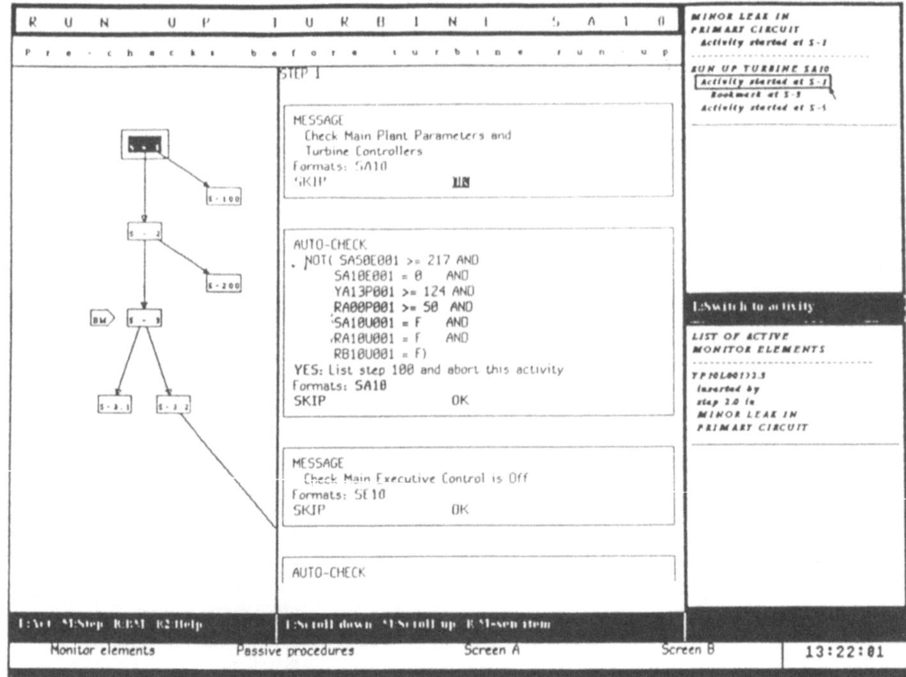

Figure 3.   Example of a COPMA screen

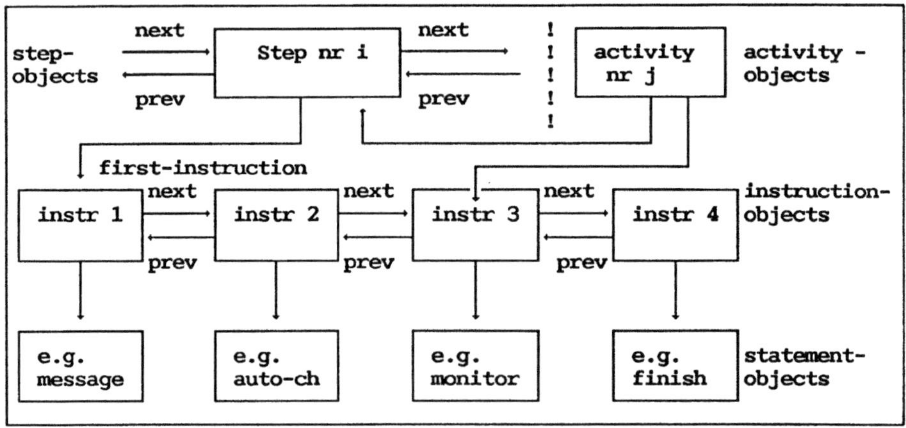

Figure 4.   Objects involved in COPMA kernel

The object structure describing the step is static, i.e. it will not
change whatever happens with the procedure. This means that it can be used
by more than one activity at a time, and hence there are no limit on how
many activities which can be associated with a particular step.

The instruction-objects are characterized by having several attributes,
i.e. they will have variables which are local to the object. In Lisp such
similarities is made explicit by defining a certain "flavor", e.g.

```
(defflavor instruction-objects(next previous statement)())
```

This construction expresses that all objects of type (or flavor if you
like) "instruction-object" will have three local variables, namely next,
previous and statement. In referring to this flavor it is possible to create
any number of instruction-objects, executing statements of the following
kind:

```
(setq instruction-1 (make-instance 'instruction-objects))
```

In addition to local variables (attributes), objects will normally also
have message handlers which are supposed to take care of incoming messages.
Handlers are function-like constructs which can make free references to the
variables inside the object.

The setting of local variables can for instance be implemented by
passing a message like the following:

```
(send instruction-1 :set-next instruction-2)
```

This message is passed to the object referenced by the variable in-
struction-1. This object receives this message and treat it according to the
handler. Handlers, also called methods, are specified according to the
syntax in the following example:

```
(defmethod (instructions :set-next)(instruction)
    (setq next instruction))
```


CONCLUSION

A prototype of a computerised procedure system has been developed based
on LISP and object-oriented programming techniques. The prototype is interfaced to the NORS PWR fullscale colour CRT based simulator system by means
of a serial link. A number of procedures will be implemented and the system
will be extensively tested both from a human factors and system designs
point of view.

REFERENCES

1. K. Haugset,"The Integrated Surveillance and Control System ISACS",
     HWR-212, OECD Halden Reactor Project, May 1987.

2. M. H. Morgenstern et al.," Study of Operating Procedures in Nuclear
     Power Plants : Practices and Problems", NUREG/CR-3968. February 1987.

INTERFACE OF AN EXPERT SYSTEM WITH ON-LINE DATA FOR

TECHNICAL SPECIFICATION MONITORING

William H. Schlegelmilch   Raymond P. Jefferis

PSE&G   Widener University
Hancocks Bridge, NJ   Chester, PA

ABSTRACT

   Compliance with the Technical Specifications (tech specs) of a Nuclear
Power Plant has been and continues to be an important issue for the utilities
operating the plants as well as for the public at large. However, the
volume and complexity of the tech specs often make compliance a tedious,
time consuming, and sometimes error prone task. The result is operating
inflexibility, increased manpower usage, and greater probability of lost
plant availability. This paper describes a system that will enable utilities
to better manage their nuclear plant operations by automating the compliance
decisions by using an expert system.

   The system runs on a microcomputer and is designed to take advantage
of data gathering from existing plant computers. Since the amount of data
needed to correctly evaluate the rules (i.e., tech specs) is very large,
manual input of all of the data is impractical. The system itself is
comprised of several programs running in a multi-tasking environment. These
subsystems include an expert system, a program to gather data from an
external computer, a user I/O system, and various support programs for the
above. The various options for running the system are discussed. These
include an examination of the extent to which PROLOG and FORTRAN components aid the expert system and optimize its running time.

INTRODUCTION

   With the many options available for evaluating compliance with the
Tech Specs of a Nuclear Power Plant, the debate must end when it comes to
deciding on how the data should be gathered for use by the system. Because
of the literally thousands of pieces of data needed for in-depth compliance
monitoring, there is no choice left to the engineer but to provide for automatic inclusion of this data. Human interface, in this capacity, must be
limited to a relatively minor role. This paper focuses on an implementation of compliance monitoring and the steps both taken and needing to be
taken for automatic inclusion of on-line plant data.

CONCEPTS

The general concepts behind the Technical Specification Compliance
Monitor are centered around three main goals. These are to (1) provide
on-line support to nuclear power plant personnel, (2) provide off-line
inquiry capabilities for planning purposes, and (3) to do this with a
microcomputer system. With the power and popularity of the newer micros,
they have become a very viable option to many applications formerly reserved
for mini and mainframe computers. In addition to the financial motivators
for this type of approach, flexibility in software selection and customiz-
ation allow the engineer a large degree of freedom in system design.

The provisions for off-line inquiry are necessary to expand the scope
of use of the compliance monitor. For although there is great benefit from
on-line monitoring; off-line inquiry gives us the capability to put the
system into the hands of those who need but have no knowledge of technical
specification requirements. Allowing them to have, in effect, an expert in
the tech specs always at their side.

The focus of this paper, however, is on the first goal of providing on-
line monitoring of tech spec compliance. In order to do a thorough job of
this, there needs to exist three major sources of data. These are (1) plant
component data, (2) plant process information, and (3) surveillance test
results for the equipment covered by the tech specs. With these sources
available for query by the expert system, human data input can be reduced
to an acceptable level and frequency.

DESIGN CONSIDERATIONS

Many nuclear plants in this country and around the world have, or are
planning to have, a computer system which will enable them to keep current
information on the many valves, circuit breakers and other components in
their plants. Certainly all plants have computers which, to varying
degrees, handle plant process information. And many plants also have
computers which process tech spec surveillance information. So, it is seen
that the bulk of the data needed for tech spec compliance monitoring is
already resident in a nuclear plant's computing environment. The only
questions remaining are how to gather the data, how to make it available
to the expert system, and how the expert system will use the data.

In the gathering of data for expert system processing, there are
several key factors which must be considered. These are as follows:

o   The level of the incoming data. By this, we mean to decide
        whether the expert system will process the data directly, or
        whether some preprocessing will need to be performed on the data.

o   The format of the incoming data. Is the data in a form that can
        be successfully utilized by the system?

o   The scope of the incoming data. Should the data be sent to the
        system in generic blocks, or should only the data that is neces-
        sary for processing be sent?

o   The frequency of the incoming data. Should the data be re-
        freshed constantly, or only when there is a pertinent change in
        it?

To decide on each of these factors, both ends of the data transfer need to be looked at in detail.

In general, the level of the incoming data should be suited to the expert system environment. As with most computer systems, the expert system needs to be geared toward the user. When performing the design from this standpoint, the level of detail that the user will require must be determined. Once this is done, and the raw data is known, the knowledge engineer can decide on the amount of preprocessing that is required. A secondary contributing factor to this decision must also be the effect of the level of the data on the system runtime. If the incoming data is in such a form that the system processing time will become unacceptably long, preprocessing may also be preferred.

The format of the data is certainly a critical issue in the design stages of the system. The expert system must be able to understand the incoming data in order for it to be of any use at all. This can sometimes be handled in a file transfer protocol; or, it preprocessing is to be done anyway, it might be more convenient to reformat the data at that time.

When deciding on the scope of the incoming data, the knowledge engineer must consider the effects of future changes to the knowledge base. If the knowledge base is expected to remain relatively constant in regard to the data necessary to evaluate the rules, then it would be preferable to have specific blocks of data transferred. If the knowledge base is expected to go through some sizeable changes throughout its lifetime, a more general approach would be called for. Another factor in this decision lies in the overall system design. If it is decided that only specific pieces of data will be sent to the expert system, then the knowledge engineer has actually given the data generating programs some of the domain knowledge. This may or may not be desired. When this route is taken, however, it must be clear that changes which affect the data needed by the application will also affect the data gatherers.

Of course, the incoming data will have to be refreshed on a periodic basis. The period is determined by the application. A good heuristic approach would be to bring in new data only when there is a change to it. There is no sense in using computing resources to replace existing data with new data that has not changed.

Analysis of each of these factors must be done while keeping both the source and the destination software and system hardware in mind. As with many engineering solutions, trade-offs must be made to obtain an optimally designed system from a total integrated system standpoint. Unduly penalizing one system at the expense of another is both a bad engineering and a bad business practice.

IMPLEMENTATION

Before discussing how the expert system will use the data, we need to look into the workings of the expert system itself. The expert system which has been prototyped consists of two separate approaches. The major portion uses a rule based shell called EXSYS by Exsys, Inc. The secondary portion makes use of predicate calculus implemented in PROLOG. An analysis of each follows.

The rule based shell was chosen as the primary means to perform the compliance decisions for the following reasons. In regard to the tech specs; (1) the rule based shell allows for close correlation to the text and logical structure of the tech specs, and (2) the large interrelated knowledge base can be easily understood in the If-Then-Else production rule format. In regard to other rule based systems, this shell provides; (1) forward and backward chaining capabilities, (2) external program calls, (3) numeric and string variable processing, and (4) all input in English text, among other valuable features. Each of these factors play an important role in the compliance monitor.

This shell's capability to make external program calls is the major factor enabling the automatic inclusion of the plant data. By utilizing this option, the expert system has been designed to invoke several generic programs which will return the data necessary for the processing of a rule. The specific program which is called will depend on both the level and type of data needed by a particular rule. Figure 1 illustrates this process.

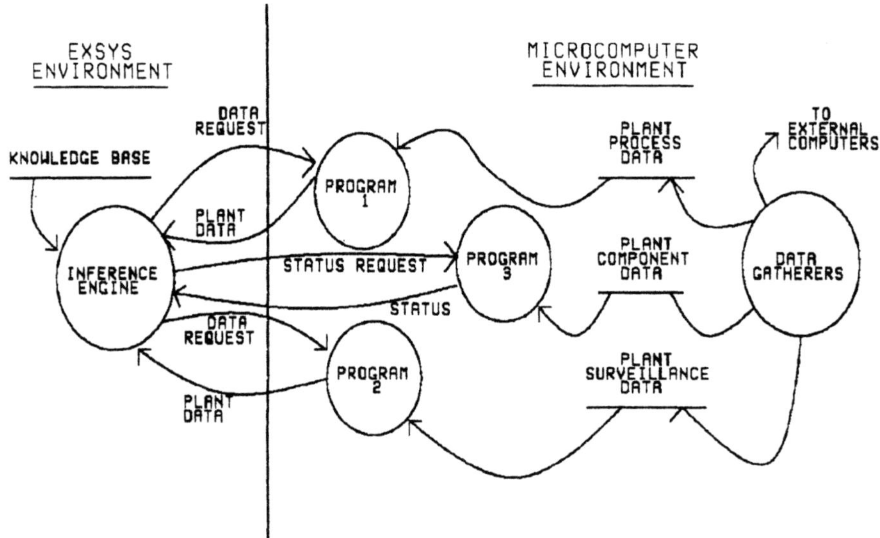

Figure 1 - Automatic Query of a Data File for Rule Evaluation

As mentioned previously, the program which is invoked to find the data and feed it into the rule based system will vary with the level and type of data required. For instance, suppose that a rule needs a value for the current volume of a Reactor Coolant System Accumulator in order to determine its operability. Let us also say that this data is resident in a file (this file will be refreshed periodically with current data) in the microcomputer in table format. By using a table search algorithm, this data can be found and sent to the rule based system to evaluate the rule. Suppose now that the status of 11 Safety Injection Pump is needed to process another rule. In this case, a different approach would be called for since the status of the pump depends on many different pieces of the raw data which is available from the plant computers. One might ask; why not write another rule which will be able to make use of the raw data? In the cases where many pieces of data need to be evaluated, the number of rules and external program calls needed would cause the system runtime to become excessively slow. This is where the expansion of the expert system into PROLOG becomes very useful.

Following up on the above example, a PROLOG program can be called
which will evaluate the status of the safety injection pump and return the
deduced status to the rule based system. Figure 2 shows this process.

From the data flow diagram, it is seen that the knowledge base area
of interest is passed to the PROLOG program. With this information, the
PROLOG knowledge base can rapidly access many pieces of raw data in various
data files to arrive at a conclusion on the status of the safety injection
pump. The status of the pump can therefore be returned to the rule based
system in a fraction of the time which would otherwise be necessary.

In order to quickly evaluate equipment status, the PROLOG program was
designed to use a method of data consultation. This method allows for the
contents of the entire data file necessary for evaluation to be read directly into accessible RAM. This, then, requires the use of only one clause
(instruction) for each piece of raw data from the file being utilized.

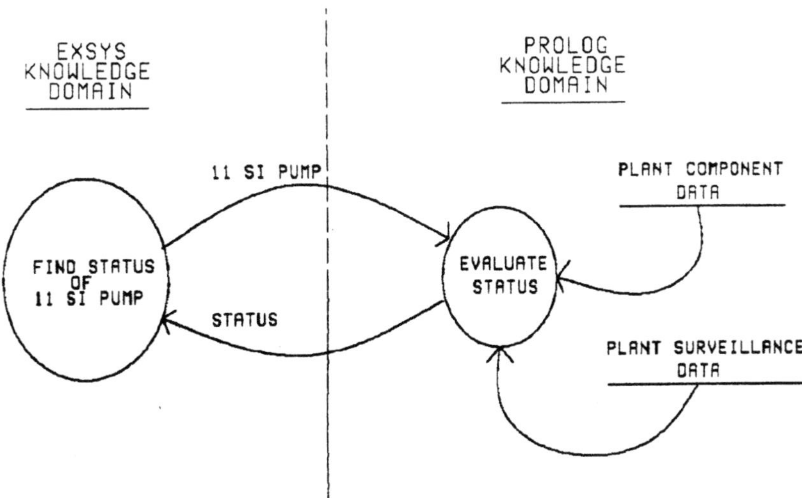

Figure 2 - Chaining the Rule Based System to a PROLOG Knowledge
Base

CONCLUSIONS

Since the expert system is intended to run on-line, there must be an
independent interface to keep the user informed of entries into and termin-
ations of action statements as well as certain query capabilities. Using
the ability of the rule-based core of the expert system to make external
program calls, the problem of tracking action statement activity has been
solved. As a part of the rules which decide on compliance with or violation
of Limiting Conditions for Operation (LCO), another generic program is
called to perform this function. The program, written in FORTRAN, creates
and purges empty files as action statements are entered and terminated. The
directory contents can therefore be utilized by the user interface program
to keep the operator appraised of current action statements in effect.

The compliance monitor's user interface consists of a main menu, from which five categories of input and output functions can be accessed. These include (1) LCO Status, (2) Surveillance Status, (3) LCO Text, (4) Reports, and (5) Plant Data. These categories provide the user with a comprehensive picture of both the current and static data which are part of or impact on the technical specifications.

The system, which has been developed as a prototype, makes use of on-line data for plant components. Expansion into the areas of plant process and equipment surveillance information as on-line inputs needs to be completed. This, in conjunction with a complete tech spec model, will provide for continuous monitoring of the impact of plant activities on the technical specifications.

INTEGRATED DISTURBANCE ANALYSIS (IDA),

A BASIS FOR EXPERT SYSTEMS IN NUCLEAR POWER PLANTS

W. Bastl

Gesellschaft für Reaktorsicherheit (GRS) mbH
Forschungsgelände
8046 Garching, F.R.G.

1. INTRODUCTION

In the framework of a project sponsored by the Bundesminister für Forschung und Technologie (BMFT) and the Rheinisch-Westfälisches Elektrizitätswerk AG (RWE) the concept of Integrated Disturbance Analysis (IDA) has been developed and applied to the Biblis Unit B Nuclear Power Plant. Rather than using the classical approach, which aims at finding the cause of a plant disturbance, the main goal was to make transparent to the operator what it really means for the process, when a set of parameters begins to deviate from its normal state. Therefore the goal was to help the operator to combine the information on the individual process parameters to the complete picture of the evolving transient, thus facilitating the decision about the importance of a disturbance for operational safety. Considering this goal the most important subjects taken care of were

-   to extend event sequence analysis by means of analog information
-   to provide adequate modelling of time dependent processes and feedback
    loops
-   to integrate IDA information effectively into the existing control room
    environment

The software used was based on the STAR-GENERIS package developed by GRS [1, 2]. The effectiveness of IDA was demonstrated by applying it to several systems of the Biblis Unit B Nuclear Power Plant.

2. METHODOLOGY

In order to meet the requirements explained above, IDA is to inform the operator on evolving disturbances at an early stage, i.e. before the conventional alarms are activated. It should assist him to understand the character of the disturbance and to estimate the further development and consequences respectively. The information focuses on disturbances which are slow enough as to allow the operator to take appropriate corrective actions in time.

Based upon these ideas IDA was applied to the condensate and feedwater systems of the Biblis NPP comprising the following items: hot-well, main condenser feeding, low pressure preheating, feedwater tank, main feedwater, high pressure preheating, steam generator. Consequently systems and process analysis was performed, which - in short - concentrated on following problems:

- The effect of disturbance propagation in the cooling water system to the condenser vacuum. When the condenser vacuum changes within certain limits the operator can try to annihilate the disturbance or to reduce the reactor power should he not succeed.

- The effect of disturbances in the main and auxiliary condensate systems on the low pressure preheaters and mass balances of condenser and feed-water tank. Within certain limits the operator has the possibility to take measures for keeping full-power operation before automatical systems are activated (automatical bypassing of the preheaters).

- Disturbance of the steam generator supply as well as undesired feedwater temperatures caused by disturbances in the feedwater system as well as in the condensate system. If the operator is able to quickly understand the disturbance (decreased feedwater flow, changed feedwater temperature) he can continue power operation by means of reconfiguring pumps and preheaters.

In the sequel the analysis methodology is explained which has been used to achieve the desired information for the operator.

2.1 Basic Analysis Models

Besides the specific models describing the aforementioned systems there are basic analysis models, which are used to process the analog signals. They are operated for every scan (actualization) of the signal.

The basic analysis module FILTER delivers representative absolute values and gradients of the analog signal. Basically it is assumed that the process to be observed can be described within a time window of five scanning points $[x_{i-4}, x_i]$ of the process variable x by means of a linear function (five-point fixed memory filter). This means IDA evaluates every analog process variable by a time window of five scanning points and calculates, under the assumption of a linear function within this interval, representative values for the zeroth and first order derivation $(x, \dot{x})$ of this function. The algorithm was selected because of its simplicity and stability. A recursive algorithm would have been more advantageous, however, systematic faults would have been of major concern. This filter also calculates the estimation values of the variances for $(x, \dot{x})$ considering the variation of the input data $[x_{i-4}, x_i]$. Therefore it is possible to define the minimal threshold value for $\dot{x}$, below which we consider the process variable x as not having changed. The possibility to define this threshold value is very important in order to avoid false alarms in practical applications.

The basic analysis modules LIMIT and VAL are used for check and validation of the actual analog value. LIMIT performs straightforward comparison of the actual analog value with alarm thresholds. VAL performs the validation by analytical redundancy. This means a functional relation between the process variable x and $x^*=f(z)$ (if there is such a variable z at all) exists, so that the actual value x can be compared with the calculated value x^* in order to get an indicator for incorrect measuring values. An example for such a relation could be the flow and the pressure which is denoted by the characteristic curve of the pump. The supervision of erroneous binary signals rests upon the already validated analog signals.

Sometimes, however, the validation turns out to be too complicated. We should bear in mind that to discard the erroneous signals we need at least three values to be compared. Therefore, frequently only messages are given on the fact that two signals which should be the same do not match. In which case, for further analysis, the value x is assumed to be in order. In proceeding that way we considered the fact that in most cases the process relevant variable x is anyhow triggering further actions which eventually are supervised by IDA.

The basic analysis model STATUS characterizes the state of a process variable x and its gradient $\dot{x}$. It serves as a means for evaluating the severity of a disturbance, but also generates the control variables for the appropriate IDA supervision (analysis) mode. We defined four bands between the nominal value x_o and x_{ua} and x_{la} (first upper and first lower alarm value = pre-warning values), which are called

- upper annunciated band
- upper pre-warning band
- (tolerance band)
- lower pre-warning band
- annunciated lower band

This is shown in fig. 1 including the classification of the status which is dependent on the deviation of the variable from the nominal value and the sign of the gradient. As a result we get four status categories (0, 1, 2, 3) which are interpreted according to table 1. In addition the basic analysis model TREND calculates changes of the gradient in case of progressively developing disturbances in the annunciated area.

Typical for IDA are the two pre-warning bands which allow supervision of the variable between the nominal value and the pre-warning values. In order to avoid misinterpretations of IDA, a tolerance band has been defined which can change according to the process status and which is set in a way as to allow for changes of the observed process variables within their operational limits. Thus a compromise can be found in between the detection and the mis-interpretation of a disturbance, which of course implies consideration of the measuring signal quality. In general the automatic controls are responsible to keep the nominal value within the tolerance band. Therefore IDA can be used to observe automatic controls for there proper operation. If the tolerance band is exceeded, the information of an evolving disturbance can be given and the functional causes can be analyzed. In many cases this means the characterization of the status of a certain automatic control.

If a further escalation of the disturbance takes place and the pre-warning values are reached, the nominal operational conditions begin to be left. At this point-in-time IDA performs a comparison of the disturbance evolution in the pre-warning band and in the annunciated band, if necessary considering the statuses of redundant trains, so that it is able to inform about a system degradation caused by a process behaviour deviating from normal.

2.2 Process and Systems Analysis

In order to derive the desired information a systematical analysis of the systems and their tasks is performed (fig. 2). As a first step the decomposition of the technological goals from the overall plant process is carried out. Therefore Functional Units (FU) with a specific technological task, e.g. the storing or delivering of coolant, are defined. As a rule these tasks can be described by means of a few process variables. They form the basic structure of the information system at the lowest analysis level, the Functional Sub-Unit (SU). In order to enable an early disturbance detection, an evaluation of the observed deviation from the nominal value

Table 1. Interpretation of the status categories
of a process variable x

STATUS	INTERPRETATION
0	x within the tolerance band or normalization of an existing disturbance
1	x stagnant within one of the pre-warning bands or an annunciated disturbance begins to disappear
2	x progressively changing within one of the pre-warning bands or an existing and annunciated disturbance remains stagnant
3	annunciated disturbance with progressive behaviour

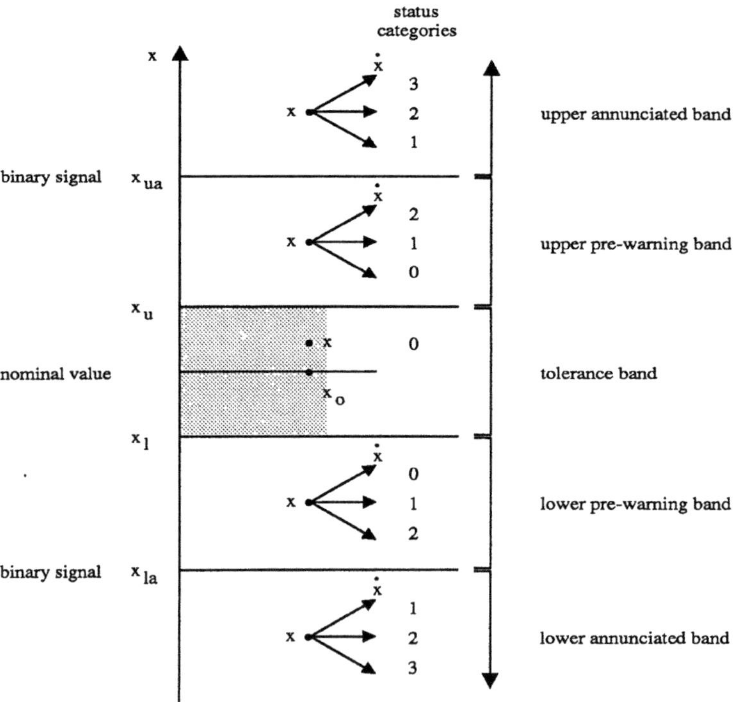

x_u, x_l upper and lower limitation of the tolerance band
(generated by IDA)

x_{ua}, x_{la} first upper and first lower alarm value
(binary signal from conventional alarm system)

Fig. 1. Characterizing the state of a process variable x
by means of the analysis model STATUS

is performed which implies the absolute value of deviation from the nominal
value and its gradient, in addition considering the momentary plant state
(i.e. normal process changes are taken into account). This means a dynamical
definition of the status of the surveyed process variable and hence the
associated disturbance behaviour (progressive, stagnant, normalizing). The
spectrum of the provided surveillance modes is status dependent, e.g. we
perform mass balances or surveillance of the controls around the nominal
value, supervision of automatic actions or trend prediction for advanced
disturbances.

In order to be able to describe realistically the status of the
Functional Unit (FU) it is necessary to take care of so-called external
influences. This is done at the second analysis level (FU-level), where in
redundant systems the Functional Unit corresponds to one train. We have to
consider external influences, e.g.

-   if the surveyed process variable at the SU-level is rather insensitive
    (the level in a tank with a large surface), so that we better take note
    of the mass flow to and from the tank, if we like to find out the level
    behaviour
-   if in associated auxiliary systems or neighboring technological units a
    situation is evolving, which influences the dynamics of the observed
    Functional Unit.

At the first analysis level the FUs are compiled within the Functional
Section (FS). At this stage the status of the redundant system is evaluated
with reference to its technological goal considering the momentary power
related requirements.

By means of the procedure described above we achieve a system/process
description consisting of stand alone models (typically hot-well, condensate
feeding, etc.). Therefore it is easy to exchange models, if the quality
should turn out not to be sufficient to achieve the information task. This
implies the clear definition of interfaces with respect to consequences to,
and/or feedback effects from, neighboring systems.

2.3 <u>Organization of Analysis Models</u>

In this chapter we discuss which methods were used to integrate the
analysis results of IDA into an overall surveillance system. For our actual
case we have to take into account the effect of disturbances within the
analyzed systems onto the secondary circuit as such, e.g. mass transfers in
the large tanks with the consequence of a turbine trip. Therefore a supervision with respect to higher level technological goals has to be considered,
which are in fact the three possible ways of heat transport in the secondary
circuit; following the operational practice according to the sequence
turbine/condenser (1), live steam station/condenser (2), roof station
(release and safety valves) (3). A pre-condition for heat transfer via the
secondary circuit is, of course, the proper functioning of the steam generators. The associated signal analysis is performed by the analysis module
PRIORITY. It provides information

-   if and how the heat transfer through the secondary loop is jeopardized
-   about the causes of the abnormal situation in heat transfer and how they
    time wise compete with each other.

I.e. PRIORITY takes care of the fact that there is a ranking of the
trains for heat transfer (e.g. if there is no turbine trip, then avoid it;
if there is a turbine trip, then avoid condensate pump emergency shut-down).
This leads already to some tasks of an SPDS-system and clearly shows the
relationship between disturbance analysis and such a system.

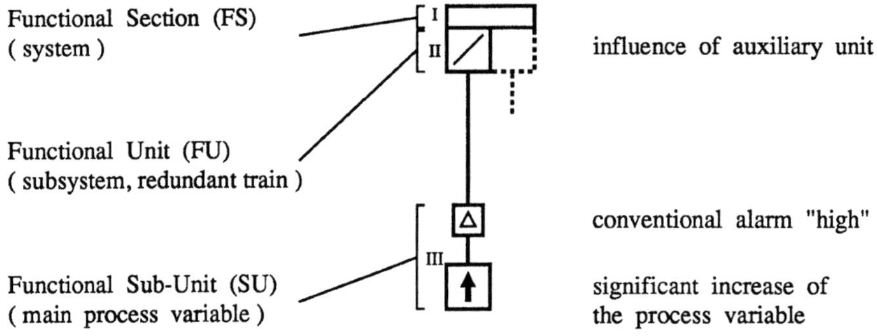

Fig. 2. Analysis and information hierarchy of IDA

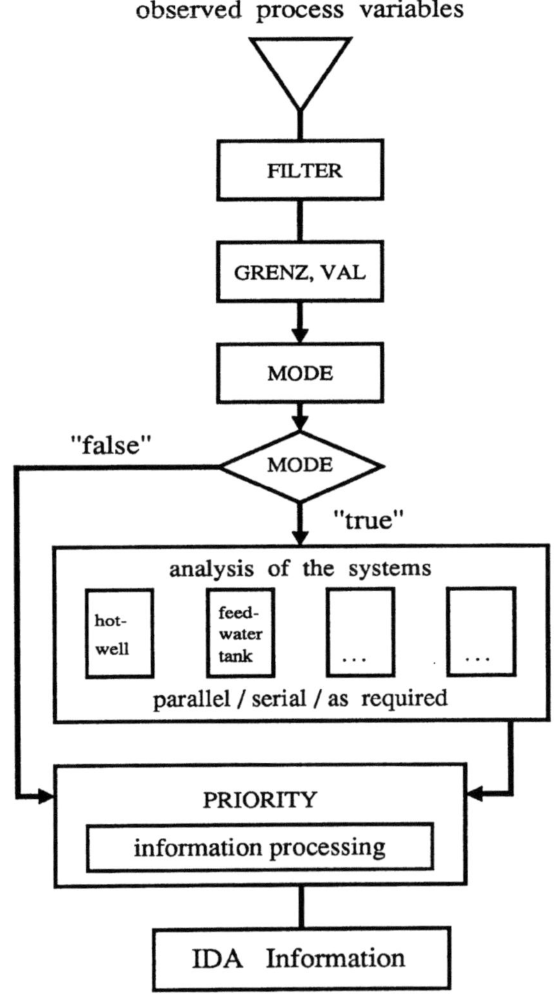

Fig. 3. Organisation of IDA analysis models

The type of information given by PRIORITY is the degree of jeopardy of
a heat transfer path in terms of residual times in case of progressive
disturbances. Residual time (RT) is the time span left until an automatic
protective action is initiated by the observed process variable x. Furthermore a mutually exclusive classification of the residual time is used:

$$
\begin{aligned}
&0 \text{ for } RT > 5 \text{ min} \\
&1 \text{ for } RT < 5 \text{ min} \\
&2 \text{ for } RT < 2 \text{ min} \\
&3 \text{ for } RT < 1/2 \text{ min}
\end{aligned}
$$

Typically the analysis module PRIORITY is used subsequently to the
analysis of the functional sections as described in the previous chapter. If
disturbances are escalating, the models used there will not be valid any
more, because they were developed to detect disturbances at a very early
stage. However, taking into account the fact that also in these cases
PRIORITY can deliver useful information, the signals are fed directly into
this model. The decision is performed automatically by the analysis module
MODE.

MODE distinguishes between normal analysis states and abnormal analysis
states (fig. 3). Normal analysis states relate to operational conditions,
where the steam generators are in order and the heat transfer path (1)
turbine/condenser is likely to be lost, if, due to mass imbalances in the
inventories of the secondary loop a trip of the main coolant pumps is likely
to occur. MODE identifies abnormal states, if there are fast transients or if
the plant is no longer in power operation. Again heat coupling between
primary and secondary loop has to exist. There is no simple and unique
criterium on heat coupling, but it certainly gets worse with decreasing
inventory of the steam generators. For that reason the individual levels of
the steam generators and the life steam pressure and their trends (gradients)
are indicated, so that mass- and energy-imbalances can be recognized by the
operator. Furthermore, feeding of the steam generators considering the status
of the main emergency feedwater pumps and associated valves and controls can
be observed. Residual times or residual distances are not indicated any more.

3. INFORMATION PRESENTATION

The information about status and status changes of the systems supervised by IDA are presented according to a hierarchical structure. Deviations
from nominal values are clearly indicated considering the trends of the
influencing variables. The essential process variables are shown with
reference to higher order process goals, in the time line including extrapolation. Within our project two mimic diagrams were realized, the IDA
overview picture and the IDA transient picture.

3.1 Presentation of IDA Results

Following the analysis as given in chapter 2 the status information
about the Functional Sub-Unit (SU), Functional Unit (FU) and Functional
Section (FS) is given as indicated in fig. 2:

With reference to the SUs following symbols are used in windows III:

blank   process variable within the tolerance band
↑   significant increase of the process variable above the
         tolerance band
↓   significant decrease of the process variable below the
         tolerance band

≙   process variable with small gradient above/below the tolerance
         band
NP   analog signal not plausible
?   analog signal out of measuring range

    Alarm values from the conventional information system are given in the
smaller quadratic fields as additional information distinguishing in between
'high', 'very high', 'low', 'very low'.

    At the FU-level information is given in windows II, considering also
auxiliary components or subsystems necessary for adequate operation. If an
influence of one of these items actually exists, it is indicated by means
of a diagonal bar and changing color of the part of the window located below.
More information about the cause of the influence can be taken from the
relevant SU-level. In addition the symbols "NP" and "?" are taken as summary
signals from the SU-level.

    In the windows I the state of the functional section is indicated depe-
ndent on the status of the functional units and on the requirements of the
plant to the Functional Section, e.g. considering how many trains of the
system are used according to the actual power of the plant. Also at this
level external influences are indicated by means of a diagonal bar. In order
to give an easy-to-comprehend picture on the status of the various surveyed
systems background colors are used for windows I and II. They mean:

             dark green   - undisturbed state
             light green   - disturbance degree 1
             yellow   - disturbance degree 2
             orange   - disturbance degree 3
             red   - failure
             grey   - no surveillance by IDA

3.2 IDA Overview Picture

    The IDA overview picture shows the relevant horizontal information which
is necessary for evaluating the analysis results from the FS-level and below
at a higher level. Amongst others it permits a fast view on how disturbances
spread and which influences they have on the essential operational goals. The
overview picture is subdivided into three information areas (fig. 4). Information areas I and II are oriented from left to right according to the heat
transport in the secondary loop whereas information area III is oriented from
right to left according to the reverse transport of the fluid (condensate
system).

    Information area I: Overview on power status (reactor left, turbine
right) with a message window for reactor scram (RESA) and turbine trip
(TUSA).

    Information area II: Overview on the statuses of the essential com-
ponents for the heat transfer in the secondary circuit. Both the left and
the center parts relate to the steam generator, whereas the right part is
provided with the display of the different ways for heat transport (turbine/
condenser, live steam station/condenser, roof station). By means of
actualized arrows the direction of flow is visualized. Furthermore the
maximal gradients of the live steam pressure and temperature are indicated.
If the gradient of the live steam pressure exceeds 4 bar/min the message
window above the lines indicates the residual time until initiation of
closure of the secondary loop (YZ60). If YZ60 has been initiated this is
shown above the forementioned message window. Moreover the information
'emergency power' can be given.

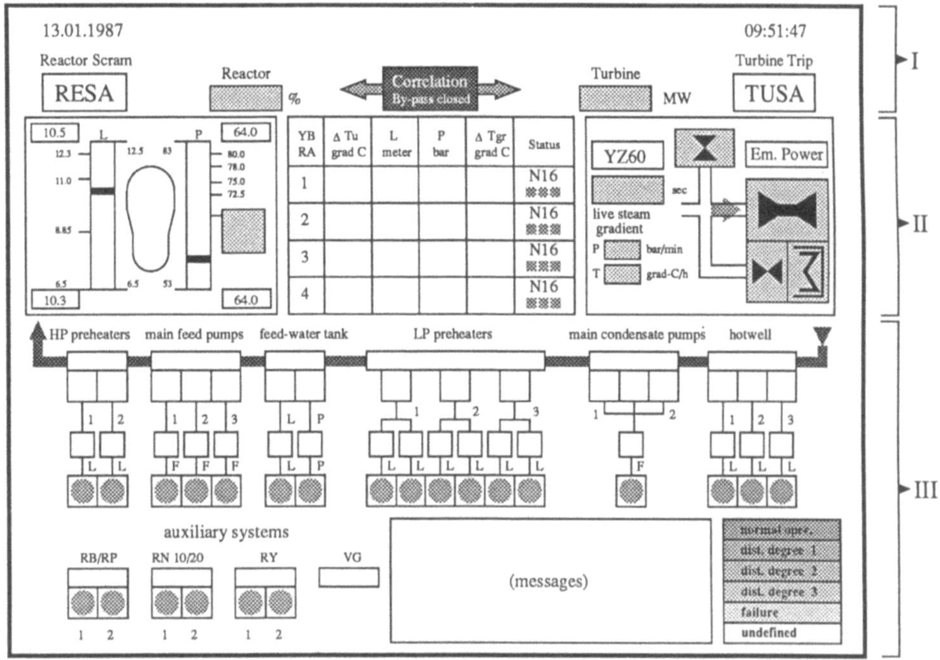

Fig. 4. IDA Overview Picture

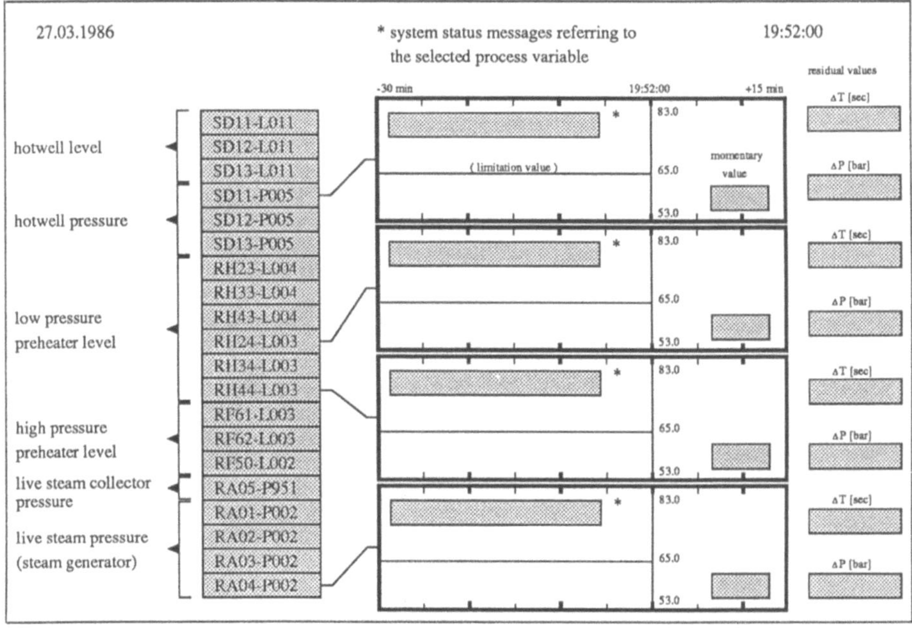

Fig. 5. IDA Transient Picture

Information area III presents an overview on statuses of the systems as described in chapter 2.

3.3 Transient Picture

In this picture residual values and residual times for the different heat transfer paths are computed and ordered for those process variables that have some effect on the current heat transfer path once they have crossed a threshold and are approaching the trip level (fig. 5). This gives the operator an information about competing or mutually influencing factors of a jeopardy of the respective heat transfer path.

Twenty relevant process variables have been selected. Depending on the status of the heat transfer path the system offers four signals for presentation. Other signal combinations can be selected by the operator. The relation between the selected signal and the transient presentation is shown by means of lines. Beginning with the actual value (fixed marker line) the transient is mapped thirty minutes into the past and an extrapolation of fifteen minutes into the future is performed. The actual measuring values and the departure of measuring values and time residuals from threshold values are given in the right-hand message windows. The background color of the message windows is oriented towards the color code described before.

4. FUTURE PERSPECTIVE

The Generis-package already has the essential properties generally attributed to knowledge-based systems (e.g. the propagation model, cause-consequence trees).

However, while the information generated and displayed is considered very useful for plant operation, it is as yet passive in the sense that the user has to take what the system provides (except, for example, he may choose other process parameters to view in the transient picture). It would, however, be very desirable to get an explanation on why this and that parameter is like it is as well as to provide a device that could perform hypothetical reasoning, i.e. to be able to ask, how a transient would affect the systems behaviour if it took a different turn.

Development as far as these questions are concerned is under way in another BMFT-project dealing with on-line real-time expert systems. First results of this project are reported in [3].

5. ACKNOWLEGEMENT

The author wishes to thank Lothar Felkel, Hans-Georg Herdtle and Heinz Polke who have significantly contributed to the success of the project.

6. REFERENCES

[1] W. Bastl et al., STAR Disturbance Analysis System, IAEA-SM-265/96, Vienna 1983.

[2] L. Felkel, The STAR Concept, Systems to Assist the Operator during Abnormal Events, Atomkernenergie/Kerntechnik, Vol. 45, No. 4, 1984.

[3] L. Felkel, An On-line Diagnostic Expert System, this conference.

DEVELOPMENT OF A KNOWLEDGE-BASED SYSTEM FOR LOOP DIAGNOSIS

Lih-Yih Liao, Hsien-Chang Tang and Sam-Suan Chen

Institute of Nuclear Energy Research
P.O. Box 3-3, Lung-Tan, Taiwan, R.O.C.

ABSTRACT

An accident diagnostic system is developed as an attempt to provide a useful aid for the operators of an experimental loop or a nuclear power plant in the case of emergency condition. Because the current practices in the system diagnosis are not satisfactory, there is an increasing demand on the establishment of various operator decision support systems. The knowledge based system is a new and promising technique which can be used to fulfill this demand. With the capability of automatic reasoning and by incorporating the information about system status, the knowledge based system can simulate the process of human thinking and serve as a good decision support system.

This knowledge based decision support system can be helpful for both a fast, violent accident and a slowly developed accident. Specifically, a fast diagnostic report can be provided for a fast and violent accident of which time is the main concern and a complete diagnostic report can be provided for a slowly developed accident of which complexity is the main concern.

Such a knowledge based decision support system also provides many other equally important advantages, such as the elimination of human error, the automatic validation of signal readings, the establishment of a simulation environment, ..., etc. A perspective usage of this system is rather broad and promising.

INTRODUCTION

There is an increasing demand on the establishment of various operator support systems for nuclear power plants. Among them, the one which receives more and more attention is the operator support system for accident diagnosis. The knowledge based system[12] is a new and promising technique which can be used to fulfill this demand. With the capability of automatic reasoning and by incorporating the information about system states, the knowledge-based system can simulate the process of human thinking and serve as a good decision support tool for accident diagnosis. It is believed that such a system can be very helpful for a fast and violent accident as well as for a complex and slowly developed accident.

NEEDS FOR THE ACCIDENT DIAGNOSIS SUPPORT SYSTEM

The accident diagnosis in a nuclear reactor is a complex and difficult job for the operator. Many different initiating events can result in quite similar system behavior. For example, an uncontrolled rod withdraw accident and a loss of heat sink in a pressurized water reactor can result in the similar primary system heat up and pressurization. The accident diagnosis is further complicated by the fact that the abnormalities tend to spread out and mutual interactions between abnormalities are very common. These are the inherent difficulties faced in the process of accident diagnosis. In the implementation of accident diagnosis, all these difficulties will be emerged in the process of key signal identification, cause-consequence analysis and root cause confirmation.

In a nuclear reactor many alarm devices are installed to generate alarm signals such as sounds and lights to alert the operators that the values of the measured parameters have been departed from their normal operational range. With the knowledge about the nuclear reactor system, the operators can identify the root cause of transients or accidents by interpretating the abnormal symptoms of alarm signals. This identification process may work well for normal operation and minor plant upset condition of which the scenario is simple, the number of abnormalities are small and the time is sufficient. As the circumstance becomes more complex and the time available becomes insufficient, the need for a reliable operator support system for accident diagnosis arises. Under such circumstance, it is expected that lots of alarm signals will be generated. When many information are presented, the operators have the difficulties in identifying the key signals immediately because of the delay associated with the process of searching for the right information. As a result of the slow accident diagnosis, either the continued reactor operation will be hampered or the safety of the reactor will be challenged. Since the operators are trained to prevent the occurrence of reactor shutdown to the maximum extent in order to avoid the penalties of shutdown, the safety of the reactor is more likely to be challenged. Without knowing the real cause of the accident, the mitigation action will also be delayed.

Depending on the type of the accident, there are some additional problems. when a rapid and violent transient or accident occurs in a nuclear power plant, the time available is rather short while the number of signals to be processed are rather large. As it is understood that the capability of a human being in simultaneous multiple tasks handling is rather limited. The overflow of the alarm signals will result in emotional pressure of the operators. The emotional pressure is known as a main source for the mistaken decision making. Along with the time pressure and the emotional pressure, the most unreliable decision probably will be made while the most correct judgement is just needed.

On the other hand, when a more severe accident is developed from the slightly offset condition, the developing process is slow and the long transient period makes the occurrence of multiple failure possible. The occurrence of multiple failure adds the additional complexity to the identification of root causes. The symptom, presented by the measured plant parameters, can be affected by the initiation event, the additional malfunction of the components, the false instrument readings, the reaction of the automatic control system, the action of the operator or their combinations. As we can see that the distinction of one cause from another is not trivial because the same symptom can be resulted from many different causes and the causes are mutually interactive.

Over years many efforts have been devoted to resolving the signal processing problems. Among them, one proposal is to reduce the amount of alarm signals. Another approach is to introduce an additional system which contains only the safety related information. The " Safety Parameter Display System (SPDS) " currently installed on the commercial nuclear power plant is the product of the latter concept. These two approaches may increase the reactor safety by limiting the operators attention to a small amount of parameters critical to the reactor safety. But on the other hand, they provides relatively little help to the identification of accident root cause and the continuation of the reactor operation. To resolve the difficulties in accident diagnosis, one proposal is made by providing a kind of symptom oriented procedure which allows the operators to take remedy action without root cause identification. The emergency operation procedure currently being implemented to the commercial power plant is the product of this approach. Although the reactor can be brought to a stable condition by applying the emergency operation procedure, in many cases the identification of root cause is still required to bring the reactor to the final cold shutdown condition.

The provision of an intelligent knowledge based system offers great potential for eliminating the emotional pressure problem, speeding up the searching process and providing a complete record of all the events occurred in the complex situation.

As a first step to develop a useful and reliable knowledge-based system for the nuclear power plant, a prototype system is developed for a thermal hydraulic experimental loop.

EXPERIMENTAL LOOP

The experimental loop is a high pressure and high temperature water loop which possesses the similar basic components used in the nuclear

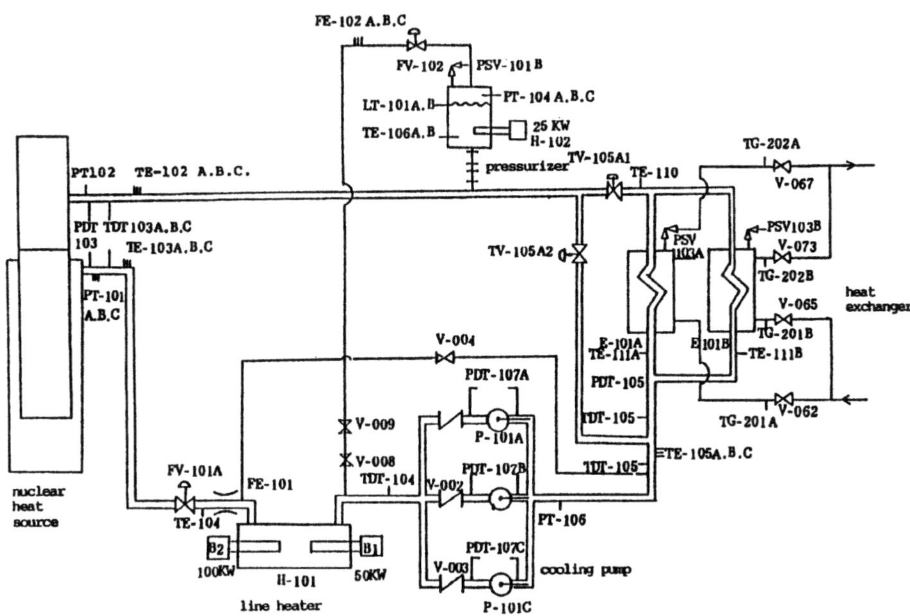

Fig.1. Schematic Diagram of the Experimental Loop

power plant such as pressurizer, heat exchangers, pumps, valves and feedwater heaters. The schematic diagram of this loop is shown in figure 1. The pressure of the system is controlled by the pressurizer heater and spray. The heat removed by the heat exchangers are controlled by the simultaneous adjustment of the parallel control valves TV-105A1 and TV-105A2. The test section inlet temperature is kept constant by controlling the power generation rate of the line heater. The loop status is represented by the readings of pressure, temperature and water level meters. The instrument readings are classified into three states: high, normal and low. The symptom of the loop is indicated by the parameters with state high or low.

SYMPTOM DESCRIPTION AND ROOT CAUSE IDENTIFICATION

The symptoms of this experimental loop are represented as the abnormalities of pressure, temperature, flow rate and water level readings. When a component failure occurs, the effect of this failure will propagate to the nearby component and this process will go on until the effects are compensated by the control systems or the effects are spreaded out to the whole system. As mentioned before, in addition to the propagation of the abnormalities, mutual interactions between abnormalities make the diagnosis complex. A line heater control failure will be used to illustrate the propagation and interaction of abnormalities. When extra power is generated by line heater H-101 due to the failure of heater controller, the line heater outlet temperature will be high. Through the flow propagation, the test section inlet temperature will also be high later. When the process goes on, the test section outlet temperature will also be high. When the temperature of the system increases, the water level in the pressurizer will and the pressure of the loop will also increase. The mutual interaction of temperature, water level and pressure makes the system behavior complex. Despite of the complexity, as long as there is no masking effects, the root cause can still be found through the trace back of the cause-consequence relationship. The masking effects can be observed when the consequence of system response overrides the consequence of root initiation event. One example of masking effect is provided here through the study of the response of line heater. In the loop, the line heater is composed of two heaters which are designed differently. Heater B1 is a small heater with adjustable power and heater B2 is a large on-off type heater. If the heater B1 fails to produce power, the temperature TE-104 will be low which will activate the line heater B2. The heat generated by heater B2 is greater than that lost by heater B1 and consequently a high TE-104 reading will be obtained. Because the consequence of root cause, TE-104 temperature low, is overrided by the response of heater B2, TE-104 temperature high, the root cause will be identified as heater B2 failure instead of heater B1 failure which is the correct root cause. Since masking effect is more complex, no attempt is made to resolve the problem of masking effect in this prototype system.

A cause-consequence relation graph is shown in figure 2 to represent the initiation component failures and symptoms resulted from these failures.

STRUCTURE OF THE KNOWLEDGE BASED SYSTEM

The system is composed of a knowledge base, an inference engine and a user interface. The fundamental thermal hydraulic concepts have been used to generate the rules. The knowledge base consists of 111 rules which describe the relationships between the causes and the consequences, 23 instrument definitions which provide the confidence levels and English

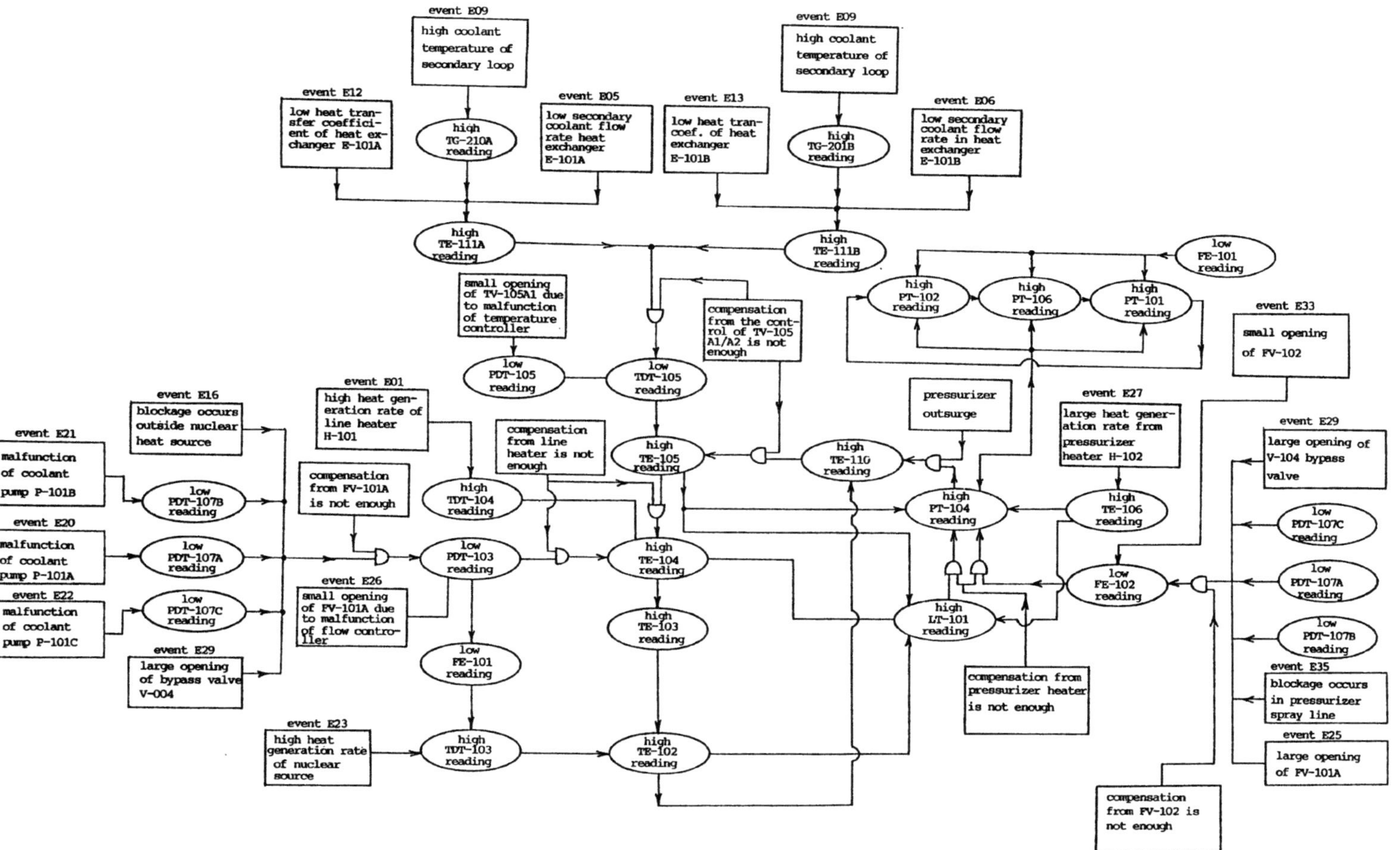

Fig.2. A cause-consequence relation graph

Table 1. Examples of Rules

```
((Rule R01 (TE-103 temperature High) <--- (TE-104 temperature High) ))
((Rule R02 (TE-103 temperature Low) <--- (TE-104 temperature Low) ))
((Rule R03 (TE-104 temperature High) <--- (PDT-103 pressure-difference
           low) (linear heater H-101 compensation not enough) ))
((Rule R04 (TE-104 temperature High) <--- (TE-105 temperature High)
           (linear heater H-101 compensation not enough) ))
((Rule R05 (TE-104 temperature High) <--- (TDT-104
           temperature-difference High) ))
((Rule R06 (TE-104 temperature Low) <--- (PDT-103 pressure-difference
           high) (linear heater H-101 compensation not enough) ))
((Rule R07 (TE-104 temperature Low) <--- (TE-105 temperature Low)
           (linear heater H-101 compensation not enough) ))
((Rule R08 (TE-104 temperature Low) <---(TDT-104 temperature-difference
           Low) ))
((Rule R09 (TDT-104 temperature-difference High) <--- ((linear heater
           H-101 failure) make (high linear heater power)) ))
((Rule R10 (TDT-104 temperature-difference Low) <--- ((linear he
           H-101 failure) make (low linear heater power)) ))
```


Table 2. Examples of Instrument Definitions

```
((definition 01 TE-102 3 (reactor-exit temperature) ))
((definition 02 PT-104 3 (pressure of "P.Z.R") ))
((definition 03 TE-105 3 (temperature after heater-exchanger E-101) ))
((definition 04 TE-103 3 (reactor-entrance temperature) ))
((definition 05 FE-101 1 (flow of primary loop) ))
((definition 06 TE-110 1 (temperature after "P.Z.R") ))
((definition 07 PT-106 1 (pressure after "P.Z.R" and heat-exchanger
                          E-101) ))
((definition 08 PDT-103 1 (pressure drop across reactor) ))
((definition 09 TE-104 1 (temperature after linear-heater E-101) ))
((definition 10 PDT-105 1 (pressure drop across heat-exchanger E-101)))
```


definitions of the instrument readings, and a list of possible causes of
the system abnormalities. Examples of rules and instrument definitions are
given in Table 1 and 2 respectively. The inference engine controls the
process of reasoning and questioning. Backward chaining is established by
using the micro-PROLOG languages. This system is installed on the IBM 16
bits, 640k 5150 series personal computer.

FEATURES OF THE SYSTEM

The major features of the system include:

1. Diagnosis of the loop

A quick and reliable diagnosis of the loop abnormalities is the main
purpose of our knowledge based system. Backward chaining is used to obtain
the root cause of the abnormal loop status. To accelerate the diagnostic
process, the order of rules has been carefully arranged so that the key
information can be collected in a short period.

2. Diagnosis of multiple failure

To make the diagnosis of the complex and slowly developed accident
possible, a multiple failure diagnostic capabilities are developed. As
long as there are no masking effects from two causes, the multiple
failures can be found one by one. Options are provided so that either
single failure or multiple failures can be selected. Because of the

limited measured information, there may be more than one root causes
for the same given system. This option also provides a way of generating
a complete list of all the possible causes.

3. Reasoning with uncertainty

Under some conditions, such as maintenance, the instruments may be
out of service and the states of instrument readings are therefore
unknown. To avoid the interruption of the reasoning process, the cause and
consequence relations are automatically investigated and a possible state
is suggested to the user during the reasoning process. To agree with the
recommended state or not is determined by the user.

4. Identification of the instrument malfunction

The plant status is judged through the instrument readings. If some
instruments fail in a loop which is in good operation condition, we may
have a misleading interpretation of the plant status. Capabilities are
established to identify the malfunction of the instruments.

5. Processing of infinite reasoning loop

Due to the characteristics of the experimental loop, the connection
of the cause-consequence relations of the whole system results in a
relation graph rather than a relation tree. A pruning technique was used
to avoid the infinite looping reasoning process of the graph.

6. Provision of a simulation environment

This system can be executed by either one of the two modes -- "auto"
and "manual". When it is executed in the "auto" mode, all the instrument
readings are supplied in one step. The root cause is found automatically.
When the system is executed by the "manual" mode, the information required
is supplied by the interactive querying process. The "manual" model can
provide a simulation environment which helps the operator in understanding
the cause and effect relations.

7. Explanation

After diagnosis, one can select the explanation option to reveal the
reasoning process. The explanation is given by English sentences. With a
number attached to each rule, this can be an effective way of finding
mistakes in the knowledge base.

8. User interface

8.1 Easy correction of the input data

The input data can be corrected in the input process. This correction
can be done without going out of the micro-PROLOG environment.

8.2 Intelligent way of questioning

There is no pre-structure of the questions. Questions are dynamically
oriented according to the answers being received and the rules being
fired. Once the knowledge base is modified, the sequence of question
asking will be changed automatically.

CONCLUSION

The provision of an intelligent knowledge based system offers great
potential in diagnosing the nuclear power plant or an experimental loop.
This knowledge based system can be helpful for a fast and violent accident
as well as for a complex and slowly developed accident. A prototype
knowledge based plant diagonstic system has been developed and preliminary
capabilities of plant diagnosis and signal validation have been
demonstrated. Extension and implementation of this system to our AI
machine will be done in the near future.

REFERENCE

1. Hoyes-Roth, et al., "Building Expert System", Addison-Wesley,
    Reading, MA, 1983.

2. T. kiguchi, et al., "A knowledge Based System for plant
    Diagnosis", 1985 International Topical Meeting on Control,
    pp.635-642, American Nuclear Society, La Grange Park, Illinois
    (1985).

3. K.L. Clark and F.G. McCabe, "micro-PROLOG : Programming in
    Logic", Logic Programming Associates Ltd., London, England (1983).

KNOWLEDGE BASED DIAGNOSTICS IN NUCLEAR POWER PLANTS

Frank Baldeweg, Uwe Fiedler,
Frank-Peter Weiss, and Mathias Werner

Akademie der Wissenschaften der DDR
Zentralinstitut für Kernforschung Rossendorf
PSF 19   Dresden   8051

ABSTRACT

In the paper to be given a special process diagnostic
system (PDS) will be presented. It must be seen the result
of a long term work on computerized process surveillance and
control; it includes a model based system for noise analysis
of mechanical vibrations, which has recently been enhanced by
using of knowledge based technique (expert systems). The paper
will discuss the process diagnostic frame concept and emphasize
the vibration analysis expertsystem.

INTRODUCTION

Availability and safety of Nuclear Power Plants (NPP)
are subjected to very demanding standards. To achieve adequate
decision capability measures for supporting the operator crew
during normal and emergency situations and predictive diagnosis
to enable preventive actions are of great importance.

But especially predictive diagnosis proves difficult for
inspections of internal parts of the plant; they are possible
only at long time intervals. Often in addition, only major
faults can be detected during maintenance measures. By the use
of noise analysis methods informations about internal processes
and states can be received 1 . Mechanical vibrations for
instance can be infered from various signals e.g. neutron
flux, body sound, pressure, coolant flow e.t.c.

The interpretation of these signals is difficult because
of the complexity of relations between components of noise
signals and the process initiating them, and the number of
these processes. Exact algorithmic solutions seem to be not
applicable, rather do semiempirical approaches that compare
the components of noise signals from different detectors with
reference data. In this case, the experience of human reactor
diagnosticians are valuable.

Only few experts apply knowledge of this kind actively and comprehensively. Expert systems are suitable to widespread this know-how in form of operator support systems.

In the paper to be given especially a diagnosis system for the analysis of mechanical vibrations of reactor components (core barrel, control elements) will be presented. Vibrations of these kinds are undesired; they are surveilled in many reactor plants.

The vibration diagnosis system (VDS) must be considered a partial system of a complex process diagnostic system PDS, furthermore integrating the disturbance analysis system SAAP-22 .

The noise analysis system has been first of all installed separatly. The coordination of the both partial systems have not been satisfying yet; especially due to the very complicate know-how of diagnosis, which has to be activated by the plant operator and due to some special problems in the computer controlled dialogue.

By using of knowledge based technique the capability of those systems are being enhanced. The advantage of such technique therefore is twofold, at least:
Firstly – the informational structure of the partial processes monitoring, interpretation of process states, diagnosis of faults, planning of recovery actions becomes more efficient.
Secondly – a suitable natural language oriented man-machine dialogue will be possible.

THE PROCESS DIAGNOSTIC SYSTEM (PDS)

<u>Aims</u>

The process diagnostic system, PDS essentially consists of the disturbance analysis system SAAP and the noise analysis system VDS. They are parts of the decentralized hierarchical informational system (HIS) (figure 1) which has been developed for monitoring and control of the Rossendorf Research Reactor 2 comprising the functions:
- surveillance
- process analysis
- optimal power control
- start-up and shut-down

It is meant to be a prototypical solution for large scale technological plant automation.

<u>Architecture</u> (figure 2)

Following the concept mentioned above, SAAP was intended to ease tasks of man by extending the scope of autonomous problem solving as well as amending man-machine-interaction. Principally, designing of man-machine interfaces means to define and coordinate automatic and interactive components.

a) automatic diagnosis: it creates and updates a process image
   in computer memory and makes it accessible for interactive
   diagnosis
b) interactive diagnosis: it makes the results of automatic
   diagnosis visible for the operator, following a top-down
   strategy from general categories to details about single
   events.

The knowledge-based extension is designed with a 3-level
architecture (figure 2):

Level 0: provides raw data for further analysis. The
scope of its tasks reaches from simple limit checking of
analogue values to complex numeric calculations to compute
symptoms necessary for advanced diagnosis.

Level 1: a separate analysis module for each physical
partial system is implemented. From symptoms given by level 0
they deduce possible consequences. Degree and dynamics of
disturbances are determined to be heuristically evaluated
later.

Level 2: heuristic evalution, which partial system should
be considered, is carried out. The required mode of operation
(operator-or menu-driven) is inferred from the actual process
states, specific information about the operator as well as
from his input, possibly given menu-or command-oriented. The
knowledge necessary for such inferences is represented as
production rules.

The knowledge based version of the process diagnostic
system PDS integrates the noise analysis system, e.g. for the
identification of mechanical vibrations of technological
components.

<u>Man-machine-aspects</u>

Efficiency of surveillance and control, especially process
diagnosis, and the capacity of man-machine-communication must
be seen in a strong relation. These tasks, e.g. diagnosis, have
to be executed in real-time. The results must be presented and
understood within the context of human diagnosis, as an integral
part of cognitive processes.

Advanced operator support systems for power plants (e.g.
NPP), including more intelligent components like expert systems
support knowledge-based problem solving by the operator staff.
The man-machine interface has to consider this fact. It seems
to be necessary to turn over from correctly widespread used
computer controlled dialogue to a user controlled dialogue
using natural language.

By the use of AI methods such an "intelligent power"
should be reached that the computer unterstand the operator
in any situation.

The interface to the human operator should be a natural
language one, but support other inputs like function keys
e.t.c., too. By means of an associative memory the original
inputs are translated into reduced one's using a restricted
word set (elimination of synonyms).

The reduced input is translated in a second step into
a knowledge representation considering the current context
and background knowledge. This step includes language
analysis which is based on the ideas of word understanding.

It is expected that the information exchange between
the man-machine interface processor and other one's occurs
via a Local Area Network (LAN). For this reason the knowledge
representation is written on a "blackboard" where other
processors can read the information after activating. In
the same way the results of computation are given back to
the man-machine interface processor of course as a knowledge
representation which is transformed into an adequate output.

A major part of the system is the self organizing data
base. Up to now we favour an associative memory designed in
software; a so-called "serial net". It allows to compress
large character strings to simple numbers and therefore
speeds up the interal work of the computer.

THE VIBRATION ANALYSIS EXPERT SYSTEM

<u>Aims</u>

The system,which may be considered a partial system of
the process diagnostic system (s. figure 2),is intended to
advise the operator in selecting test procedures in regard to
their potential decision capability, application constraints
and minimization of costs. It bases on the assumption that all
significant reactivity and transmission effects become visible
in neutron flux fluctuations. Special emphasis is denoted to
core barrel motion and vibration of control elements of the
reactor. The conventional system is well established. By using
the expert system technique the following special expert
experiences are tried to be integrated.

- probability of different possible causes for given faults
- capability, constraints, and costs of test procedures
- experimentally developed vibration models for reactor
  components
- empirical interpretation of signal features in terms of
  diagnostic and therapeutic concepts.

<u>Architecture</u>

The architecture of the noise analysis expert system may
be seen from figure 2. It comprises the three levels
- signal acquisition (0. level)
- signal analysis (1. level)
- diagnosis strategy (2. level)

0. level: the basic information is sampled: neutron flux

fluctuations, accelerator or body sound, pressure and coolant
flow.

1. level: the first step is to achieve significant signal
functions and parameters (auto-, cross correlationfunction,
power density, coherence and phase shift relations). Figure 4
e.g. shows the autospectral density of different neutron flux
signals, (in the frequency regions between 2...3 Hz there are
significant resonances, caused by flow induced motion of the
control elements); the next step is to get information about
the system by interpreting the estimated signal functions.
There e.g. direct (mathematical and physical models) or indirect methods (pattern recognition) can be used to find the
relations between process parameters and the specific noise
sources on the one side and noise sources and the signal
functions on the other side.

2. level: the computations of the 2. level are supervised
by heuristically developed rules, which determine a sequence
of tests suitable to find a correct interpretation for the
actual situation; in more detail from initial facts production
rules are fired in forward chaining mode to establish general
hypotheses, e.g. vibration of the reactor barrel. These hypotheses can be verified and refined by backward chaining in
successive steps. For this purpose, additional sources of
information (pressure, sound, flow signals) are exploited
(1., 2. level). In general, the acquisition of process parameters comprises the usage of rather complicated numerical
approaches and physical experiments.

<u>Instrumentation and implementation</u>

The instrumentation of the noise analysis partial system
is shown in figure 3. The physical values are normalized and
transformed. The time averaged part u(t) and the noise part
u(t) are selected and suitably processed either off-line or
on-line. A monitor may be used to supervise special known
noise sources.

It is implemented on a 16-bit microcomputer system an the
high level part of the distributed control system HIS (see
figure 1).

CONCLUSIONS

The current version of the vibration analysis expert
system is of pilot character. It is dedicated to introduce
plant operators in using computerized diagnosis tools to
motivate them for further cooperation. Further training will
incrementally enhance their expertise in handling the system
and to contribute to the acquisition of knowledge, which would
enable the system to make suitable decisions in a wide range
of situations.

The system is a partial system of the overall process
diagnosis system; the AI components are punctual solutions,
yet. Some effort must be done to integrate them into the
overall system to expand it to an operator assistent with
multiexpert properties.

REFERENCES

1. A. Grabner, et.al., "Decomposition of Noise Signals composed of many similar Components" *Progress in Nuclear Energy*, 1 (1977) 615

2. U. Fiedler, et.al., "An AI Approach towards Disturbance Analysis", *Kernenergie*, 29 (1986) 302

CHAPTER 12

PROBABILISTIC RISK ASSESSMENT

AN EXPERT SYSTEM FOR FAULT TREE ANALYSIS

Bjorn Frogner

Expert-EASE Systems, Inc.
Belmont, CA 94002

Abstract

This paper summarizes work on the development a Fault Tree Construction Assistant (FTCA).[1,2] This system can help the analyst produce high quality fault trees by the use of expert system and workstation technologies. Introducing a workstation environment with an interactive set of integrated software tools on a local computer promotes improved continuity in the thought process of a fault tree development. The expert system technology simplifies the way fault trees are specified and it allows the incorporation of various kinds of prior knowledge and experience. The key contribution demonstrated in this paper is the ability of very effectively resolving logical loops. In our experience, an expert system approach is effective in promoting consistency throughout the tree and giving "expert" assistance at various points in the fault tree development.

Introduction

It is common experience in probabilistic risk analysis of nuclear power plants that one of the most tedious and time consuming aspects of the work is to produce the fault trees.[3-5] There are four major areas of difficulty:

o The drawing and input of the initial set of fault trees is time-consuming.

o As the risk analysis proceeds, more and more effort is required to update the fault trees and to ascertain that they remain consistent.

o There is a problem of coordination when more than one person is involved in the drawing and updating of the fault trees.

o As fault trees for subsystems are assembled into an overall plant fault tree, logical loops frequently occur. These loops must be broken before the trees can be analyzed by conventional fault tree codes.

Development Environment

The Knowledge Engineering Environment (KEE) software package from IntelliCorp was used as the expert system shell[6-7]. KEE provided the environment for the plant component data structures, fault tree representation, and expert reasoning. The features in KEE that were critical to the effective implementation of the FTCA are: object-oriented programming environment, frame based data structures, inheritance capabilities, and backward-chaining inferencing.

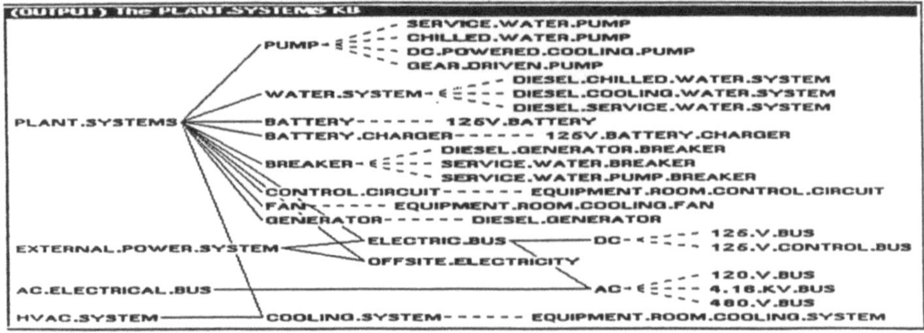

Figure 1. Representation of the knowledge base for the example system

InterLisp was used to develop the user interface, quantitative evaluation of the fault tree, and a few other special functions. The Xerox 1186 Lisp machine served as the workstation for all the software.

Fault Tree Construction Using Object-Oriented Representation

Figure 1 shows the objects involved in the analysis and their hierarchical relationships for the example problem discussed in this paper. The PLANT.SYSTEMS were divided into several sub-classes to compartmentalize the failure modes and other attributes associated with the systems and components in the plant. The inheritance capabilities available in KEE have been utilized in several areas:

o A logical loop in the fault tree is detected by KEE since it automatically will identify circular inheritance. Figure 2 shows a screen dump of one page of an example fault tree and the appearance of a logical loop.

o All the logic gates, associated object names, and inheritance classifications, are easily obtained for the branch in the fault tree that contains the loop (called loop path) by taking the intersection between the descendants of the top of the loop path and the ancestors of the bottom node.

o Once the loop path is known, the FTCA checks the attributes of the gates along the loop path.

Each of the objects in Figure 1 can have more than one parent. To support the resolution of logical loops, the following three classifications were introduced: EXTERNAL.POWER.SYSTEMS, AC.ELECTRICAL.BUS, and HVAC.SYSTEM.

By giving these three classes special attritbutes to facilitate dealing with logical loops, the presence of, for example, a 480.V.BUS will immediately be flagged as being a member of the external.power.system classification.

All the attributes of the individual objects are contained in separate frames. Early definition of the objects provides two benefits. First, the analyst can specify the aspects of the plant, the associated failure modes, and other attributes that he wants the fault tree to deal with. To a large extent, this corresponds to the abstraction process described above. Second, the subsequent construction of the actual fault tree can then access the information in the frames in an easy and uniform manner. This speeds up the generation of the fault tree and improves its consistency.

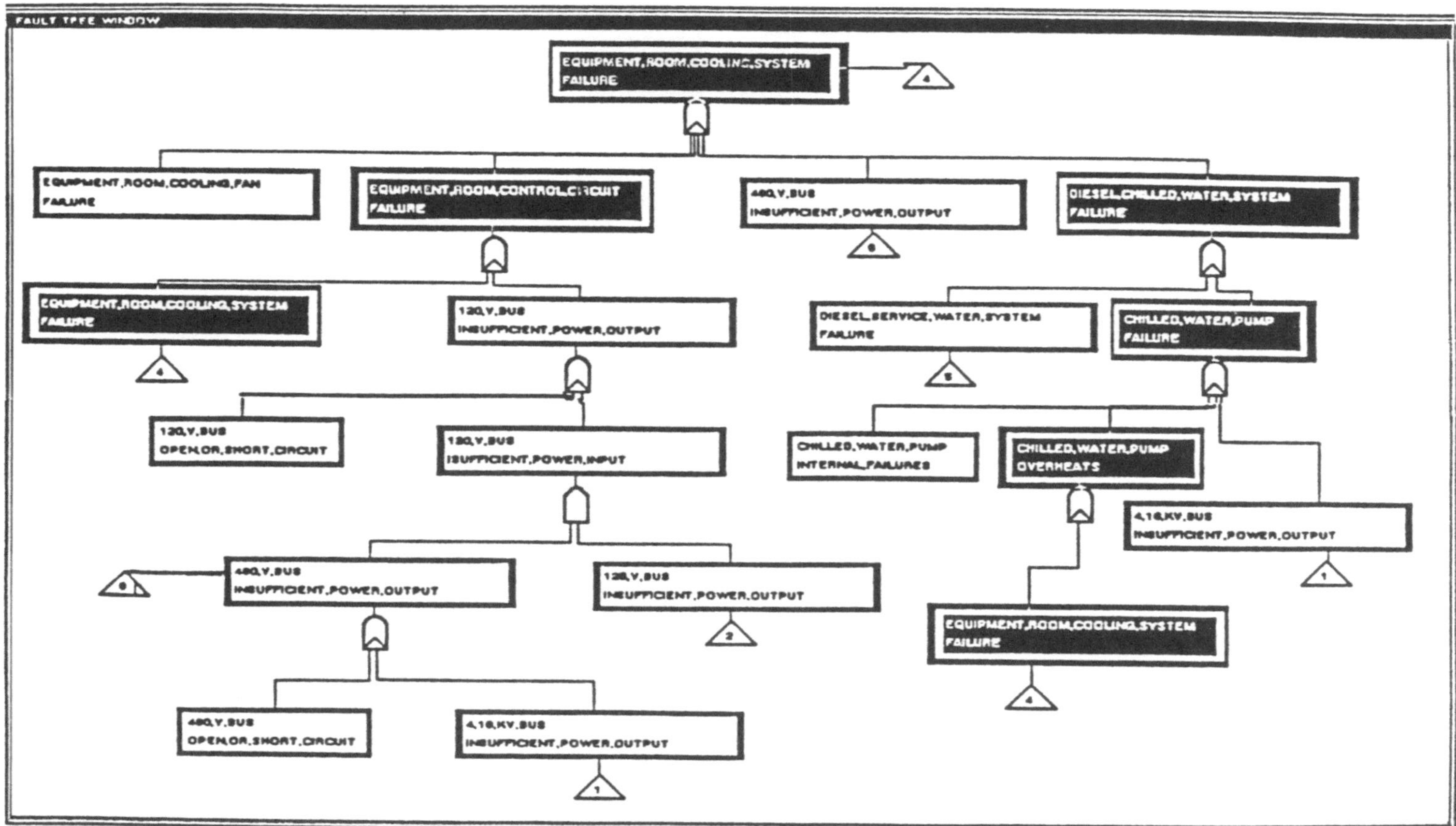

Figure 2. One page of a fault tree for example problem

References

1. B. Frogner, J.M. Holzer, R. Dourandish, and H. Lambert, "Use of Workstation and Expert System Technologies to Facilitate Development and Evaluation of Fault Trees," SBIR Report, NRC-04-85-134, April 1986.
2. B. Frogner and R. Dourandish, "Use of an Advanced Expert System Tool for Fault Tree Analysis in Nuclear Power Plants," Proceedings of COMPCON Spring 87, pp 454-457, Feb. 1987.
3. "PRA Procedures Guide," Final Report, NUREG/CR-2300, 1983.
4. W.E. Vesely et. al., "Fault Tree Handbook," NUREG-0492, 1981.
5. "ANS/ENS International Topical Meeting on Probabilistic Safety Methods and Applications," Volumes 1,2 and 3, San Francisco, Feb. 1985.
6. "The Knowledge Engineering Environment," Technical Brochure Developed by Knowledge Systems Division at IntelliCorp, Menlo Park, California, 1984.
7. J. Kunz et. al., "Applications Development Using a Hybrid AI Development System," The AI Magazine, pp. 41-54, Fall 1984.
8. D. Carlson, et. al., "Interim Reliability Evaluation Program Proceedures Guide," Sandia National Laboratories NUREG/CR-2728, 1983.

REASONING AND REPRESENTATION IN RiTSE

Steven A. Epstein

Management Analysis Company
12671 High Bluff Drive
San Diego, CA 92126

ABSTRACT

RiTSE, the Reactor Trip Simulation Environment, is a frame and rule-
base artificial intelligence (AI) application which aids nuclear power
plant operators in determining if a proposed change in a state of a component or process will cause safety systems to automatically shut the
plant down. This paper stresses the software techniques used to implement the system.

We demonstrate that prototypes and industrial strength systems can
be built quickly and economically without "expert system shells", and that
integration of AI and database techniques create a novel programming paradigm.

   Topic Areas: Representation and Reasoning, Nuclear Engineering,
                 Simulation, General Tools

INTRODUCTION

RiTSE, the Reactor Trip Simulation Environment, is a frame and rule-
base AI application which aids nuclear power plant operators in determining if a proposed change in a state of a component or process will cause
safety systems to automatically shut the plant down, or SCRAM. We have
employed RiTSE at a major East coast utility and described its success in
detail (1-8).

This paper stresses the software techniques for implementing such a
system, the use of standard programming languages and environments for
both prototyping and implementation, the integration of database and AI
programming techniques, and the attainment of extremely fast execution
speed on an IBM/PC-AT.

We conclude that both prototypes and installed systems need neither
special hardware nor software to be successful, and that novel integration
of AI and database techniques provide excellent results.

120V DC BUS
BKR # 6
CLV # 14
T/M
120V DC BUS
BKR # 3
CLV # 13
T/M
BKR # 13
125V DC BUS
1/2
Controller can only open valve
C 2
operator can only close valve
V2OP
MOTOR STARTER # 30
480V AC BUS
INSTRUMENT AIR
T/M
VALVE V2 (AOV)

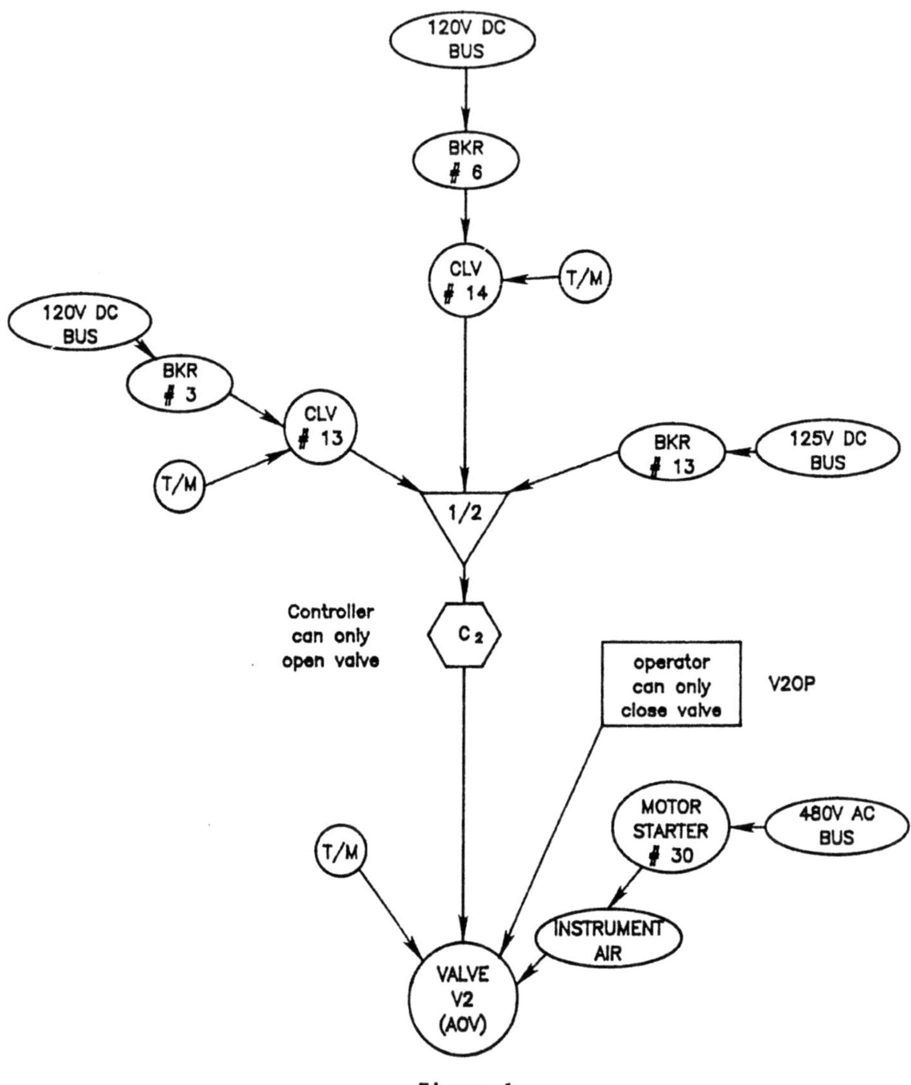

Figure 1

We have tried to allude to Wirth's formulation in our title. Just as "algorithms + data structures = programs", we believe that "reasoning + representations = AI programs". Judicious choice of data representations and algorithms for manipulating them are key to the success and use of any computer system.

BACKGROUND

During the spring of 1985, while investigating the causes of inadvertent actuation of reactor trip signals at a major East coast utility, we developed RiTSE, the Reactor Trip Simulation Environment.

One of the principal causes of unplanned outages and poor emergency response is human error resulting from the subtle and complex interactions of the simultaneous activities of maintenance, testing and surveillance at the plant. These activities often affect systems or components at a very deep level.

Typically, operations and maintenance personnel are not trained to analysize the potential interactions and ramifications of their activities. Those responsible for authorizing such activities are often quite knowledgeable and may understand the plant to the extent of hundreds of subsystems and components. However, it is quite impossible for any individual to remain aware of the tens of thousands of components, their potential interactions, and the impact of the next activity on the system as a whole.

Therefore, our engineers wanted a system that could (1) allow the plant staff to predict if an action or set of actions would cause a reactor trip, and (2) should a trip occur, aid in determination of the components and systems which directly contributed to the trip.

THE PROTOTYPE

Two members of the engineering division and two members of the information systems division formed the rapid prototyping team: manager, two experts, one knowledge engineer. Figure 1 is a schematic of the problem as presented by the experts: if the state or operating condition of any component in this valve system were to change, how would that affect the current state of the valve, V1?

We proposed the following representation for the problem. We supposed a database of the valve system where every record was a component/process or a rule for determining the state of a component/process. Rules were directly connected to their referents by pointer. Those components which did not have rules associated with them would represent the atoms of the model. For example:

```
ObjectId:        V1              ObjectId:        RULE1
isA:             Device          isA:             Rule
Class:           Valve           RefersTo:        V1
SubClass:        MotorOperated    English:         If V1 is closed, it open
StartUpValue:    Open                              if MS351 and 480DC are
CurrentValue:    Open                              operating and either the
Operability:     Operable                          operator or controller
EvaluationRule:  RULE1                             signals for open
                                 ThenCompute:     (Special rule language)
                                 UsingHeuristic:  H1
                                 IfTrueDo:        Return open
                                 IfFalseDo:       Nil
```

Figure 2

As is evident from Figure 2, a rule is much more than an "if-then" pair; rules also determine actions to take, values to return, and heuristics for execution order, i.e., conflict resolution. Two languages were designed. An LR1 grammar for the "if part" of the rule, and another for describing the consequent actions to take. These languages were lexically scanned into tokens at datum creation, and interpreted by the rule executor at run time.

Simple "what if" facilities were provided by allowing the user to create a list of components with designated states. This list of values would then override both current values and rule execution when the state of a list component were requested.

A back-chaining inference engine was written to walk the interconnectivity of the plant database. The algorithm can be outlined in pseudocode:

```
getstate(object)
  case seen_object_before
    return current_state
  case object on what_if_list
    return what_if_value
  case object isa process
    return eval(object)
  case object isa component
    does_object_have_a_rule ?
      if yes then return getstate(rule)
      if no then return current_state
  case object isa rule
    execute_rule so_that
      for_each_object_in_rule_do
        getstate(object)
      compute_rule_value
      return rule_value
```

Explanation facilities were more than a trace of which rules were executed and their values. By keeping track of which components just changed state per "what if" run, how they changed state (from component interaction, breakdown, designation, etc.), and joining this with information in the component database, a forward-chaining program which indicates which components to look at next was constructed.

We employed a variant of the PICK operating system as a programming environment because we had an extensive software library of user interface tools written for PICK and the user interface was crucial. Remember, this was no research project, we had only four months to deliver a system AND a correct solution to the client. Data entry alone was estimated at eight man-months (which proved to be correct).

When using the model depicted in Figure 1, we built a prototype in four man-days using an IBM/PC-AT. We used data template screeners, data file managers, menu drivers, and query facilities which were part of our software libraries. Only three new systems had to be written: the backchainer, the rule evaluator, and the explainer. To everyone's delight, it ran correctly.

Why did we have such an easy time with the prototype? First the problem was begging for such a solution. It was an example of what we began

to call Von Neumann's Thesis: anything that can be exhaustively and
unambiguously described ... is ... realizable by a suitable finite neural
network (9). We exhaustively and unambiguously described the topology of
a nuclear plant. We realized the plant as an implicit graph implemented
as a rule/database.

We also worked in a programming environment with which our software
engineers were intimately familiar; they were able to add new capabilities
to our software repertoire and learn new programming techniques.

Lastly, we were lucky that the natural way nuclear engineers represent
system interactions is by Boolean algebra; the mapping to rules should be
obvious. More importantly, it provided a common level of abstraction, an
agreement as to what a rule "means", and gateways to other representation
schemata (i.e., fault trees, digraphs, event trees, min-terms) for extending the system.

EXTENDING THE PROTOTYPE

After our initial successes with several small systems exemplified by
Figure 1, our engineers started to work on our client's problem. Our scope
of work for our client was to model one reactor trip function, automatic
shutdown due to insufficient flow to the feedwater system. Plant drawings,
walkdowns, and interviews were used to identify components, create rules,
and break the plant down into logical systems and sub-systems. Data entry
was accomplished in two months. The database consisted of 4,000 system/
components, 1,200 rules, and two processes.

Run time became a major concern. It took nearly two hours to traverse
the complete database in a back-chaining fashion. The engineers wanted
response time in seconds.

We realized that if we could initialize the database, then a data
driven type of rule execution could be used to traverse a minimum path
through the database. Any subsequent change in a component or system
could affect other objects in the database if, and only if, that subsequent
change caused a change in some other rule. For example, if a breaker is
opened, only those components whole rules mention that breaker in their
antecedents can change state. Those components which do change state, of
course, can cascade that change upward through the database, until either
the system reaches another "steady state", or a designated top event, such
as reactor trip, changes state.

Analogically, the database is a run time table, records are variables,
and rules are executable statements; changes in a variable's value causes
execution of statements only if that variable is mentioned in the statement.

To implement this data driven method of rule propagation, we reached
back into standard database techniques. We used file inversion on the
antecedent of each rule to create an inverted item of rule names mentioned
per each system object. For example, if breaker1 was menioned in the
antecedent of rule1 and rule2, then an inverted item with the primary key
of "breaker1" would be created with two values: rule1 and rule2. Therefore, if breaker1 were to change state, rule1 and rule2 would be executed
to see if they in turn change state ... and so on. In pseudocode:

```
datadrive(event)
  read_rulelist_for_event
  case no_rulelist
    this_is_a_top_event_which_has_changed
    display_change
    changed  ◄--- 1
  case rulelist_exists
    for_each_rule_i_in_rulelist_do
      execute_rule
    if_newstate_for_rule_is_different_then
      get_object_rule_refers_to
    if_operability_is_ok_then
      changed(i) ◄--- datadrive(object)
    end else changed(i)◄--- 0
    end else changed (i)◄-- 0
    next
    if_there_isa_1_in_changed_then
      return
    end else return 0
```

Run times improved dramatically now that a minimal path through the database was followed instead of back-chaining blindly from top events. The average run time from initiating event to response using the data driven method was between 10 and 20 seconds on an IBM/PC-AT, which the engineers considered to be quick enough for daily use.

Using RiTSE as a tool for exploring the plant connectivity, our engineers discovered an air valve that when routinely taken out of service, would cause reactor trip on low steam generator level. It is extremely doubtful whether this system weakness could have been uncovered using other methods.

The great drawback using the data driven method, however, is that the database of component and system state values must constantly be kept updated. During normal operating conditions of the plant this may not prove difficult. But during emergency situations, not only are components changing state rapidly, operators have little time to act as data entry clerks. Too few components and systems are monitored at deep enough levels by data acquisition devices to make this option realizable.

In reliability studies of power plants, use is made of fault trees and event trees. Faults and events are propositions which can take truth values of "true" and "false". Trees are constructed from faults and events by logical combination. To find out, _a priori_, which faults or events in what combinations can cause a top event to occur, reliability engineers have traditionally generated the minimum set of components in disjunctive normal form which can cause a designated top event. These are called cut-sets, derived from calling the "or" connective "cut". These cut-sets are then reduced by Boolean identities to form the so called minimum cut-sets.

We, therefore, used the component rules to generate Boolean expressions and then created minimum cut-sets from the Boolean representation of the rule space.

GENERATING CUT-SETS

The algorithms for transforming a set of Boolean equations into their most reduced form are usually classified as NP-complete. This means that the length of time necessary for executing the algorithm grows exponentially

with respect to some parameter. In our case, we find that execution time
grows exponentially with respect to literal terms; and literal terms are
plant components and their associated states. Therefore, doubling the
number of plant components squares the solution time.

The term"simplest" also poses problems. Given the two equations:
"ab + ac" and "a(b + c)", which expression is more simple? Under what
criteria? We have addressed this problem by transforming Boolean equations into their corresponding min-term forms. Min-terms, or cut-sets,
are AND gates; the database of the plant is one very large OR gate with
AND gated inputs.

The simplest, most reduced form for an equation is therefore the
smallest set of min-terms which are logically implied by the equation.
We call this reduction and simplification the minimum cut-set problem.

Historically, three methods have been used to solve the minimum
cut-set problem: Monte Carlo methods, the Quine-McClusky method, and
Nelson's method (10, 11, 12). Our method is a radical departure: we
have created an environment for the engineer to interact with the problem
so as to allow timely solution sp-ed even on personal computers.

Monte Carlo methods have been shown to be powerful aids in such
applications as solution of the Boltzmann Equation, PRAs and other
probabilistic analyses where large uncertainties abound. The minimum
cut-set problem, however, is deterministic. With no equivocation, using
a Monte Carlo method here is equivalent to randomingly assigning states
to components, doing a RiTSE run, and writing down the result as a
Boolean equation. After doing a million or so runs, we could conclude
that the smallest set of failed states which we discovered was actually
the simplest most reduced form for which we were searching. This is
similar to locating a lake by shooting cannonballs and watching for
splashes with binoculars.

The Quine-McClusky method is the oldest algorithm for the reduction
of min-term form Boolean equations. It suffers from two drawbacks.
First, each min-term must be represented as a vector: the min-term "XY-Z"
(read: "x" and "Y" and not "Z") would represent the same min-term as
"X*Y*-Z" with "*" meaning logical AND, avoiding once again, representa-
tional mismatches. Second, each min-term must be put into expanded
normal form, with each literal occurring exactly one time. In a plant
with three components: X, Y and Z, the equation "XY + -Z" expands into
"XYZ + XY-Z + X-Y-Z + -XY-Z + -X-Y-Z". The expansion process is exponential in time with respect to the number of literals. Moreover, in our
example the reduction process, also exponential, would return the original
equation.

The Nelson method is a response to the inelegance, as well as the
cost, of expanding each min-term into normal form. Nelson notes that the
complexity of the Quine-McClusky method arises from mimicking the Boolean
identity "AB + -AB = b". Nelson circumvents this bottleneck by transforming min-terms to max-terms, applying the Boolean identity "A +AB = A",
transforming back into min-term and applying the identity one more time.
Of course, one must pay a price twice for the min-max max-min transformations. This price, however, is polynomial in time with respect to the
number of min-terms and max-terms.

Our method uses symbolic manipulation and heuristic methods for
reducing search spaces. Each min-term is conceived as a record in a database. Its primary key is exactly the min-term; "XYZ" is written "X*Y*Z".

In the same manner in which MACSYMA reduces mathematical equations by
symbolic manipulation, RiTSE reduces Boolean algebraic equations. RiTSE
applies Boolean identities to the min-terms, reducing them if so needed.
RiTSE does not wander about the entire search space like a lost puppy.
While expanding the Boolean representation of the rules, the program keeps
track of where in the database of min-terms Boolean reductions have a
change of occurring. It then only looks into those spaces when applying
the costly reduction algorithms. We have given the engineer full control
over the nearly one hundred programs and subroutines used to perform simplification and reduction.

CONCLUSION

     The importance of our work lies in the incorporation of AI software
techniques with database techniques to deliver an application in a short
amount of time. Moreover, the development and delivery vehicle was a
personal computer. We believe that writing our own inference engines
proved an invaluable experience for our staff.

     Essentially, we moved simulation logic out of the program and into
data therefore allowing the engineers to change the model as well as the
states of components. Programs are data, as Von Neumann would say.

     Many developers when presented with this type of problem immediately
see the solution as an instance of diagnostics. They begin by asking
about symptoms, using medical diagnosis as the paradigm.

     However, medical diagnosis infers conclusions from symptoms and an
etiology: facts, perhaps observables, about the system right now, AND
deeper knowledge of the causative links within the system.

     We began at the bottom, so to speak. We pretended no knowledge of
the system as a whole, and therefore constructed a database of every
component and system in the plant, and the rules for their interconnectivity.

     We have developed a deep representation of the plant. Our current
work combines logical inference and deep representation with analogical
inference and surface symptoms (the coolant temperature is too high,
white steam is coming from the turbine, etc.). It is our belief, however,
that the deep model must come first, that any reasoning from high-level
observables must have a deep representation of the causative links of the
domain.

ACKNOWLEDGEMENTS

     We cannot thank enough Dr. Raymond Nelson, Dr. Harry Winston Mergler,
and Dr. Michael Frank, without whose teaching, patience and brilliance
this work could never have been accomplished.

REFERENCES

1.   S. A. Epstein and M. V. Frank, "RiTSE: The Reactor Trip Simulation
        Environment", AAAI Workshop on Simulation, AAAI-86, August 1986.
2.   S. A. Epstein and M. V. Frank, "An Online Approach to Reducing SCRAMS",
        International Conference on SCRAM Reduction, OECD and NEA, April
        1986.

3.   S. A. Epstein and M. V. Frank, "RiTSE/KCGE: An AI Approach to Improve Plant Availability", Nuclear Power Thermohydraulics Proc., Atomic Energy Society of Japan, April 1986.

4.   S. A. Epstein, "Knowledge Representations and Process Plants", Halden Project Workshop, OECD sponsored March 1986, available from author.

5.   S. A. Epstein, "Artificial Intelligence and Rapid Prototyping", Proc. of Multi-Conference 86, Society for Computer Simulation, January 1986.

6.   S. A. Epstein and M. V. Frank, "Artificial Intelligence and SCRAM Reduction", Proc. Computer Applications for Nuclear Power Plant Operation and Control, American Nuclear Society, October 1985.

7.   S. A. Epstein and M. V. Frank, "Current Work in AI", delivered to EPRI Workshop on AI Application to Nuclear Power, June 1985, available from authors.

8.   Simulation Staff, "Artificial Intelligence to Improve Nuclear Plant Availability", Simulation Magazine, March 1986.

9.   J. Bernstein, "Science Observed: Essays Out of My Mind", New York Basic Books, 1982, pg. 68.

10. E. J. Henely, H. Kumanoto, "Reliability Engineering and Risk Assessment", Prentice Hall, 1981, Englewood Cliffs, NJ.

11. W. V. Quine, "The Problem of Simplifying Truth Functions", American Mathematical Monthly, Vol. 59, 1952.

12. R. Nelson, "Simplest Normal Truth Functions", The Journal of Symbolic Logic, Vol. 20, June 1954.

EXPRESS: AN EXPERT SYSTEM TO PERFORM SYSTEM SAFETY STUDIES

C. Ancelin, P. Le, S. De Saint-Quentin, and N. Villatte

EDF-DER
1, avenue du Général de Gaulle
92141 Clamart, France

ABSTRACT

     This paper presents EXPRESS, an expert system developed for the
automation of reliability studies. The first part consists in the
description of the method for static thermohydraulic systems. In this
step, we define the knowledge representation based on the two inference
engines -ALOUETTE and LRC developed by EDF. We explain all the process to
construct a fault tree from a topological and functional description of
the system. Numerous examples are exhibited in illustration of the method.
This is followed by the lessons derived from the studies performed on
some safety systems of the PALUEL nuclear power plant. The development of
the same approach for electric power systems is described, insisting on
the difference resulting from the sequential nature of these systems.
Finally we show the main advantages identified during the studies.

Keywords : Expert System, Knowledge Representation,
           Reliability, Fault Tree,
           Nuclear Power Plant.

I. INTRODUCTION

The main objective of the EXPRESS software is the automation of systems reliability studies by means of expert systems.

Automation was first applied to the construction of the fault trees used to assess the reliability of static thermohydraulic systems. "Static" means that no changes in the system configuration such as switch over to a backup system, initially on standby, take place.

A second phase is now being implemented. It consists in building a fault tree for electric power systems - essentially sequential systems with configuration changes - and trying to maintain the same approach (knowledge representation and reasoning) as for thermohydraulic systems.

II. KNOWLEDGE REPRESENTATION

The EXPRESS software uses two types of representation : one in connection with ALOUETTE [1,2] a first-order inference engine and the other with the knowledge representation language, i.e LRC [3]. These two tools have been developed by the Direction des Etudes et Recherches at EDF.

II.1 ALOUETTE

ALOUETTE is an inference engine in forward chaining with a knowledge representation based on production rules <if....then....> in first order logic and facts written as triplets <object, relation, object>.

II.2 LRC

LRC is a language used to represent knowledge in propositional logic - order 0 - In the EXPRESS software, only the LRC2 demonstrator is used to build a fault tree by means of backward chaining, starting from an undesirable event.

III. PRESENTATION OF THE METHOD DEVISED FOR STATIC THERMOHYDRAULIC SYSTEMS

III.1 General Principles

The method implemented by the EXPRESS software is issued from a basic observation !

According to their consequences on the system, failures relating to the components of thermohydraulic system can always be grouped under a few large categories. The main categories are :
- fluid flow interruption called "blockage"
- loss of fluid outwards also called "external leak"

Based on the above observations, the EXPRESS software revolves two main phases (Appendix, figure 1) performed by the first-order ALOUETTE engine.

The first phase is aimed at grouping the components of a system into larger component categories - or macrocomponents - according to the consequences of their failures (blockage or leakage).
The second phase deduces the failure consequences, for each component, in terms of path losses, according to the macrocomponent(s) to which the said component belongs. Moreover, the possible causes of these failures are stated according to the configuration and the mission studied.

762

These two phases lead to the generation of a simple order 0 rule base (LRC-AUTO) written in LRC language.
  If a hand-written rule base (LRC-MANU) concerning the undesirable
event and the boundary conditions mainly is added to (LRC-AUTO), all the
data required to build a fault tree are then available.

  The fault tree is then obtained by applying a demonstrator (LRC2) to
the entire rule base. This demonstrator presents the results in a form
which can be handled by a conventional code computing minimal cut-sets.

  In the discussion below the examples will concern the thermohydraulic system presented in the appendix (figure 2).

III.2 <u>The Initial Data Base</u>

  All the data required for the first phase are contained in the initial data base.
  The base includes two types of facts :

        . <u>topological facts</u> which describe the system topology, i.e. :
                      . The sources,
                      . The goals,
                      . The sequential relations between the components.
  Thus, for the system presented, we have the following facts :
                      (TANK, nature, source)
                      (SG1, nature goal)
                      (5VD, next item, motor-driven pump 21PO)

COMMENT

  Topological facts are obtained simply by reading a system diagram.
They do not, therefore, depend on the mission or the configuration studied.
        . <u>functional facts</u> which :
           - on the one hand, specify the failure <u>consequences</u> (blockage,
        leakage) and especially indicate :
                      . possible paths between a source and a goal,
                      . the components having an isolation function
                        * real (normally-closed valve during the system mis-
                          sion)
                        * potential (valve that can be closed during the sys-
                          tem operation)
                        * upon a pressure drop (presence of membranes...),
           - on the other hand, state the failure <u>causes</u> (blockage, leaka-
        ge) and indicate, for instance :
                      . the component nature
                      . their initial position
                      . their position during operation
                      . whether their position is displayed in the control
                        room.....

  <u>Examples of functional facts</u>

        (TURBINE DRIVEN PUMP 31PO, nature pump),
        (PATH1, starting from, TANK),
        (PATH2, leading to, SG1)
        (PAHT1, via, MOTOR DRIVEN PUMP 21PO),
        (31VD, nature, motor operated valve),
        (31VD, initial position, open).

COMMENT

Most functional facts depend on the studied mission or configuration.
Moreover, the way they are formulated often depends on the functional assumptions adopted (Can a given component be considered an isolating device? Is there enough time to send someone on the spot?...)

III.3 <u>Topological Rule Base</u>

The topological rule base builds all the concepts needed to express
an undesirable event in terms of path losses, i.e. :

- The arcs connecting a source to a node, a node to another...

- The paths running from a source to a goal,

- The "leakage" blocks which are built, starting from a node, by
  incoporating the components, one after the other, up to the
  first isolating devices met whatever actually the nature of
  the isolation -whether one-sided in the case of check valves,
  or two-sided in that of valves.

By applying the topological rule-base to the initial data base, all
the above-described entities can be constructed and added to this data base. The knowledge of the system is then contained in a base called "intermediate data base" which will be used in the second phase
to generate order 0 rules (LRC-AUTO)

III.4 <u>Generating Rule Base</u>

It includes two types of rules :

<u>topological rules</u> which use the results obtained during the first pha-
se and express the component failure consequences in terms of "path
losses", according to the "leakage" and or "blockage" block(s) to
which they belong.

<u>functional rules</u> which :
. on the one hand, state the causes of the blockage or leakage; thus,
  for an initially shut down pump, the failure modes - maintenance outage, primary failure - are automatically generated.
. on the other hand, contribute to the automatic processing of certain specific problems.

- either to facilitate the work : when a given problem must be repeated for several components, it is sometimes convenient to have an
  automated system for taking the problem into account. Then as a ge-
  neratic rule is available, manual drafting of numerous rules can be
  avoided.

- or because those who study the systems think that they are likely
  to meet the same specific problems when studying other systems.

<u>An Example of Generating Rule</u>

. <u>topological rule</u>

Blockage <u>rule</u> :

```
IF          nature (BB) - block-blockage
            nature (P)  - path
            item   (P)  - (BB)
            nature (X)  - component
            item   (BB) - (X)

KNOWING THAT name (C)  - PATH1
THEN          write        (in LRC Syntax)
              "IF          X - BLOCKAGE - YES
              "THEN     PATH1 - LOSS    - YES
```

<u>functional rule</u>

Pump maintenance rule :

```
IF          nature (PV) = pump
THEN        write
            "IF          PV - MAINTENANCE = YES
            "THEN        PV - BLOCKAGE    = YES
```

III.5 <u>The LRC-AUTO Base</u>

By applying the generating rules to the intermediate data base developed in the first phase, a set of very simple rules written in LRC language is obtained. These rules give the lost paths.

An example of LRC rule is given below :

```
RULE NUMBER 11
IF     TANK - LEAKAGE - EXTERNAL = YES
THEN   PATH 1 - LOSS = YES
```

III.6 <u>The LRC-MANU Base</u>

Like the initial data base, the LRC-MANU base is entered manually by the user and is concatenated to the LRC-AUTO base.
It essentially contains :

- the undesirable event,

- the boundary conditions (electric power supplies, cooling systems, controls..)

- certain specific problems.

III.7 <u>Teaching Derived from the Studies Performed</u>

The method described above has been validated on the AFS[*] and RHRS[*] systems of the PALUEL nuclear-power plant. The main merits identified during these two studies are as follows :

- . <u>short execution time</u> for most of the reliability study. The reliability specialist can then devote more time to the tricky problems in his system reliability study,
- . greater <u>exhaustivity</u> as most of the failures and failure combinations are automatically generated,
- . <u>transparency</u> and greater <u>flexibility</u> of the tool because the reliability specialist can rapidly find out which failures have already been taken into account and introduce (or ask those operating the tool to introduce) the failures he wishes.

IV. DEVELOPMENT STAGE OF THE STUDIES ON ELECTRIC POWER SYSTEMS

The same approach - generation of simple order 0 rules by a first order engine - was adopted for these systems. However, due to the important differences resulting from configuration changes (sequential nature of the systems) and from upstream short-circuit propagation, difficulties have arisen, thus preventing the same generating rule-base to be used. Development work is now underway in order to build a generating rule-base peculiar to electric power systems while retaining a representation of the systems similar to that of the static thermohydraulic systems in the initial data base.

Moreover, the sequential feature which has to be taken into account when quantifying minimal cut-sets will be addressed by the code GSI^4 (Inferential Sequence Generation Code).

V. CONCLUSIONS

The EXPRESS software has been successfully used to study the reliability of the AFS in the PALUEL nuclear power plant. it will be applied, in the framework of 1300 MWe nuclear power plant PSA, to other static thermohydraulic systems.

The processing of electric power systems is now being developed using, as a reference, the electric power supplies of the PALUEL 1300 MWe nuclear power plant.
Note that, with the proposed approach, all the assumptions required to conduct reliability studies have to be clearly formulated.

Consequently :

- . the approach is a good tool for <u>dialogues</u> with the systems designed, operators....
- . it ensures the <u>consistency</u> of the reliability studies both from the quality and presentation point of view.
- . it will, therefore, facilitate the <u>management</u> of these studies as applied to other problems (sequences, modes common to several systems) while minimizing the error risk.

---

*Auxiliary feedwater system
Residual heat removal system

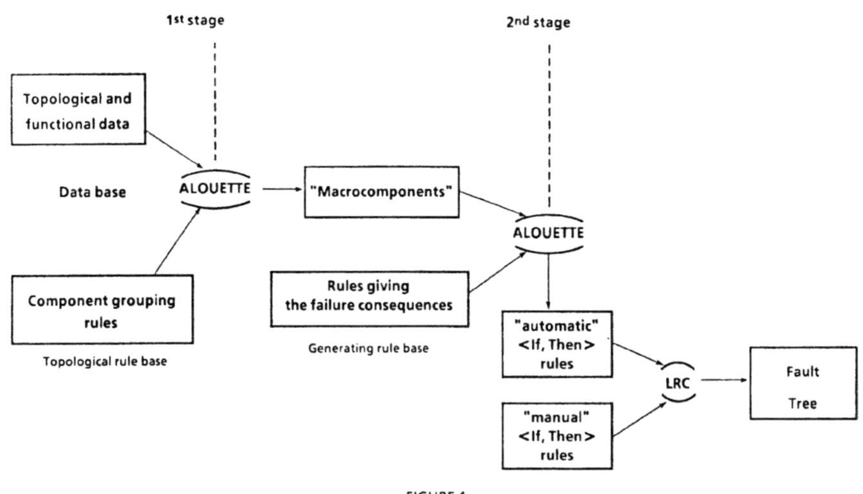

FIGURE 1

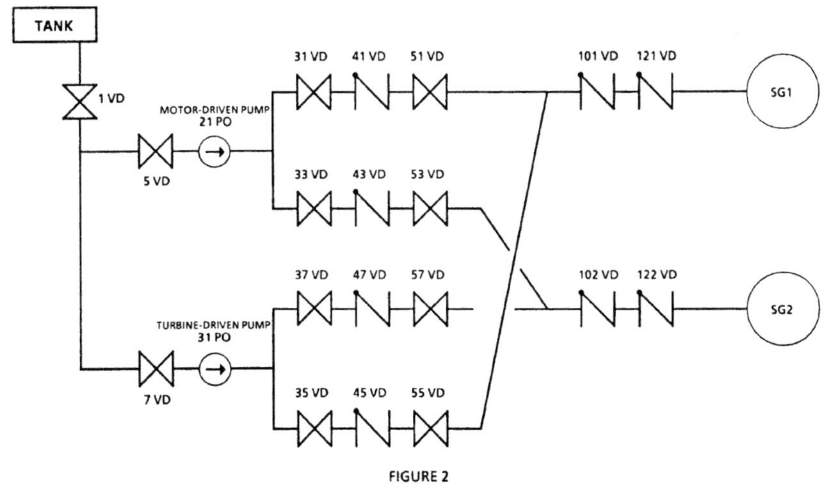

FIGURE 2

REFERENCES

/1/   M. Gondran and D. Mulet-Marquis
      "Un langage pour les Systèmes Experts: ALOUETTE"
      EDF-DER - Note HI/4773 - 1984

/2/   D. Mulet-Marquis and P. Benhamou
      "Manuel d'utilisation du langage ALOUETTE"
      EDF-DER

/3/   J.F. Hery and J.C. Laleuf
      "Notice d'utilisation et analyse des logiciels L.R.C."
      EDF-DER - Note HI/5097-02 - 1985

/4/   M. Bouissou and N. Villatte
      "GSI2, a simple link between the FMEA and the quantification for
      complex sequential systems; automatic generation of FMEA by an
      expert system

A paraître IASTED International conference
      Quality control and reliability

      June 24-26, 1987 Paris

THE DEVELOPMENT OF AN EXPERT SYSTEM FOR FINDING FRAGILITY CURVES OF BUILDING STRUCTURAL SYSTEMS IN THE PRELIMINARY DESIGN STAGE

Lone-Yee Yee and David Okrent

Mechanical, Aerospace and Nuclear Engineering Department
University of California
Los Angeles, California 90024-1597

INTRODUCTION

This research is a starting point for the development of an expert system for determining seismic fragility curves of structural systems in a nuclear power plant or conventional building at the preliminary design stage. The resulting system will assist an engineer with moderate engineering background and limited reliability knowledge to analyze the failure functions of building structures. It simulates the performance of an expert in identifying the potential failure modes and their variabilities for a structure of interest. Then, by combining the stress results of crude analyses, the method of linear statistical model, and the methodology developed by Kennedy [Kennedy et al., 1980, 1984], the fragility curves of the structure will be estimated.

On reviewing the methodology of seismic fragility evaluation for existing building structures in the nuclear power plant industry, one finds that the investigation process starts with the identification of critical components or substructures, whose failures will result in the functional failure of safety related equipment or the failure of structural integrity itself, and follows with complicated numerical analyses to estimate the capacity functions associated with the limit states of these components or substructures [Campbell et al., 1981a, 1981b; Kennedy et al., 1980, 1984; Wesley et al., 1980, 1981, 1984]. The identification of critical structures is conducted by a complex decision process in the expert's head with the help of plant specific data. The estimation of the quantitative value of the capacity (or fragility) often relies on expert judgment to get meaningful results. However, to estimate the fragility curve in the preliminary design stage, some modification of this procedure is necessary because none of the detailed plant specific data and complicated numerical analyses are available.

PROBLEM ANALYSIS

Although each building structure is a complex and unique system, it generally is designed inside the design envelop. In other words, the behavior of a building structure under earthquakes is predictable with some uncertainties. Hence, some common aspects will exist for buildings designed with the same structural conceptions. Based on these common

aspects, the attributes to describe the characteristics of the structural
behavior under seismic conditions may be deduced. Once a set of values to
describe each attribute has been defined, a building structure may be
represented as the hierarchy of pattern spaces configured from these
attributes and their values. Under this representation, the complex
decision-making process for the identification of critical components or
substructures can be decomposed into several simpler hierarchical
subproblems; for this reason, they may be identified by diagnosing the
values of attributes without performing any numerical analysis.

The local responses, however, are needed in order to estimate the
quantitative value of the capacity function for the selected components or
substructures. For getting these local responses, some crude stress analysis schemes will be employed. These crude schemes are the same as the
tools used in the preliminary design stage by senior engineers for their
everyday job. However, there are some limitations of these methods;
therefore, obtaining reasonable results for the local responses will rely
on engineer's judgment. In fact, the median values and variabilities of
the response factor and its subfactors [Kennedy et al., 1980, 1984] cannot
be evaluated by these methods. Fortunately, some common aspects exist
because these factors or subfactors describe the global behavior of a
structural system under seismic conditions. Therefore, meaningful results
of these parameters may be found based on expert diagnosis of the attributes and their values, corresponding to these common aspects.

KNOWLEDGE ACQUISITION

A difficult and time consuming part of developing an expert system is
how to transfer the knowledge in the expert's heat into a computer-
compatible form [Efstathion et al., 1986]. In most existing expert
systems, their achievements relied on the performance of knowledge engineers. Some techniques have been developed for supporting communications
between experts and knowledge engineers in order to elicit the experts'
knowledge easily and quickly. One available technique is induction, which
uses the attributes and their values to describe the essential features of
the problem [Michalski, 1975, 1980]. Figure 1 depicts a hierarchical
diagram for seismic fragility investigation.

By this hierarchical representation, not only the related attributes
could be defined but also the procedure of a systematic way to represent
the knowledge would be developed. The procedure associated with critical
components identifications may be written as a sequence of nine steps:

(a) Classify the potential structural configurations into different struc-
    tural system categories.

(b) Define the attributes and their values to describe the essential
    features of structural configurations.

(c) Use expert opinions to develop rules to assign buildings into the
    corresponding structural system catagories.

(d) Define the attributes and their values for each structural system to
    describe the dynamic characteristics of this system.

(e) Use expert opinions to develop rules to describe the dynamic behavior
    of each structural system.

(f) Use expert opinions to develop rules to identify the critical stick or
    sticks and weak link or links between sticks.

(g) Define the attributes and their values to describe the capacity
    distribution per each stick or each link.

(h) Use expert opinions to develop rules to identify the weakest component
    associated with critical stick or link.

(i) Use expert opinions to develop rules to identify the potential failure
    modes associated with soil-foundation system.

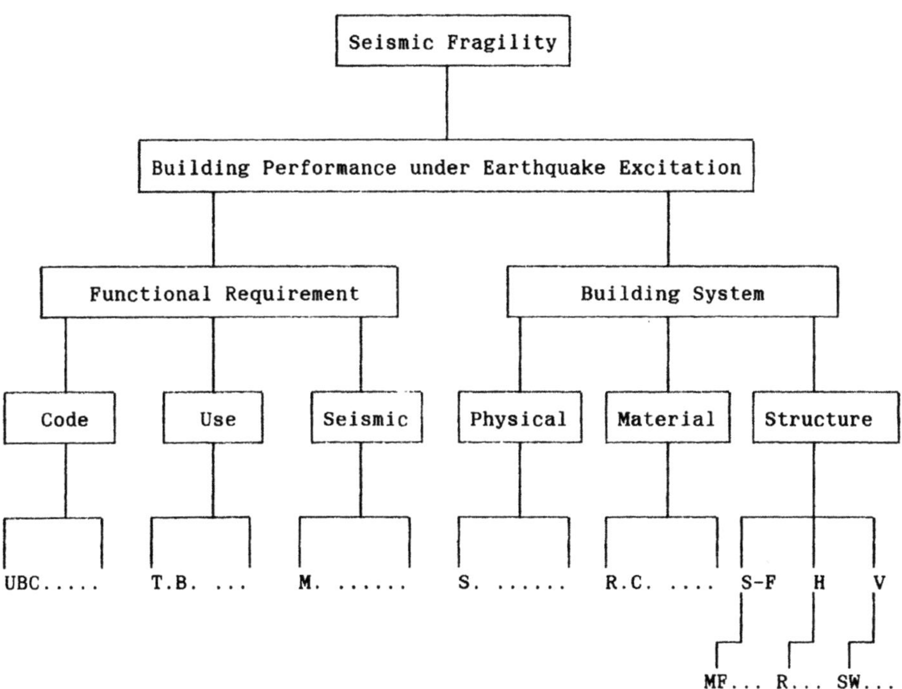

where: UBC : Uniform Building Code;   T.B.: Turbine Building
        M   : Magnitude;   S   : Symmetry
        R.C.: Reinforced Concrete;   S-F : Soil Foundation System
        H   : Horizontal System;   V   : Vertical System
        MF : Mat Foundation;   R   : Rigid
        SW : Shear Wall

Figure 1: A Hierarchical Diagram for Seismic Fragility Investigation.

Similarly, the procedure for estimating these parameters associated fragility evaluation may be written as follows:

(a) Collect the generic statistical data for material properties, size
    variability and model uncertainty.

(b) Use expert opinions to develop rules to translate the generic data
    into a plant specific data.

(c) Review the literatures and reports; for example, the existing reports
    for probabilistic risk assessments.

(d) Use expert opinions to develop rules for using the results in items
    (a) and (c) to estimate the uncertainty or randomness of these factors
    or their subfactors.

(e) Use expert opinions to define the potential loading distribution under
    seismic conditions.

(f) Use expert opinions to choose crude numerial schemes to estimate the
    local respone of a selected component.

(g) Use expert opinions and information of design criteria and the result
    in (f) to estimate the possible configuration of the selected
    component.

(h) Use expert opinions to estimate the mean values and variabilities of
    these parameters.

(i) Apply Kennedy's model to estimate the capacity function.

<u>EX. 1:</u> **Development of a rule set to describe the rigidity of a diaphragm.**

    (a) Assume the rigidity of a diaphragm can be divided into four cate-
    gories, such as

        (i)   rigid   (ii)   semi-rigid

        (iii)   semi-flexible   (iv)   flexible

    (b) The physical configuration of a diaphragm may be described as
    follows:

        (i)   metal deck diaphragm with structural concrete topping,

        (ii)   cata-in-place concrete, etc.

    (c) The associated rules may be deduced as follows:

        <u>Rule 00001</u>: If it is a metal deck diaphragm with structural
    concrete topping and without shear studs, then the
    rigidity of it is (rigid/0. + semi-rigid/1.0 +
    semi-flexible/0.3 + flexible/0.) with CF = 1.0.

        <u>Rule 00002</u>: If it is a cast-in-place concrete diaphragm without
    large opening, then the rigidity of it is (rigid/1.0
    + semi-rigid/0.5 + semi-flexible/0. + flexible/0.)
    with CF = 0.8.

    etc.

    Where CF = confident factor associated with the rule.

    Alternatively, the rigidity of a diaphragm may be defined by the
    relative stiffness ratio kr, that is

$$kr = \frac{\text{stiffness of diaphragm}}{\text{stiffness of adjoint vertical system}}$$

Thus, the rule set based on the value of kr may be written as
follows:

Rule 01001: It is a rigid diaphragm of kr ⩾ 2 with CF = 0.9.

Rule 01002: It is a semi-rigid diaphragm if 0.4 ⩽ kr < 2 with
             CF = 0.9.

Rule 01003: It is a semi-flexible diaphragm if 0.1 ⩽ kr < 0.4
             with CF = 0.9.

Rule 01004: It is a flexible diaphragm if kr < 0.1 with
             CF = 0.9.

SUMMARY

Because the set of attributes and their values is incomplete and the
numerical results estimated by crude analysis schemes are inexact, an
inexact reasoning inference network will be built into this developing
expert system. Therefore, each rule or stress result will attach a confidence factor to describe the truth value of it. In order to conduct the
evaluation of the truth value, fuzzy set theory [Yao et al., 1985, 1986;
Zadeh, 1965, 1973] will be employed. In fact, the crude analysis schemes
are implemented into the system; as a result, it will be a hybrid system.

Although a final result of this study is not available yet, it has
defined the research field for applying expertise, fuzzy set theory, the
theory of evidence, and crude numerical schemes to estimate the fragility
functions of structural systems in the preliminary design stage. Until
now, about 120 rules have been developed. After completing the whole rule
set, a prototype system will be built.

ACKNOWLEDGMENT

This work was supported in part by Department of Energy Contract No.
DE-FG03-865F-16610.

REFERENCES

Campbell, R.D. and Wesley, D.A., 1981a, Preliminary Failure Mode Predica-
    tions for the SSMRP Reference Plant (Zion 1), U.S. NRC, NUREG/CR-1703.
Campbell, R.D. and Wesley, D.A., 1981b, Potential Seismic Structural
    Failure Modes Associated with the Zion Nuclear Plant, U.S. NRC,
    NUREG/CR-1704.
Efstathion, J., and Mandani, E.H., 1986, "Expert Systems and How they are
    Applied to Industrial Making," Elsevier Science Publ., B.V.
Kennedy, R.P., et al., 1980, Probabilistic Seismic Safety Study of the
    Existing Nuclear Power Plant, Nucl. Engr. and Design, 59:315-338.
Kennedy, R.P. and Ravindra, 1984, Seismic Fragilities for Nuclear Power
    Plant Risk Studies, Nucl. Engr. and Design, 79:47-68.
Michalski, R.S., 1975, Variable-Valued Logic and Its Applications to Pat-
    tern Recognition and Machine Learning, chapter in monograph:
    "Multiple-Valued Logic and Science, D. Rine, ed.
Michalski, R.S., 1980, Pattern Recognition as Rule-Guided Inductive
    Inference, IEEE Trans. on Pattern Analysis and Machine Intelligence,
    PAMI-1.

Wesley, D.A. et al., 1980, Conditional Probabilities of Seismic Induced
     Failures for Structures and Components for the Zion Nuclear Generating
     Station, <u>Zion Safety Study</u>.
Wesley, D.A. and Hashimoto, 1981. Seismic Structural Fragility Investiga-
     tion for the Zion Nuclear Power Plant, <u>SSMRP Report</u>, U.S. NRC,
     NUREG/CR-2320.
Wesley, D.A., et al., 1984, "Seismic Fragilities of Structures and Com-
     ponents at the Millstone 3 Nuclear Power Station, <u>Millstone Safety
     Study</u>.
Yao, J.T.P. et al., 1985, "Fuzzy Logic Approach in Metals Fatigue,
     <u>CE-STR-35-35</u>, Purdue University.
Yao, J.T.P. and Zhang, X.J., 1986 The Development of SPERIL Expert Systems
     for Damage Assessment, <u>CE-STR-86-29</u>, Purdue University.
Zadeh, L.A., 1965, Fuzzy Sets, <u>Info. and Cntrl.</u>, 8:338-353.
Zadeh, L.A., 1973, Outline of a New Approach to the Analysis of Complex
     Systems and Decision Processes, <u>IEEE Trans. Syst. Man and Cybrntcs.</u>,
     SMC-3:28-44.

PRA-BASED DATA ANALYSIS FOR USE IN COMPUTERIZED DECISION AIDS

Shimshon Arueti and George Apostolakis

Mechanical, Aerospace, and Nuclear Engineering Department
5532 Boelter Hall, UCLA
Los Angeles, CA 90024

1. INTRODUCTION

1.1 Scope

The purpose of this paper is to develop a methodology for the analysis
of data that could be used in a knowledge based software tool for maintenance scheduling. This tool would assist plant management in planning and
scheduling preventive and corrective maintenance in advance, based on
deterministic constraints (refueling outages, technical specifications,
etc.) and probabilistic decision rules and data (failure rates, maintenance
cost and duration, etc.) related to various objects. The objects that the
tool deals with are components and systems of a (nuclear) plant. One type
of relationship among those objects is a TYPE-OF relationship (e.g. between
the objects PUMP and MOTOR-DRIVEN-PUMP), which is conveniently modeled in a
frame based system.

We propose a data base organization, which fits well into this structure and which can also be used for purposes other than maintenance, such
as probabilistic risk assessment (PRA) and availability and reliability
programs data base. This organization, coupled with an updating proces, can
be instrumental in maintaining a consistent data base for those purposes,
which is based on all (both complete and incomplete) data available.

1.2 Justification

Section 4.8 of reference [1] lists several reasons for the limited
applicability of PRA. Among the most important ones are the lack of a uni-
versally accepted statistical data base, and the subjective judgement involved
in every step, which strongly influence the results. Both are related to the
large uncertainty on parameters such as failure rates, maintenance duration
and cost, manpower and spare parts availability, etc. The same drawbacks
apply to maintenance scheduling. As for now, preventive maintenance is
based on subjective and very general manufacturer recommendations. Decisions on corrective maintenance involve judgment and very often ad-hoc
solutions, which lack optimality and are prone to human errors.

A partial solution to those problems is offered by the following
methodology, which consists of a hierarchically structured data base (such
as in Figure 1 below) and a technique for propagating event frequency

distributions among the related objects in the structure. The proposed
methodology allows such features as utilization of incomplete data,
detaching dependencies among probability distributions (e.g. due to common
source of prior distributions), and correcting errors introduced by expert
opinion based prior distributions. It reduces the necessity to use
excessive subjective opinions by offering a "natural prior" system, in
which each level of generalization can be used as a natuarl prior distribution for its more specific descendents. The result is an organized
knowledge base, that as such avoids both ambiguity and overlapping of
objects on one hand and incompleteness on the other hand.

In terms of computational efficiency, the method allows to store most
of the information regarding an object with its more generic ancestor in
the hierarchy, thus avoiding unnecessary repetition.

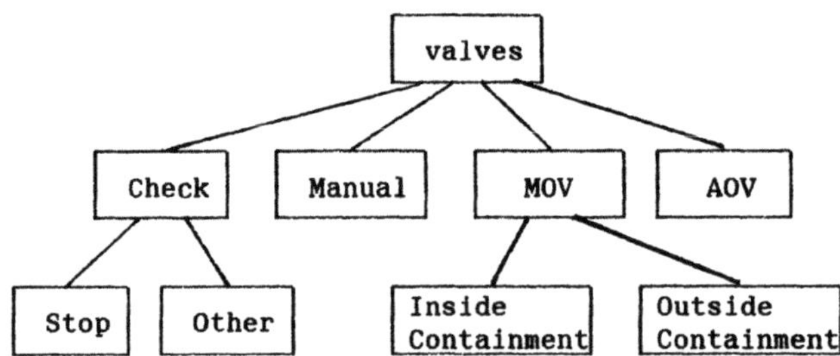

Figure 1. Hierarchical Structure for Valves

1.3 Bayes' Theorem

Bayes' Theorem is the main tool used for inference in the data analysis process for PRA of nuclear plants. Good illustrations of its use for
data analysis are given in references [3-5]. References [6-8] concentrate
on two special cases: one is the use of discrete probability distributions
(DPD) in conjuction with Bayes' Theorem, while the second is a development
of two stage approach to Bayesian updating.

To clarify the subject of this paper one application of Bayes' Theorem
is briefly described. The basic idea behind the methodology is that one
wants to use all the available sources of information. The process is
started by putting together a prior distribution $\phi(\lambda)$ based on judgement.
This distribution is then updated, using equation (1), with the likelihood
function $L(Eg/\lambda)$ based on generic evidence Eg:

$$\phi(\lambda/Eg) = \frac{\phi(\lambda)L(Eg/\lambda)}{\int \phi(\lambda)L(Eg/\lambda)d\lambda} \tag{1}$$

The third step applies equation (2) for specializing the distribution
with likelihood function $L(Ep/\lambda,Eg)$, based on plant specific evidence Ep.

$$\phi(\lambda/Eg,Ep) = \frac{\phi(\lambda/Eg)L(Ep/\lambda,Eg)}{\int \phi(\lambda/Eg)L(Ep/\lambda,Eg)d\lambda} \tag{2}$$

In the following discussion we refer to this method as the "lumped data"
method, as opposed to the "variability method" that is developed here.

2. UNCERTAINTY IN THE UPDATING PROCESS

The decision making process for maintenance involves many sources of
uncertainty, among which are task length, spare parts availability, supply

time and failure rate, operating equipment failure rate, repair time and
cost for corrective and preventive maintenance.

In respect to PRA, it was pointed out[5,10] that "the overwhelming evidence from research on uncertain quantities is that people's probability
distribution tends to be too tight." Since any addition of evidence tends
to narrow down the posterior distribution, we may suspect that the
posterior distribution too represents much less uncertainty then the actual
state of knowledge requires. To compensate for this defficiency, several
methods are discussed in the literature for broadening the originally submitted expert failure rate distribution[5,11,12,13].

But even if the right state of knowledge prior distribution is used, it
is shown below that the current methodology tends to produce results that
are tighter than suggested by the evidence. To illustrate this point let us
repeat example #1 of reference [3]. This example updates the WASH-1400
diesel-generator (D.G.) failure rate distribution, which is Lognormal with
5th percentile 0.01 and 95th percentile 0.1, by evidence from tests made in
a specific plant. To simplify the calculations we replace the original
Lognormal prior distribution with a Beta(n=47, r=2) distribution. Applying
the evidence r = 5 failures in n = 227 tests, the posterior distribution is
Beta(n = 274, r = 7), which has a mean = 0.0283 and a variance = 8.7E-5.

Let us assume now a hypothetical case, that the evidence comes from 4
different D.G.s, which were tested as follows :

D.G. #	Tests	Failures
1	40	0
2	80	1
3	70	2
4	37	2

Although the hypothetical evidence shows that one out of the four D.G.s
had 0 failures, and the second one had only 1 in 80 tests (1.25E-2), the
posterior allows only for 5% probability of having $\lambda \leqslant 1.2E-2$. On the
other extreme we have one D.G. which had 2 out of 37 failures (5.41E-2).
Nevertheless the distribution allows only 5% above 4.3E-2. In other words,
the distribution is very narrow compared to our real state of knowledge.

The technique which is proposed in the next sections modifies the way
that evidence is treated, rather than modifying the prior distribution. In
particular it is proposed to deal with a variability curve of the evidence,
rather than to lump it to a single result for a given component type. In
some cases this approach may eliminate the need to "massage" the prior
distribution. In the general case it is a complementary method to the
various prior distribution broadening methods described above.

3. EXTENSION TO THE POPULATION VARIABILITY CONCEPT

In order to express our real state of knowledge, we propose the population variability concept, which is, in this application, an extension to
the plant variability concept that is used in references [3,5].

To illustrate this process assume the hierarchical data-base structure
in Figure 1. One can describe any object in Figure 1 as specialization of
a higher level. Let the object VALVES represent a generic valves failure
rate distribution, and the object MOV represent the failure rate distribution of motor operated valves. Thus Bayes theorem can be applied, for
example, to VALVES, such that VALVES distribution will be used as generic
prior distributions, MOV specific events will be used as evidence for specialization, and the result will be MOV specific posterior distributions.

This application is an extention of the concept of "generic" and "plant specific" distribution[3,8], and is especially important in cases where very little data are available at the lower level.

References [7,8] suggest equation (3) for combining the prior distribution and the population data into a plant population distribution. The technique is based on the variability of the distribution parameters. If a two parameter distribution is used (e.g. Lognormal), and the parameters are μ, σ, then the population distribution is:

$$\bar{F}(\lambda) = \int F(\lambda/\mu,\sigma)g''(\mu,\sigma)d\mu d\sigma \tag{3}$$

Thus the parameter distribution $g''(\mu,\sigma)$ is used as a weighting function on the various distributions $F(\lambda/\mu,\sigma)$ given the parameters μ,σ in order to get an expected distribution $\bar{F}(\lambda)$ for all possible parameters. The uncertainty, is expressed in terms of uncertainty on the parameters μ,σ, which in turn is updated by Bayes' Theorem as given by equation (4):

$$g''(\mu,\sigma) = \frac{1}{C} L(E/\mu,\sigma)g'(\mu,\sigma) \tag{4}$$

In this paper we use a different technique, in which the distribution $F(\lambda)$ itself is directly derived, rather then it's parameters. By doing that we detach from one assumption on which the other technique is based, namely, that the family of distributions $F(\lambda)$ preserves the distribution type (e.g. Lognormal) after updating, and only the parameters are changed.

4. CONSTRUCTING A POPULATION VARIABILITY DISTRIBUTION

Returning to the D.G.s example, we now present the distribution in various levels as follows:

$F(\lambda)$ =Prior distribution based on expert opinion, which is identical to $\phi(\lambda)$ in equation (1) for the lumped data method.

$F(\lambda/E)$ =Population variability distribution, updated by the evidence for the whole population.

$F(\lambda/E,Si)$=Sample specific distributions, updated from the prior distribution $F(\lambda)$ by that part of the evidence E that represents sample Si. For the purpose of developing a variability distribution, a sample Si consists of all the data collected on object #i, which is in this example D.G. #i.

The equivalent equation to (3) for our method will be equation (5):

$$F(\lambda/E)= \sum_{\substack{\text{all samples} \\ \text{Si in E}}} F(\lambda/E,Si)*p(Si) \tag{5}$$

Where p(Si) represents the probability that Si is the "right" sample representing the specific population that will actually be challenged. Note that reference [7] treats each specific machine as a random sample of the generic distribution. In the current treatment, on the other hand, the sample probability p(Si) is chosen based on the specific population and the specific decision that the distribution is derived for. Assessing p(Si) is itself an important issue, which is discussed in reference [14]. For a given sample we use the Bayes' equation (6) for updating:

$$F(\lambda/E,Si)= \frac{F(\lambda)L(E,Si/\lambda)}{\int F(\lambda)L(E,Si/\lambda)d\lambda} \tag{6}$$

Combining equations (5), (6) we finally get:

$$F(\lambda/E)= \sum_{i} \frac{F(\lambda)L(E,Si/\lambda)*p(Si)}{\int F(\lambda)L(E,Si/\lambda)d\lambda} \tag{7}$$

Which is the updated population variability distribution. One can now apply
equation (8) to the example in the previous section. It will be assumed
that $p(S1) = p(S2) = p(S3) = p(S4) = 0.25$, where Si is the sample for D.G.
#i, and $p(Si)$ is the probability that D.G. #i will be challenged for a real
purpose (as opposed to test)[*]

$$F(\lambda)=Beta(47,2)$$

$$
E= \left|
\begin{array}{lll}
S1 : & n=40, & r=0 \\
S2 : & n=80, & r=1 \\
S4 : & n=70, & r=2 \\
S3 : & n=37, & r=2
\end{array}
\right|
$$

Using the convenience of the conjugate family we get:

$F(\lambda/E,S1) = Beta(87,2)$	$m_1 = 2.30-2$	$Var_1 = 2.55-4$
$F(\lambda/E,S2) = Beta(127,3)$	$m_2 = 2.36-2$	$Var_2 = 1.80-4$
$F(\lambda/E,S3) = Beta(117,4)$	$m_3 = 3.42-2$	$Var_3 = 2.80-4$
$F(\lambda/E,S4) = Beta(84,4)$	$m_4 = 4.76-2$	$Var_4 = 5.34-4$

And finally from equation (5):

$$F(\lambda/E)= \frac{1}{4} \left\{ Beta(87,2)+Beta(127,3)+Beta(117,4)+Beta(84,4) \right\}$$

Figure 2 is a plot of $F(\lambda)$, $\phi(\lambda/E)$ using the lumped data method and
$F(\lambda/E)$ using the population variability method. In addition to a better
expression of uncertainty[**], the variability distributions technique has
another advantage, in that it removes in part the concern expressed above
that "peoples' probability distributions are too tight", since this method,
as opposed to the lumped data method, allows for broadening the distribution if the evidence requires such broadening.

5. A SYSTEMATIC HIERARCHICAL STRUCTURE

Assume that from the four D.G.s, #1, #2, #3 are locaterd in dry
environment and #4 is in a humid room. This information may now be used in
order to construct separate distributions for the first group and a different one for D.G. #4, using the same methodology. We end up with a
hierarchical structure as described in Figure 3.

6.USE OF VARIABILITY DISTRIBUTION AS PRIOR DISTRIBUTION

An example for the application of this concept to maintenance related
decision making is a maintenance program which takes into account environmental conditions. Consider the TYPEOF relations of Figure 1. Current data
bases and maintenance programs do not distinguish between MOVs that are
located inside the containment and those located outside the containment.
It is quite possible that data specialization will give rise to a more

---

[*] Although the 4 samples include different sample sizes, we assume the same
   $p(si)$ for all. The curve will be used for calculating the risk during real
   challenge, the probability of which is independent of the test history.

[**] Variance (Variability method) = $4.12*10-4$
   Variance (Lumped data method) = $8.70*10-5$
   Variance (Raw data   ) = $5.44*10-4$

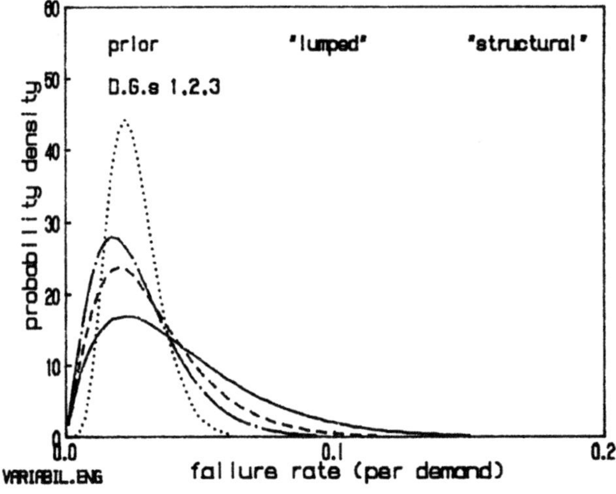

Figure 2. Four Diesel Generators Variability Distribution

efficient program in which those components that are exposed to severe conditions will be tested, calibrated, fixed and replaced more frequently
then other components of the same type. Furthermore, this type of information may help in development of improved environmental conditions such as
chemical treatment of coolant, humidity and temperature in various
controlled areas, etc.

Another application of this concept can be for control of spare parts.
Some probabilistic factors in maintaining a proper spare parts shop are the
frequency in which they are required, the rate of their degradation in the
shop, storage cost (direct and indirect), and the cost of their abscence
when required.

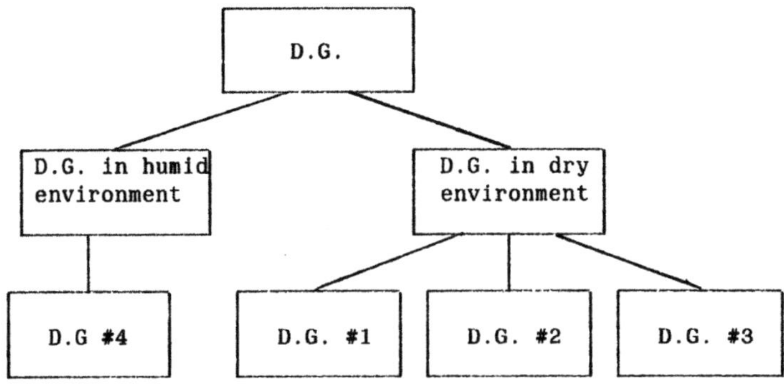

Figure 3. Hierarchical Structure for Diesel Generators

A third application is possible development of separate data bases for normal operation and for accident conditions.

Several questions come to mind while examining this process. The first question is: "what is the point in developing a generic distribution from the population data, and then using the same distribution as a prior for part of that population?" To answer this question we have to compare the prior distribution to the population variability distribution for the D.G.s. Both represent the same type of information: Our knowledge about the generic behavior of D.G.s. The only differance between the two is that the first one is based on some vague idea, while the second one utilized "harder" information. Both are, therefore, legitimate prior distributions for specialization. We don't want, though, to use the vague one while the information on hand enables us to use a better prior. In order to see the conceptual similarity between the two distributions, note that as a person becomes more experienced, he will give as his expert opinion a distribution which is closer to the population variability distribution. The only difference in using our methodology is that we use the experiance explicitly.[*]

Under the obove reasoning, we come to the question that was raised in reference [8], namely: should we exclude from the population evidence that part which is included in the specific assesment? Since we view the population distribution as a prior for updating, we should not exclude that part. The resultant variability distribution must represent the belief in each distribution $\phi(\lambda/E,Si)$ being the "true distribution", with the corresponding probability $p(Si)$. In the development of that distribution, all the specific points were considered for deriving a "general behavior" curve, and this general behavior must consider every available data.

In most cases this question is practically unimportant.In extreme cases, though, (such as D.G. #4 in our example) it is important not to eliminate the specific data from the generic curve. If we used curve #4 in Figure 2 (which is based on D.G.s 1,2,3 only) as a prior distribution, it would have such a low probability density around the #4 D.G. data, that even the updating process will not be able to raise it to a "true" value. Curve #4 expresses a false prior distribution that forsees almost 0 probability of having failure rate above 0.06, while our real expectation is higher.

Finally comes the important point of how this top down updating process can be used. The following algorithm is suggested :

1) Use expert opinion as a prior distribution for the top most item in the hierarchical structure. (example: VALVES in Figure 1).
2) Use equation (7) to construct an updated population variability distribution, where Si are all the lowest level items.
3) For every item i that has an updated disribution, and that has below it in the structure an item j whose distribution has not yet been updated, use the distribution of i as a prior and the data for j as evidence to update the item j distribution. Repeat this process until all the distributions in the structure are updated.

---

[*] Note that conceptually the variability curve is not our prior. The subjectivistic approach treats probabilities as degrees of belief. The prior distribution is the analysts prior estimate based on the generic curve (= variability curve). The simplest approach is to adopt the curve as is and to use it as a prior distribution.

7. CONCLUSIONS

After surveying the current methodology for data analysis we found a
weakness in the way that uncertainty is expressed. In particular, there is
a tendency to express much less uncertainty than our intuition expects.

We suggest a different methodology which utilizes population variability curves at various levels. This methodology eliminates that weakness,
and enables to construct a structure of relationships for data analysis,
and to apply the population variability method for propagating failure
distributions within the structure.

ACKNOWLEDGEMENT

The work in this study is supported in part by DOE contract No.
DE-FG03-86-sf-16610.

REFERENCES

1. W.S. Faught, "Functional Specifications for AI Software Tools for
   Electric Power Applications," EPRI NP-4141, prepared by Intellicorp,
   August 1985.
2. Commonwealth Edison, "Zion Probabilistic Safety Study, Section 1.5,"
   September 1981.
3. G. Apostolakis, S. Kaplan, B.J. Garrick, and R.J. Duphily, "Data
   Specialization for Plant Specific Risk Studies," Nuclear Engineering
   and Design 56(1980) 321-329.
4. S. Kaplan, B.J. Garrick, and P.P. Bieniarz, "On the Use of Bayes'
   Theorem in Assessing the Frequency of Anticipated Transients," ANS/ENS
   Topical Meeting, April 6-9, 1980.
5. G. Apostolakis, "Data Analysis in Risk Assessment," Nuclear Engineering
   and Design 71(1982) 375-381.
6. S. Kaplan, "On The Methodology of Discrete Probability Distributions in
   Risk and Reliability Calculations - Application to Seismic Risk
   Assessment," Risk Analysis,Vol. 1., No. 3. 1981, 189-196.
7. S. Kaplan, "On A 'Two Stage' Bayesian Procedure for Determining Failure
   Rates From Experiential Data," IEEE Transactions on Power Apparatus
   and Systems, Vol. PAS-102, No. 1, January 1983.
8. I.A. Papazoglou, "A Methodology for Assessing Uncertainties In the
   Plant-Specific Frequencies for Initiating Events In the Presence of
   Population Variability," International ANS Meeting on Thermal Nuclear
   Reactor Safety, Chicago, IL, August 29 - September 2, 1982.
9. A.D. Swain and H.E. Guttmann, "Handbook of Human Reliability Analysis
   with Emphasis on Nuclear Power Plant Applications," NUREG/CR-1278,
   final report, August 1983.
10. S. Lichtenstein, B. Fischhoff, and L.D. Phillips, "Calibration of
    Probabilities: the State of the Art," in : J.Jungerman and G. de Zeeuw,
    Editors, "Decision Making and Change in Human Affairs," D.Reidel,
    Dordrecht, Holland, 1977.
11. S. Ahmed, "Pitfalls in Rare Event Analysis and Some Suggestions," Proc.
    Workshop Low-Probability/High-Consequence Risk Analysis, Arlington,
    Virginia, June 15-17, 1982 (Plenum Press, New York, 1985).
12. H.F. Martz, "On Broadening Failure Rate Distributions in PRA Uncer-
    tainty Analysis," Risk Analysis, 4(1984)1.
13. G.Apostolakis, "The Broadening of Failure Rate Distributions in Risk
    Analysis: How Good Are the Experts?," Letter to the Editor; H.F. Martz,
    Response to the preceding Letter, Risk Analysis 5(1985)89.
14. J.K. Shutlis, W. Buranapan, and N.D Eckhoff, "Properties of Parameter
    Estimation Techniques for a Beta-Binomial Failure Model," NUREG/CR-2372,
    December 1981.

AN INTELLIGENT MAN-MACHINE INTERFACE FOR DIAGNOSIS

Eugene Filshtein and Peter Rzasa

Nuclear Services, Combustion Engineering, Inc.
1000 Prospect Hill Road
Windsor, Connecticut 06095

ABSTRACT

This paper describes a computer-based intelligent information acquisition
tool for diagnosing in emergency response situations. This tool is an
integral part of Combustion Engineering's Generic Diagnostic Shell (GDS),
a LISP based Expert System for performing power plant and process control
diagnostics. The objective of the intelligent information acquisition
tool is to optimize the man/machine interface by providing the capability
to enter off-line information into the system in a way which minimizes
diagnostic processing time. The methodology chosen is based on a minimum
entropy algorithm. This approach queries the user for information at a
point in the search space which has been calculated to have a minimum
entropy value. When the query is satisfied, it results in a search space
with the least amount of uncertainty. This system addresses the basic
issues of when and where to post requests for additional information when
diagnosing in emergency response situations. The paper focuses on how we
integrated the entropy-based algorithm into our Generic Diagnostic Shell.
The search technique use in GDS for processing on-line information is
described. An application example is given.

INTRODUCTION

A Generic Diagnostic Shell (GDS) package was developed by Combustion
Engineering to facilitate building problem specific diagnostic expert
systems for complex plant processes. The overall design and specific
features of GDS have been described in references [1,2,3,4]. This paper
focuses on the man/ machine interface module in GDS. The goal of this
software interface tool is to allow a user to enter off-line information
into the GDS system in a way which minimizes the time it takes to
diagnose the problem.

Additional information from an operator must be supplied in cases where
on-line information is insufficient for diagnosis. To optimize the
diagnosis, the system must have the capability to request this
information in an intelligent way. An unsophisticated approach would
simply compile a list of all possible off-line instruments and post
requests for their readings in an arbitrary order. Perhaps the first
request is the crucial one or perhaps the last request in the queue is
the one which will complete the diagnosis. A better approach calls for a

system which posts its request in a way which will optimize the time it takes to diagnose the problem. The intelligent request modules' objective is to guide the user in providing the data to the system in a way which will speed up completion of the diagnosis. The selection criterion uses an algorithm which maximizes expected information entropy reduction.

Information entropy expresses a measure of uncertainty as to which particular component(s) has failed, zero entropy constitutes certainty (exact knowledge) and an entropy greater then zero is related to the level of uncertainty. In other words at zero entropy maximum knowledge about the system state exists. Therefore the strategy is to reduce the entropy, i.e., place an inquiry such that the information obtained produces a maximum entropy reduction. To better understand how we applied the minimum entropy algorithm it is important to have an overview of the basic strategy use by GDS to process available (on-line) information.

GENERIC DIAGNOSTIC SHELL (GDS) OVERVIEW

GDS utilizes an Expert System architecture i.e., the problem specific knowledge is incorporated into a knowledge base while the generic problem solving portion is built into a separate control and inference scheme. (See Figure 1). The inference techniques include an enhanced forward chaining algorithm and a collection of logic reasoning rules. The knowledge base consists of two forms of information, production rules and a logic model based on a tree structure. Information enters the system either directly from on-line sensors or by manual user entry.

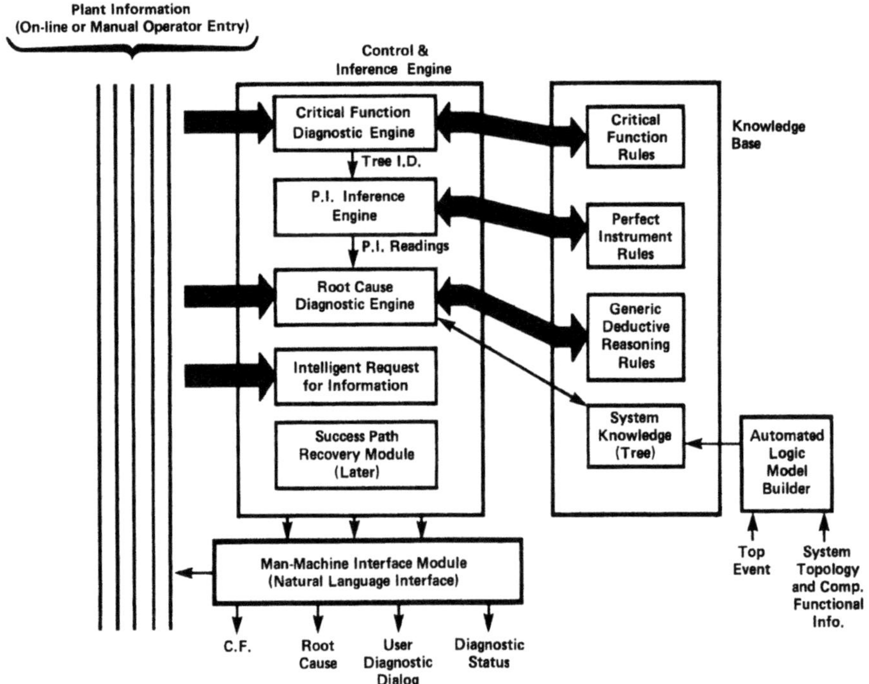

Figure 1. Generic Diagnostic Expert System Block Diagram

The critical function module polls the on-line sensor information. At
predefined time intervals the critical function module chains thru the
production rule base determining if a critical function is violated. If
a failed function is identified, then a fault tree logic model specific
to that critical function is processed in the diagnostic modules. The
critical function is the top event of a fault tree. The next step in the
diagnosis process is the root cause determination of the system failure.

The root cause diagnosis algorithm consists of two modules, a perfect
instruments (PI) module and a deductive reasoning module. "Perfect
Instruments" are introduced as imaginary instruments which are scattered
in a predetermined fashion throughout the fault tree and are capable of
directly assessing the operating states of events. Those events with
perfect instruments will be referred to as identifiable events. It
should be noted that not all events have perfect instruments. They
eliminate the need to have specific instrumentation information in the
generic diagnostic tool. They also provide information to the deductive
reasoning module which allows the diagnosis to be accomplished with
substantially less information than normally required. A transition from
the Generic Diagnostic System to a specific application is accomplished
by providing the production rules which convert real instrument readings
into indications of perfect instruments. As a result of processing the
perfect instrument rules we now have binary indications (operating or
failed) of those events on the fault tree which have perfect instruments
associated with them.

This acquired information (i.e., the PI readings) is passed to the
deductive reasoning module. This modules controls the execution of 24
deductive reasoning rules in a forward chaining process. Those rules
propagate the existing information acquired from processing the PI rules
by applying deductive reasoning logic. The system converges on the
diagnosis by eliminating branches on the tree which do not according to
the logic contain failures and by drawing inferences about cause-effect
relationship between events on the tree. Details of the root cause
module are discussed in Reference 2.

The control program identified as 'intelligent request for information'
module is activated when the root cause process cannot reach a conclusion
due to a lack of sufficient on-line information. The entropy algorithm
which is the focus of this paper is invoked at this point.

MAN-MACHINE INTERFACE

To optimize the diagnosis, the system must have the capability to request
information in an intelligent way. How should the system post a request
for additional information in a way which minimizes the number of
requests it will take to determine the failure? An unsophisticated
approach would simply compile a list of all the events and post requests
for their values in an arbitrary order. Perhaps the first request is the
crucial one or perhaps the last request in the queue is the one which
will complete the diagnosis. There is no way to assure optimal
performance. A better approach calls for a system which posts its
request in a way which will minimize the time it takes to diagnose the
problem. The objective is to guide the user in providing the data to the
system in a way which will expedite the completion of the diagnosis. The
selection criterion for placing queries uses an algorithm which maximizes
expected information entropy reduction.

Picture the diagnostic system in a state where all possible branches to
be pruned are pruned, all possible paths to be selected have been
selected by processing on-line information, and there is still a large

tree-structured search space (although reduced in size) to be considered.
If it were known which nodes (gates or basic events) had information to
offer and, in addition, which node would offer information resulting in
the smallest remaining search space (the smallest subsequent version of
the reduced tree), the decision as to where to post a request would be
any easy one to make. This is exactly the type of assistance the
implementation of the entropy principle provides to the GDS system.

Additional information about component failure rates should be provided.
For every basic event defined in the fault tree a failure rate can be
assigned. However if they are unknown, the system, by default, assumes
them to be equivalent. A calculation is performed to determine failure
rates for the intermediate events on the tree and conditional failure
probabilities are determined for all the events in the tree.

An algorithm has been derived (Ref. 4) which determines which event
corresponds to the minimum entropy value without having to apply a
time-consuming entropy calculation. It involves a new construct, a
transformed fault tree. In a transformed fault tree all the 'AND' gates
are removed and the tree is re-organized by selecting the one path
through the 'AND' gate with a minimum entropy value. It has been
demonstrated that after the conditional probabilities have been
evaluated, the event on a transformed fault tree with a conditional
probability closest to .5 is the event which corresponds to the minimum
entropy value given that the time it takes to obtain the information for
each query is approximately the same. (A method for handling situation
where acquisition times vary can be applied.)

When the on-line information processing algorithm cannot reduce the tree
any further the entropy algorithm executes as follows:

1.   Calculate failure rates and conditioned probabilities for all the
      reduced tree events.

2.   Construct a Transformed Fault-Tree.

3.   Among identifiable events, i.e., those that have instruments
      associated with them, find the one with closest to .5 conditional
      failure probability. (For situation where acquisition times are the
      same.)

4.   Post an inquiry regarding the state of this event and ascertain its
      state via the operator.

5.   Decompose the Transformed Fault-Tree according to a given set of
      rules.

6.   Return to step 3 unless the diagnostic search is complete.

With the transformed knowledge structure the search strategy can now
shift from a breadth-first search (which is what is going on in the
on-line instrument processing) to a depth first search strategy. Such a
strategy change is justified by the fact that the off-line information
acquisition via the operator is a relatively slow process. It makes
sense to focus on a single path of reasoning (i.e., depth-first) when the
time consuming question and answering session is invoked.

An example of the process described above follows.

The tree shown in Figure 2 is a reduced tree from a diagnostic process
which determined (through processing on-line data) that the feedwater

level in a given process failed Low (event E2 on the tree). The actual malfunction in this example is "steam non-return valve open in excess" (labelled MF5-3 on the tree). Because there are no 'AND' gates in this tree, the reduced tree is the transformed tree. After the first pass of calculating probabilities (the results of which are shown in Table 1), it is determined that Event 4 has a probability closest to .5 (Pr[E4] = 0.4444). Event 4 is selected. A query regarding the state of this event results in a determination that Event 4 is not failed. The reduced tree resulting from this process shown in Figure 3. The subsequent probability calculation is shown in Table 2. This process is repeated again. Because Event 6 and Event 5 have equal distances from .5 we can choose either at random for the next inquiry.

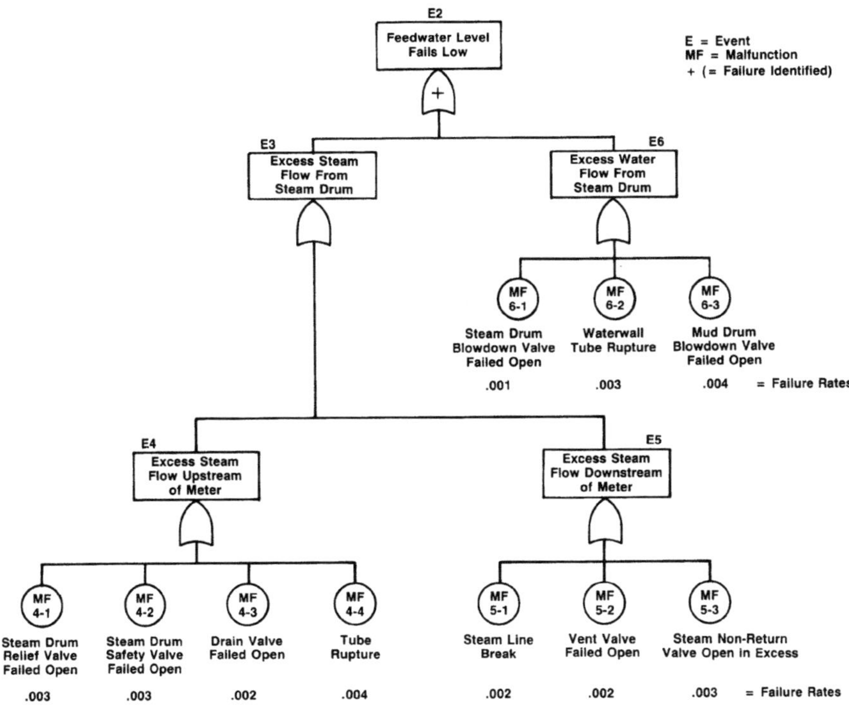

Figure 2.   Reduced Tree for Malfunctin "SNRV Opened in Excess"

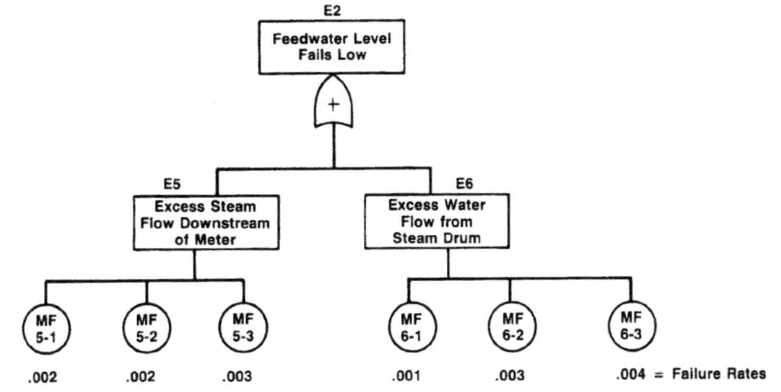

Figure 3.   Reduced Tree after E4 is Determined to be Operating

The outcome of either query will be a reduced tree similar to the one in
Figure 4. Event MF5-3 will be selected as the event closest to .5. The
results of the operator inquiry will reveal that this is the malfunction,
and the diagnosis is completed.

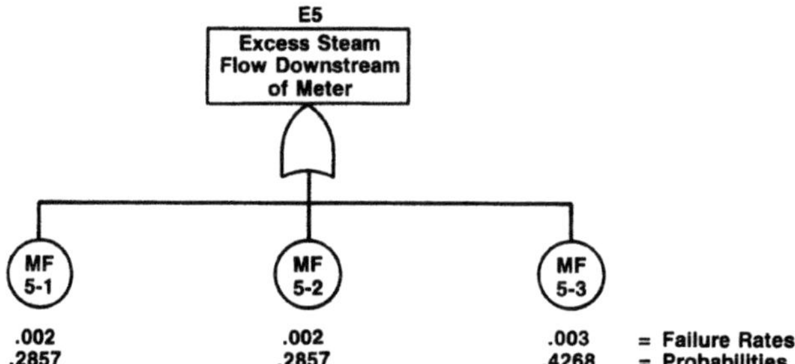

Figure 4. Final Reduced Tree

The man-machine interface module can perform in either of two operational
modes. The first mode is pursued when the operator is uncertain as to
which way to proceed with diagnosis in a particular emergency situation.
This mode of operation is driven by the entropy algorithm and has been
described above. The second mode of operation is activated by the
operator when he believes that he has some relevant off-line information,
e.g. a conspicuous instrument reading. We refer to this second mode of
operation as an "override" mode. In this mode the operator interrupts
the system and supplies this particular information. The system will
accept this input and will determine what additional information is
required to proceed with the diagnosis. This is accomplished through the
instrument rules which maps real instruments to events on the fault tree.
In the cases when no instrument rules exist to incorporate this
information, the system will log the event for future reference. By
doing this the system will help to refine the instrument rule set. In
this situation the operator would typically return to the first mode of
operation.

Table 1

Failure Rates and Probabilities of Reduced Tree
Given Event 2 is Failed

Event ID	Failure Rate	Probability
E6	.008	.3963
E4	.012	.4444
E5	.007	.2593
E3	.019	.7037
MF6-1	.001	.037
MF6-2	.003	.111
MF6-3	.004	.1483
MF5-1	.002	.0741
MF5-2	.002	.0741
MF5-3	.003	.1111
MF4-1	.003	.1111
MF4-2	.003	.1111
MF4-3	.002	.0741
MF4-4	.004	.1481

Table 2

Failure Rates and Probabilities of Reduced Tree
After E4 is Determined to be Operating

Event ID	Failure Rate	Probability
E5	.007	.4667
E6	.008	.5333
MF5-1	.002	.1333
MF5-2	.002	.1333
MF5-3	.003	.2001
MF6-1	.001	.0666
MF6-2	.003	.2
MF6-3	.004	.2667

Figure 5 illustrates how an operator would interface with the diagnostic
software. The vertical lines identify the boundary between the deductive
reasoning components the man-machine interface components of GDS. The
deductive reasoning module will pass a reduced tree to the entropy
reduction process if it cannot reach a diagnosis by processing on-line
information. The entropy reduction module will determine which event
information should be provided for. The real instruments for that event
are posted to the operator. The operator provides the readings for these
instruments. The system determines the event state and passes the
information back to the deductive reasoning components. The deductive
reasoning module can either reach a diagnosis with this new information
or further reduce the tree which starts the process just described over
again.

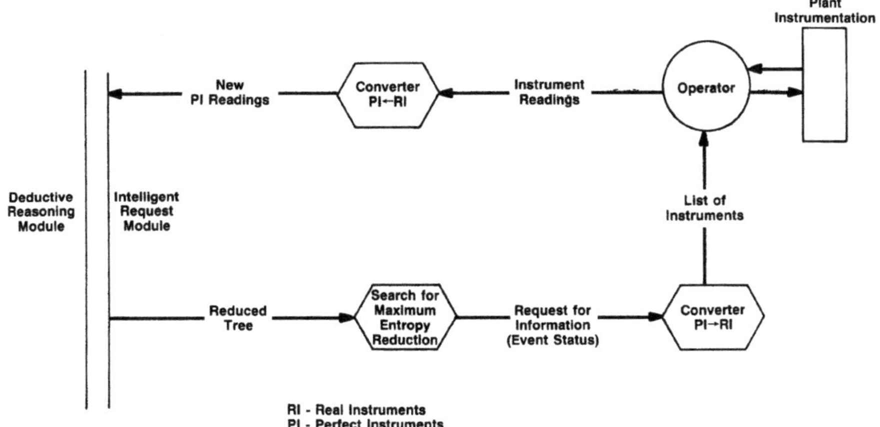

Figure 5. Flow of Information in Operator - GDS Interface

SUMMARY

An algorithm for processing off-line information efficiently in
diagnosing has been presented. This algorithm is incorporated in
Combustion Engineering's Generic Diagnostic Shell (GDS). It is based on
the concept of entropy. A description of the strategy for processing
on-line information was included as background information. An example
of how the entropy algorithm is applied was given. A general discussion
of GDS was provided to describe how the algorithms were applied.

REFERENCES

1. C. H. Neuschaefer, P. W. Rzasa, "GDS - An Intelligent Diagnostic Software Shell", ISA/86 International Conference, Houston, Texas, October 1986.

2. E. L. Filshtein, P. W. Rzasa, "Application of Automated Deductive Reasoning to Power Plant Diagnosis", ANS/ENS International Topical Meeting on Advances in Human Factors in Nuclear Power Systems, Knoxville, TN, April 1986.

3. C. H. Neuschaefer, P. W. Rzasa, Dr. E. L. Filshtein, R. L. Burrington, R. Donais, "Application of C-E's Generic Diagnostic System to Power Plant Diagnostics", EPRI Seminar: Expert Systems Applications in Power Plants, Boston, MA, 1987.

4. E. L. Filshtein, "An Efficient Search Strategy via Entropy Reduction" ANS/ENS Topical Meeting on AI Applications in the Nuclear Industry: Present and Future, Snowbird, Utah, 1987.

TIME-DEPENDENT RELIABILITY ANALYSIS OF NUCLEAR

REACTOR OPERATORS USING PROBABILISTIC NETWORK MODELS

Yoshiaki Oka, Kenji Miyata[1], Hideki Kodaira[2],

Sakae Murakami, Shunsuke Kondo[3] and Yasumasa Togo

Department of Nuclear Engineering,
University of Tokyo
Hongo, Bunkyo-ku, Tokyo, 113, Japan

INTRODUCTION

Human factors are very important for the reliability of a nuclear
power plant. Human behavior has essentially time-dependent nature. The
details of thinking and decision making processes are important for
detailed analysis of human reliability. They have, however, not been
well considered by the conventional methods of human reliability
analysis. The present paper describes the models for the time-dependent
and detailed human reliability analysis. Recovery by an operator is
taken into account and two-operators models are also presented.

BASIC MODEL

GERT(Graphical evaluation and review technique) is used for modeling the reactor operator behavior. GERT is a type of network modeling
technique, which combines PERT-type networks, flowgraphs and stochastic
networks. The network is described in terms of nodes and branches as
shown in Fig.1. There are two types of nodes, DETERMINISTIC and PROB-
ABILISTIC. The branch is characterised by the probability of being included in the network and by the time required to complete the activity
represented by the branch. A set of simulation runs by Monte Carlo is
traced for obtaining the results. A simulation program, GERTS3Q was used
for the study. Nine types of probability distributions of the time to
perform the activity are available in the program, but only constant
distribution was used in the study. The details of GERT and GERTS3Q are
described in the book (Whitehouse, 1973). They are the well established
method for systems analysis and design.

----

present address : (1) Kansai Electric Power Co. Osaka, (2) Japan Atomic
                  Power Co. Tokyo, (3) Nuclear Engineering Research Lab.
                  University of Tokyo, Tokai-mura, Ibaraki

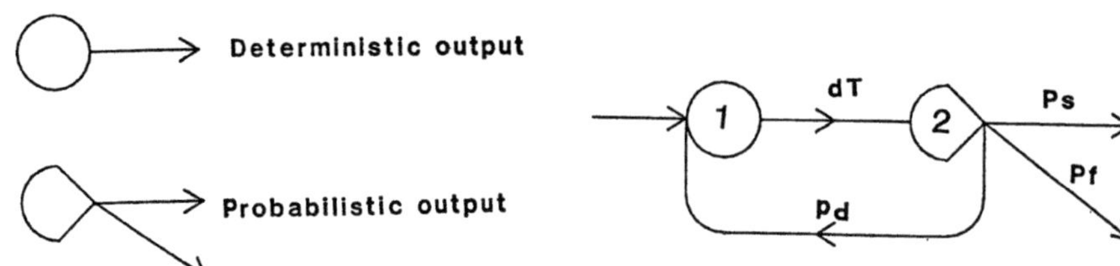

Fig.1 Types of nodes   Fig.2 Model for "Stimuli"

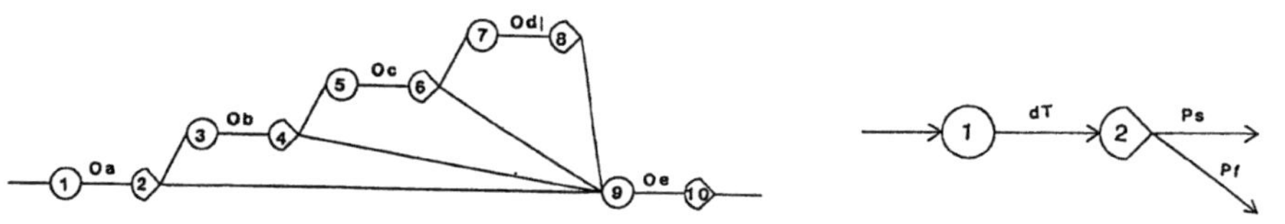

Fig.3 Model for "Organismic"   Fig.4 Model for "Responses"

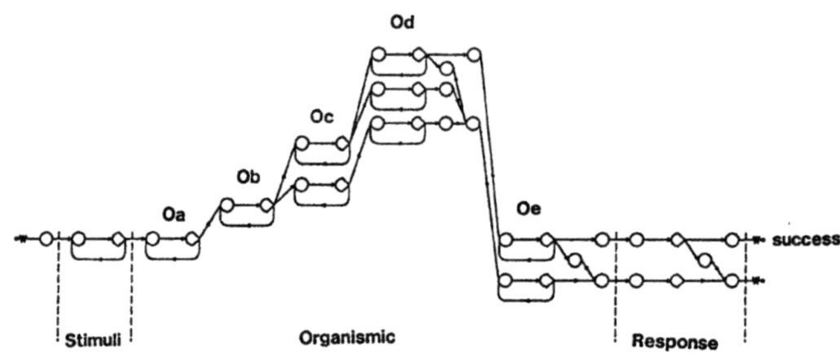

Fig.5 Model including the knowledge base process

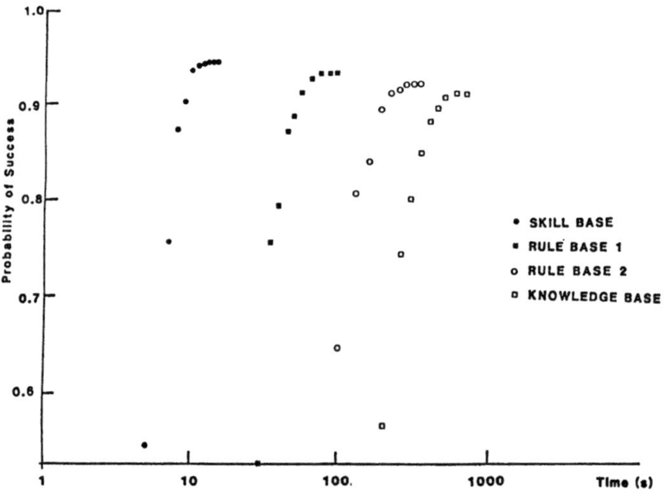

Fig.6 Success probabilities of skill base, rule base and knowledge
base operator behavior

The human behavior is described by the following three components (Swain and Guttman, 1980) ;
(1) Stimuli, (discriminations or perceptions)
(2) Organismic, (thinking and decision making process)
(3) Responses, (motor responses)
The human is expressed as a component that receives inputs, acts on these inputs, and produces outputs. In the present study, each component is expressed by a network model. The model for "Stimuli" is depicted in Fig.2. Ps is the success probability of perceptions during the time, dT. Pf is the failure probability during dT and Pd is the feedback probability during dT.

The model for "Organismic" is shown in Fig.3. It describes "skill base", "rule base" and "knowledge base" thinking and decision making processes which were proposed by Rusmussen. The processes are expressed using the following five stages.
Oa : activation, Ob : data collection, Oc : identification of system state, Od : interpretation of situation and Oe : procedure selection.
The "skill base" process is expressed by Oa-Oe.
The "rule base" process is by either Oa-Ob-Oe or Oa-Ob-Oc-Oe.
The "knowledge base" process is by Oa-Ob-Oc-Od-Oe.

The model for "Responses"is depicted in Fig.4. Ps and Pf are the success and failure probability of responses during the time, dT. The models in Fig 2, 3 and 4 are connected in series and used for describing human behavior. As an example, the model including the knowldege base process is shown in Fig.5. If only the skill base process is included, the nodes of Ob, Oc, Od of Fig.5 is omitted.

Probability of activity realization and time required to complete the activity are the input data of the model. These two parameters are generally not independent each other. However, the following assumptions is made in the present study. The sucess and the failure probabilities, Ps and Pf converge to the conventional (time-independent) values which are described in the Swain's Handbook(1980). The distribution of time is assumed to be constant during dT. Unfortunately the data base of the time distribution of human behavior has not yet been established well. It is possible to accumulate the data from simulation experiments etc.

Time dependent success probabilities were calculated for the skill base, rule base and knowledge base operator behavior. They are depicted in Fig.6. The operator behavior under normal condition may be skill base or rule base, while under transient and accidental condition it may be rule base or knowledge base behavior. It takes much time to complete the knowledge base task.

Recovery of errors is one of the features of human behavior. It has, however, not been well considered by the conventional operator models which can not describe the time-dependent behavior. The time delay associated with recovery should be taken into account. The present network model has its capability because probability and time are the inputs of the model. There are two ways for expressing the recovery in the present network models. One is the use of an internal feedback loop which is added to each stages of the thinking processes. The other is the use of an external feedback loop. It describes the recovery by an operator who has noticed his error by watching the state of the reactor system. The time dependent success probabilities were calculated using the networks with and without the external feedback loop. Figure 7 shows the model with the external feedback loop. The result are depicted in Fig.8. The success probability increases after 40 seconds due to the recovery.

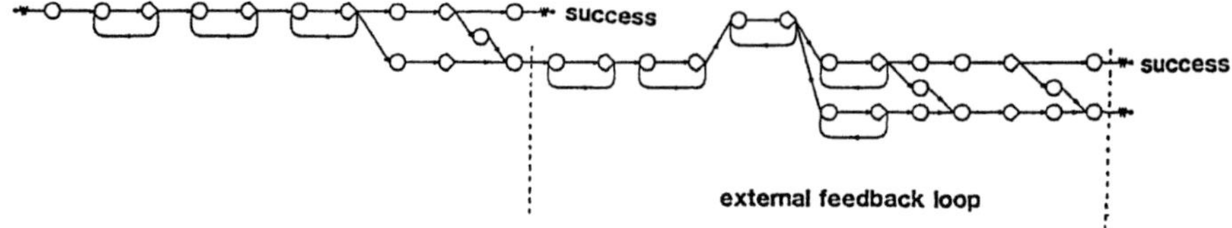

Fig.7 Model with an external feedback loop showing recovery

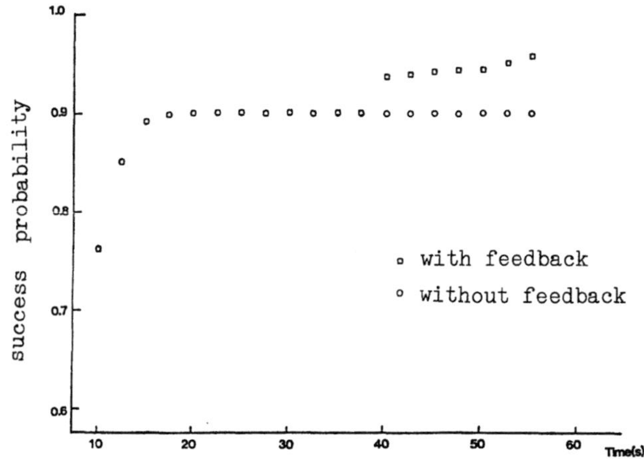

Fig.8 Success probabilities with and without a feedback loop showing recovery

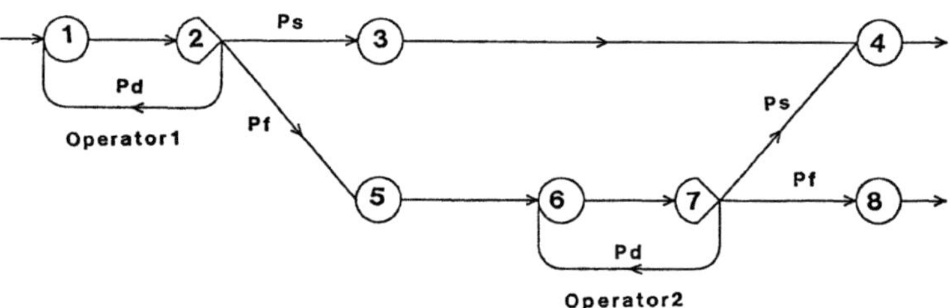

Fig.9 Supervisor model

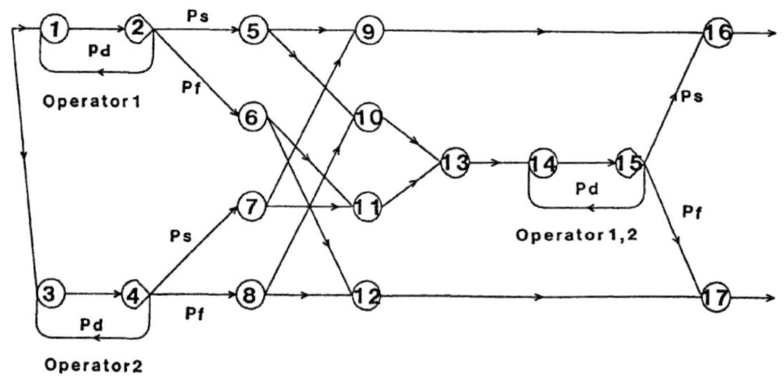

Fig.10 Consultation model

TWO-OPERATORS MODEL

The control of a nuclear power plant is usually carried out by multiple operators. Two types of models are proposed for describing multiple operator behavior. The one is "supervisor model" and the other is "consultation model". They are shown in Fig.9 and Fig.10 respectively. The supervisor model describes the case that one operator supervises the other. The error of the operator is recovered by the supervisor. This model is frequently used for discribing the multiple operator behavior. The consultation model describes the case that two operators check their own decision each other. If they do not agree, they consider the reactor state again. This model will be used for the simulation of the sophisticated thinking and decision making process.

One-operator and two-operators behaviors are compared in Fig.11. The supervisor model or the consultation model was used for the decision making process of the two-operators model. The ultimate success probability of the consultation model is higher than that of the supervisor model. The rise of the success probability is, however, low for the consultation model. The supervisor model and the one operator model show identical rise of the probability between 20 and 30 seconds.

ANALYSIS OF SG WATER LEVEL CONTROL OF A PWR

Control operation of steam generator(SG) water level of a PWR was analysed in order to demonstrate the ability of the network models. The assumed incident was as follows : The automatic control circuit of the main feedwater control valve of a SG failed. The valve was fixed open. This increased the feedwater flow and the water level of the SG. The operator noticed the annunciating indicator, tried to find the cause and stabilized the incident.

The sequence of the operator behavior is as follows :
A. perceive the annunciating indicator showing the water level deviation
B. notice the increase in water level
C. notice the increase in feedwater flow
D. notice that the rift of the feedwater valve is large
E. decide that the incident is attributed to the abnormal feedwater controller
F. decide the procedure using the instruction manual
G. change the positioner of the feedwater controller
H. notice the change
I. change the feedwater controller
J. manually close the valve
K. know the result
The task A corresponds to "Stimuli". The tasks from B to F correspond to "Organismic" and the tasks from G to K to "Responses".

The operator behavior was analysed by the one-operator model and the two-operators model. The one-operator model is depicted in Fig.12. The nodes 3,4,5 expresses correct perception of annunciator. The failure of perception is expressed by the node 301. The nodes 6 and 7 correspond to Oa (activation) of the thinking process. The nodes 8,9,10 express the success in perceiving the water level change (the operator behavior B in the above sequence). The correspondence between the node numbers (numerals) and the operator tasks (alphabets) listed above are as follows. 11,12,13-C, 14,15,16-D, 16,17-E, 19,20,21-F, 22,23,24-G, 25,26-activation of thinking process, 27,28,29-H, 30-31-decision of the next procedure, 32,33-I, 35,36,37-J, 38-39-thinking process, and 40,41,42-K. The nodes from 301 to 309 express the failure, but some of

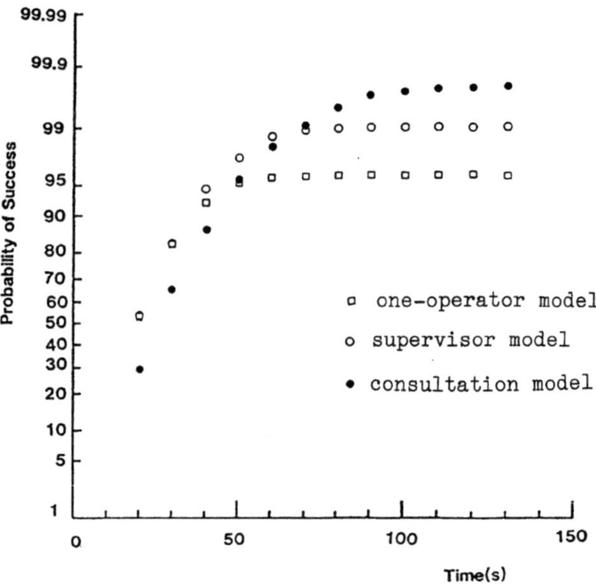

Fig.11 Comparison of the success probabilities of the decision making
process between one-operator model, the supervisor model and the
consultation model

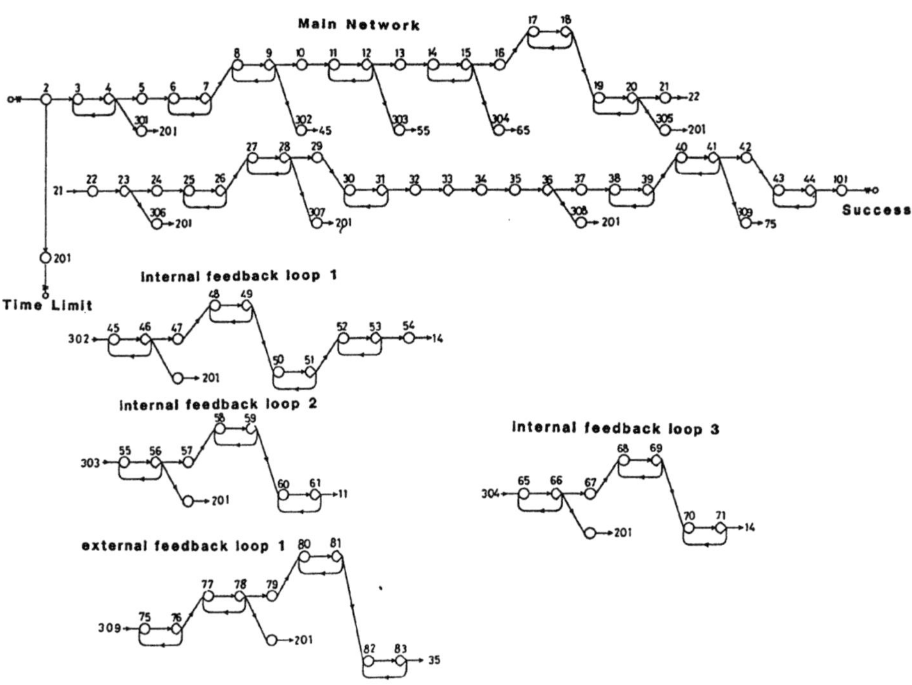

Fig.12 Network model for the analysis of SG feedwater level control
(one-operator model)

796

Table I. Comparison of the success probabilities
by the present and the conventional method

	case 1	case 2	case 3	case 4	case 5
network model (present)	0.9896	0.9614	0.9725	0.7455	0.9915
tree diagram (conventional)	0.9873	0.8898	0.9419	0.5128	0.9903

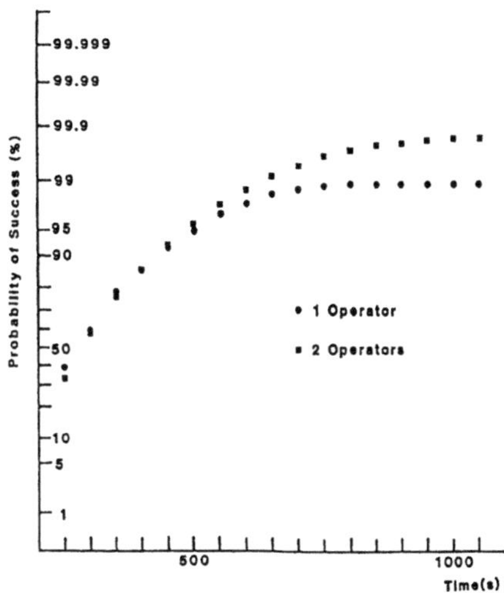

Fig.13 Success probabilities of the one-operator and the two-operators model for the SG feedwater control

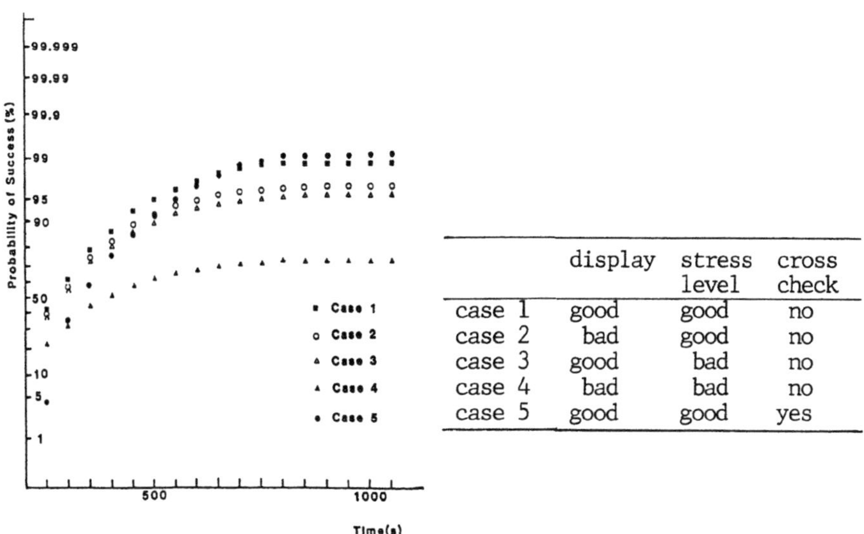

	display	stress level	cross check
case 1	good	good	no
case 2	bad	good	no
case 3	good	bad	no
case 4	bad	bad	no
case 5	good	good	yes

Fig.14 Effect of performance shaping factors and cross-check

them are connected to the internal and the external feedback loops which express the recovery of the operator.

The operator behavior was also analysed by the two-operators model. The model was constructed by incorporating the supervisor model and the consultation model to the thinking and decision making processes of the single operator model. The supervisor model was incorporated into the following nodes, 3-4-5-301, 14-15-16-304, 19-20-21-305, 22-23-24-306, 27-28-29-307, 35-36-37-308 and 40-41-42-309. The consultation model was into the followings, 8-9-10-302, 11-12-13-303, 55-56-57-401, 65-66-67-402 and 77-78-79-403. The result of the analysis is depicted in Fig.13. The success probability of the two-operators model is lower at first, but becomes higher afterwards than that of the one-operator model. The slow rise of the success probability of the two-operators model is attributed to the use of the consultation model.

The effect of the performance shaping factors (PSF) on the success probability was analysed. Visual accessibility of the display and stress level were chosen as the PSFs. Combination of good and bad case for both display and stress level generated four cases to be studied. They are shown in Fig.14 together with the results. These results were calculated by the single operator model. The results of the case 4 shows the lowest success probability due to the bad display and bad stress level. The case 5 in the figure is the result of the two-operators model for the case that cross checking of the decisions was carried out by the operators themselves. The consultation model was used for the purpose, but no supervisor model was included in the two-operators model. The values of the case 5 are low at first due to consultation, but become high afterwards.

The comparison with the conventional tree deagram analysis of human reliability was carried out for the above five cases. The result is shown in Table I. The values of the network model is higher than that of the conventional method. This is attributed to effect of the recovery which was included in the present model. The conventional method cannot treat the failure probability of the decision making process in detail.

SUMMARY
The network models have been presented for the human reliability analysis. The models are based on the GERT simulation technique which was well established. Time-dependent human reliability analysis is possible using the model. Details of thinking and decision making processes are expressed by multi-stage networks describing skill base, rule base and knowledge base operator behavior. Recovery is taken into account in the model by adding feedback loops to the network. The supervisor model and the consultation model have been proposed for describing multiple operator behavior. The analysis of the SG water level control has been carried out by the one-operator model and the two-operators model. The reliability of the two-operators case is low at first, but becomes higher than that of the one-operator model. The analysis has demonstrated the capability of the network model.

REFERENCES
Swain A.D and Guttman H.G. "Handbook of human reliability analysis with emphasis on nuclear power plant applications" NUREG/CR-1278(1980).
Whitehouse, G.E., 1973 "Systems analysis and design using network techniques" Prentice-Hall INC. Englewood Cliffs, New Jersey.

USING RISK BASED TOOLS IN EMERGENCY RESPONSE[*]

Brent W. Dixon and Karen G. Ferns

EG&G Idaho, Inc.
Idaho Falls, ID 83415 USA

ABSTRACT

Probabilistic Risk Assessment (PRA) techniques are used by the nuclear industry to model the potential response of a reactor subjected to
unusual conditions. The knowledge contained in these models can aid in
emergency response decision making. This paper presents requirements for
a PRA based emergency response support program. The authors have developed several components of this system to date. A brief discussion of
published work provides background for a detailed description of recent
developments. A rapid deep assessment capability for specific portions
of full plant models is presented. The program uses a screening rule
base to control search space expansion in a combinational algorithm.

INTRODUCTION

This paper addresses application of Probabilistic Risk Assessment
(PRA) techniques in the area of nuclear power plant emergency response.
The first portion of the paper contains a description of current issues
in emergency response which can be managed with PRA based information.
A specification for computer support requirements is presented.

The body of the paper is a description of work in progress on an
Artificial Intelligence (AI) based emergency response support program.
The project overview includes brief descriptions and references for previously completed components. This is followed by a detailed description of recent, unpublished work on real time analysis algorithms.

The particular algorithm described supports rapid, thorough assessments of specified portions of full plant PRA models. The user specifies a small group of basic events of interest. The algorithm extracts
all model solutions which include at least one of those basic events.
Several cutoff parameters are supported to constrain the search.

---

[*]This research is sponsored by the U.S. Department of Energy under DOE
Contract No. DE-AC07-76ID01570.

EMERGENCY RESPONSE ISSUES SUPPORTABLE BY PRA

Nuclear power plants are designed to automatically manage most single component failures. While the defense-in-depth design of these plants supports survival when confronted with multiple independent failures, these cases may exceed the capabilities of the automated controls. Thus, an emergency operator response is often required if several independent failures occur together.

Operations personnel respond to an emergency situation by following predefined procedures to regain plant control. These procedures are based on prioritized methods to achieve basic safety functions such as core cooling. The procedures detail the steps required to bring preferred cooling sources and paths on line. The greater the number of failures involved, the harder it is for operations personnel to identify which sources and paths are available.

Another problem confronting the emergency response personnel is the tendency to focus on solutions to near term issues. Occasionally this results in failure to identify negative longer term consequences of the situation. For example, severe situations may occur when subsequent failures confuse recovery actions.

Probabilistic Risk Assessment (PRA) models can assist the operations personnel in both of these areas. First, the PRA models contain information on the resources available to the operator and how those resources can be lost. Given specific information on current plant conditions, an on-line PRA could quickly identify which cooling sources and flow paths are available or unavailable. This would allow the operator to quickly identify which resources to use based on the priority list.

In addition to identification of current success paths, a PRA model can effectively identify the weak links in those paths. Thus a PRA model can aid in planning over the longer term of an incident.

There are two aspects of longer term support. First, the current status of components may naturally and predictably degrade over a period of several hours. For example, if ventilation systems are currently failed, components of other systems will overheat and fail over time. Such situations are easy to overlook.

The second problem is a subsequent independent failure. The plant may be on a good recovery trend with backup success paths available. However, due to previous failures and unavailable equipment all the success paths may depend on a single piece of equipment. If this situation is identified, the key component can be watched more closely and contingencies developed. Identification of any single failure conditions supports rapid comprehension of sudden changes in plant trends.

PRA SUPPORT REQUIREMENTS

Based on the applications of PRA outlined above, requirements for PRA support tools can be established. First, a means of rapid model modification is needed for specializing the plant model to the current situation. This primarily involves toggling of event or system availabilities given known conditions. It could also include model pruning to remove large nonapplicable portions. In addition, the detail of the remaining model portions could be enhanced using model generation methods.

A second necessary support capability is rapid assessment of the modified model. Total automation of the assessment is highly desirable. While a thorough assessment is always beneficial, a scoping assessment which identified the top few dozen failure modes would usually be sufficient. If any changes in status occurred, a new assessment would be performed. Thus, the need for thousands of cutsets covering contingencies is removed.

The final basic support requirement is that tools interface easily with operations and emergency response personnel. This means that the results should be self explanatory and not require a PRA specialist to interpret. In addition, the system input should be simple but flexible, allowing the user to explore multiple alternatives quickly. The ability to simulate subsequent failures and recoveries should be supported. In brief, the full power of PRA techniques should be made available in an understandable and usable fashion.

THE FORESTER CONCEPT

The authors are currently developing an integrated set of computer programs to address the requirements of PRA applications to emergency response problems. This set of programs, titled FORESTER, utilize AI methods to develop, cultivate, and harvest information from models composed of interrelated fault trees, event trees, response trees, and other forms.

Existing components of FORESTER include the Integrated Fault/event TRee Engineering Environment (IFTREE[1]), the SeQUence IMPortance calculator (SQUIMP[2]) and the System UnMaintainability Assessment Code (SUMAC). IFTREE supports the graphical construction and assessment of fault trees and event trees coupled to fault trees. The capabilities of IFTREE allow for full plant sized models to be developed and analyzed in a scoping fashion. IFTREE currently does not support distributions for basic event probabilities and does not perform sensitivity or mission time calculations. It can produce event and sequence importances in an automated manner after the initial modeling activity is completed. IFTREE is designed as an interactive model development environment.

SQUIMP is a specialized code partially integrated with IFTREE. SQUIMP supports development and assessment of large event tree networks like those developed in many Pickard Lowe and Garrick PRAs.[3] Assessment in SQUIMP is based on split fraction rules for each event. This contrasts with IFTREE where each event in the event trees is associated directly with a fault tree. SQUIMP supports multiple tiers of event trees with "pinch points" at the tier interfaces. SQUIMP produces importances across full plant models for initiators, events, sequences, and individual trees.

SUMAC is a postprocessor which accepts cutsets and event probability and repair time distributions as input. SUMAC then produces event, cutset, and system unmaintainabilities and unmaintainability importances for indicated repair periods. Unmaintainability as defined in SUMAC is the probability that the system, subsystem, or component of interest cannot be repaired in time T given that it is currently failed. This supports emergency response by indicating the likelihood of regaining lost systems or components within a particular time limit.

THE SUBSET ASSESSMENT CAPABILITY

The most recent development in FORESTER, and the topic which will be
detailed in the rest of this paper, is a specialized analysis capability
being integrated into IFTREE. This capability, called "subset assessment"
allows for the rapid location of a select group of cutsets in a very large
risk model. The common denominator between cutsets in the group is that
all of them will contain at least one event from a specified subset of
the events in the model. Thus a subset assessment of a model involves
specification of the events of interest and any cutoff parameters. The
assessment results include all cutsets from the model which meet the
cutoff criteria and involve one or more of the events of interest.

There are many applications for subset assessment, including identification of risk associated with localized fires, terrorist attacks,
maintenance errors, and any other problem involving a small group of
components in a large, complex facility. In the area of emergency
response, subset assessment can identify the particular failure modes
associated with equipment which is confirmed unavailable. It can also
support "what if" activities.

Effective use of the subset assessment capability in supporting
emergency response activities requires comparison of its results against
a base line risk derived from a small cutset cache. The cache would be
based on a prior general assessment of the full model. The base line
comparison would indicate the relative change in risk presented by the
current situation.

FINDING A SUBSET ASSESSMENT ALGORITHM

In developing the subset assessment algorithm we were faced with a
dilemma. The assessment required that the complete tree be examined in
detail, but only specific solutions would be of interest. Each standard
solution examined had problems with a combination of problem size,
solution speed, or correctness.

One way to accomplish subset assessment is to perform a general
solution, then discard all cutsets which do not include events of
interest. This is very expensive and may affect the depth of analysis
possible. Thus, this method would not support a thorough subset
assessment on a large model.

A second approach would involve discarding any tree branches that do
not include events of interest, then solving the pruned tree. Unfortunately this method is guaranteed to produce erroneous results. The
errors occur due to AND gates high in the tree. If any part of an input
to an AND gate is pruned, the solution at the gate is compromised.

The method we finally used is somewhat unconventional; a combinational algorithm. Combinations of basic events are examined as possible
solutions to the tree equation. This contrasts with most algorithms
which involve expanding and reducing the equation on a gate by gate basis
in either a top-down or bottom-up fashion.

CONTROLLING THE COMBINATORIAL EXPLOSION

A major problem with the combinational algorithm is the "combinatorial explosion" in the search space. The total number of combinations
possible is 2^N-1 where N is the number of basic events in the tree.

For an average plant model containing 1000 events, there are approximately $1.0*10^{300}$ combinations possible, a bit too many for reasonable computerized solution.

The IFTREE combinational algorithm uses AI methods to control search space growth. The predominant method is rule-based screening of all potential additions to the search space. In addition, special ordering is maintained and rule-based search space pruning is supported. The specialization of the combinational algorithm for subset assessment required only minor changes in event ordering and the rule bases.

The first change was to modify the probabilistic ordering of the event list. In the general algorithm, the events in the fault tree are ordered in descending probability. The first event is placed in the search space. The following events are then placed at the front of the search space and also combined with any existing entries to create potential new entries (see Fig. 1). Before addition to the search space, each entry is tested against the tree equation. If it solves the equation, the entry is a cutset and is placed in the solutions list instead of the search space. The probability ordering allows for probabilistic pruning of the search space, since any events examined later will have lower probabilities.

For the subset assessment, the special events (events in the subset) are placed at the front of the event list, regardless of their probabilities. The rest of the list remains probabilistically ordered. The general rule of adding subsequent events to the search space is amended so that only the special events are added as singles. This way, all sets in the search space contain at least one of the special events (see Fig. 2). Thus the basic goal of the subset assessment is met.

SUPPORT FOR CUTOFF PARAMETERS

Support of cutoff parameters required other minor modifications to the rules. The basic methods supporting order cutoff and solution progress pruning did not require modification. However, rules for probability pruning were affected by the event list reordering.

The general rule for probability screening of new search space entries is "the probability of the new set must be greater than the current probability cutoff divided by the current event's probability."

The reasoning behind this rule is that each search space entry is a subminimal set, requiring at least one more event to make it a cutset.

```
Event List  --  (a b c d . . .)

Event      Search space after considering event

           NIL

a          (a)

b          (b) (a) (b a)

c          (c) (b) (c b) (a) (c a) (b a) (c b a)

d          (d) (c) (d c) (b) (d b) (c b) (d c b) (a)
           (d a) (c a) (d c a) (b a) (d b a) (c b a)
           (d c b a)
```

Fig. 1. Unpruned search space ordering in general algorithm.

```
Subset to be assessed  --  (b c)
Event List  --  (b c a d . . .)

Event      Search space after considering event

           NIL

b          (b)

c          (c) (b) (c b)

a          (c) (a c) (b) (a b) (c b) (a c b)

d          (c) (d c) (a c) (d a c) (b) (d b) (a b)
           (d a b) (c b) (d c b) (a c b) (d a c b)
```

Fig. 2. Unpruned search space ordering in subset assessment algorithm.

Since the events are ordered in descending probability, the next event
will have a probability equal to or lower than the current event. If
the next event completes the set, the resulting cutset's probability will
be less than or equal to the product of the set probability and the current event probability. The number derived by dividing the cutoff by the
current event probability is called the "effective probability cutoff."

   The effective probability cutoff is also used for the probability
based search space pruning rule. Basically, this rule states that when
the cutoff rises (when using a dynamic cutoff), current entries in the
search space may no longer meet the requirements of the probability
screening rule. If so, those entries can be deleted. Fig. 3 provides
an example of probability based search space entry screening and search
space pruning. Note the reduction in search space growth.


```
Event List  --  ( (a .1) (b .04) (c .01) (d .002) . . .)
Probability cutoff -- .00001

Event      Effective cutoff      Search space

           .00001                NIL

a          .0001                 (a .1)

b          .00025                (b .04) (a .1) (b a .004)

c          .001                  (c .01) (b .04) (a .1)
                                 (b a .004) (c a .001)

d          .005                  (c .01) (b .04) (a .1)
```

Fig. 3. Search space ordering when constrained by a fixed probability
         cutoff.

   The modification of the probability screening rule to support subset
assessment required a "pending probability" to be established. The pending probability is the probability of the first nonspecial event in the
event list. The rule is modified to divide the cutoff by the greater of
the current event's probability or the pending probability. This correctly identifies the highest probability yet to come in the search. The
search space pruning rule also was modified to use the pending
probability (see Fig. 4).

```
Event List   --  ( (b .04) (c .01) (a .1) (d .002) . . .)
Probability cutoff -- .00001

Event       Effective cutoff     Search space

            .00001               NIL

b           .0001*               (b .04)

c           .0001*               (c .01) (b .04) (c b .0004)

a           .0001                (c .01) (a c .001) (b .04)
                                 (a b .004) (c b .0004)

d           .005                 (c .01) (b .04)


* Based on "pending probability" value
```

Fig. 4.   Subset assessment search space ordering with a fixed cutoff.

THE MINIMAL CUTSET ISSUE

One final modification required in the algorithm involved whether
an identified cutset was truly minimal. In the general algorithm, all
new solutions (cutsets) are subjected to a constrained test against
other solutions identified during the current event search. Since any
"more minimal" solutions would have already been identified, only the
existing solutions need to be checked. A strict method of search space
ordering ensures the completeness of this approach.

The problem with the minimal solution search in subset assessment
is that it depends on earlier identified solutions. Those solutions are
constrained to contain at least one special event. If the current
solution is nonminimal due to a minimal cutset which does not contain a
special event, the constrained test would fail.

We handled this problem by also testing one subset of the new solution composed of all nonspecial events in the cutset. If the subset is
also a cutset, then the cutset including special events is not minimal
and is ignored.

ALGORITHM EFFICIENCY

The described subset assessment algorithm performs reasonably well
on large problems when a small subset of 1 to 5 events is used and a
basement fixed cutoff is allowed. The larger the subset, the closer the
assessment comes to a general solution. In such a case, only scoping
assessments can be completed on large problems in the time required to
support emergency response activities (i.e. less than an hour).

The bottom line on efficiency is that reasonable cutoff values must
be available for the screening and pruning rules to control the combinatorial explosion. The general solution algorithm of IFTREE supports a
dynamic probability cutoff to automate location of reasonable cutoff
values. The initial basement cutoff is derived from the concept of a
"screening probability." The screening probability is the probability
of the most significant (i.e. highest probability) cutset for a gate.
This value can be approximated rapidly without generating cutsets.

Unfortunately, the screening probability does not apply when
performing a subset assessment. The highest probability cutset with a
special event is not guaranteed to equal or exceed the screening
probability lower bound. Therefore the screening probability cannot be

used as the basis for an initial basement cutoff value. Currently this
means that the user must specify either a fixed probability cutoff or a
reasonable basement value for the dynamic probability cutoff to work from.

The subset assessment capability has been tested and verified. An
informal benchmarking of the algorithm indicates that the basic concept is
efficient enough to support emergency response applications. Fine tuning
has been delayed to permit extensions useful for these applications.

The main extension is to add a means of comparing subset assessment
results to a background risk derived from a cache of cutsets prepared
earlier in a general assessment. This will allow the user to see both the
failure modes affected by a particular group of equipment and note whether
those modes are significant when compared to the normal facility risk.

A second extension is support for graphical display of the tree
model with failed portions and unaffected portions properly altered or
removed. The idea is to focus the user's attention on what has been
affected by the current situation, while removing much of the clutter of
unaffected areas from the display.

CONCLUSION

In this paper we have presented a philosophy for design of PRA-based
emergency response aids. Several problem areas in emergency response were
presented as possible targets for a PRA-based tool. The support requirements of that tool include easy model specialization, rapid scoping analysis capabilities, and an interface designed for operations personnel.

Next, we described an ongoing effort to realize the described tool,
including descriptions of existing components. These include fault tree
and event tree graphical modeling and analysis capabilities. The basic
requirement in their design has been support for full plant sized models.
The analysis capabilities are unique in this respect, since most tools are
designed for system sized models. The need for fully automated analysis
in emergency response aids is best realized by full size model support.

Finally, a recently developed analysis capability was described in
detail. This "subset assessment" capability locates specialized groups
of cutsets in an efficient and potentially real-time manner. The method
supports thorough assessment for specific goals across full models.
The basis of the method is intelligent control of a combinational algorithm. Search and pruning methods from the field of Artificial Intelligence are used to impede the potentially exponential expansion of the
search space.

REFERENCES

1.   B. W. Dixon, "An Integrated Fault Tree Development Environment,"
     International ANS/ENS Topical Meeting on Operability of Nuclear Power
     Systems in Normal and Adverse Environments, Sept. 29 - Oct. 3, 1986

2.   B. W. Dixon and M. F. Hinton, "Event Tree Analysis Using Artificial
     Intelligence Techniques," Proceedings of the American Nuclear
     Society Topical Meeting on Computer Applications for Nuclear Power
     Plant Operation and Control, Pasco, Washington, Oct. 1985.

3.   Pickard, Lowe and Garrick, Inc., Seabrook Station Probabilistic
     Safety Assessment, Dec. 1983.

CHAPTER 13

ADVANCED CONCEPTS AND NON-NUCLEAR APPLICATIONS

HEURISTIC SIMULATION OF NUCLEAR SYSTEMS ON A SUPERCOMPUTER USING

THE HAL-1987 GENERAL-PURPOSE PRODUCTION-RULE ANALYSIS SYSTEM

Magdi Ragheb[*], Dennis Gvillo[*], and Henry Makowitz[**]

[*]Department of Nuclear Engineering
 and Computer Science Department
 University of Illinois at Urbana-Champaign
 103 S. Goodwin Ave., Urbana, Illinois 61801
[**]Idaho National Engineering Laboratory (INEL)
 EG&G, P. O. Box 2814, Idaho Falls, Idaho 83401

ABSTRACT

HAL-1987 is a general-purpose tool for the construction of
production-rule analysis systems. It uses the rule-based paradigm from
the part of artificial intelligence concerned with knowledge engineering. It uses backward-chaining and forward-chaining in an
antecedent-consequent logic, and is programmed in Portable Standard Lisp
(PSL). The inference engine is flexible and accomodates general additions and modifications to the knowledge base. The system is used in
coupled symbolic-procedural programming adaptive methodologies for
stochastic simulations. In Monte Carlo simulations of particle
transport, the system considers the pre-processing of the input data to
the simulation and adaptively controls the variance reduction process as
the simulation progresses. This is accomplished through the use of a
knowledge base of rules which encompass the user's expertise in the
variance reduction process. It is also applied to the construction of
model-based systems for monitoring, fault-diagnosis and crisis-alert in
engineering devices, particularly in the field of nuclear reactor safety
analysis. Research is carried out on decision-making methodologies
using possibilistic and probabilistic methods, and the study of anticipatory systems.

INTRODUCTION

As a pre-fifth-generation effort aiming at implementing Artificial
Intelligence (AI) methodologies on Supercomputers, we describe the development of the HAL-1987 Production-Rule Analysis System and its implementation on a Cray X-MP high-speed multiprocessor system for scientific
applications.

The HAL-1987 Production-Rule Analysis System was developed as a tool for the generation of Model-Based systems on a supercomputer. It is programmed in Portable Standard Lisp (PSL). Versions of the system currently operate under the Cray Operating System (COS), as well as under the Cray Time-Sharing System (CTSS).

DESCRIPTION

The system is built within the framework of an antecedent-consequent logic, and is based on the Rule-Based System paradigm. The system uses backward-chaining and forward-chaining.

The structure of the if-then rules in the Knowledge-Base is shown in Fig. 1 as well as an example of the hypotheses used in the backward-chaining mode. These correspond to the physical configuration shown in Fig. 2, and to the Goal-Tree representation shown in Fig. 3.

The functioning of the Model-Based system for both forward- and backward-chaining is demonstrated in the following cases, running in the interactive mode. For forward-chaining, a set of facts about the data representing the basic events in the tree structure is chosen by the user, and the Model-Based System induces the accumulated facts about the state of the individual system components and provides a recommendation according to the constructed model. In backward-chaining, the inference engine asks the user to provide facts as input to its deduction process. Typical cases of forward-chaining and backward-chaining are shown in Figs. 4 and 5 respectively.

DISCUSSION

On a general-purpose supercomputer, an advantage of such an application is the gain in speed, even though a tagged-address architecture like on the current Lisp machines is not used. Benchmarks by Stanford University and Los Alamos National Laboratory have shown that Lisp on the Cray X-MP reaches 15 to 30 times the speed on a Symbolics 3600 Lisp machine. A more significant justification for the presented approach is that procedural-programming modelling of engineering systems can presently be efficiently performed on general-purpose supercomputers. When parallelism is used, the associated speedup will allow faster-than-real-time simulations of fast-occurring phenomena, such as accident situations, in the near future. Computer developments will produce four to sixteen central processing units, general purpose parallel vector pipeline hardware, with one to four nanosecond clock times in the next few years. Present hardware systems, such as the Cray X-MP/48 and Cray-2, and the future Cray Y-MP and Cray-3 are examples of this trend which allows the development of Model-Based systems using symbolic-procedural applications which can prognosticate the future behavior of devices and provide crisis-alert and situation-assessment systems allowing optimal recovery from faults and mitigation of accident consequences. More advanced concepts are under investigation using this system. This includes coupled symbolic-numerical adaptive methodologies for stochastic and numerical simulations, particularly for particle transport and fluid flow. It also includes decision-making methodologies using possibilistic and probabilistic notions, and research on anticipatory systems.

```lisp
(SETQ HYPOTHESES '(
        (STATE IS RECOMMENDATION-IS-NO-ACTION-NEEDED)
        (STATE IS RECOMMENDATION-IS-SHUT-DOWN-REACTOR)
        (STATE IS RECOMMENDATION-IS-DECLARE-EMERGENCY)))

(SETQ RULES '(
        (RULE SAFETY1
                (IF (STATE IS WATER-SOURCE-IS-AVAILABLE)
                    (STATE IS FLOW-IS-POSSIBLE))
                (THEN (STATE IS SYSTEM-STATUS-IS-NORMAL-OPERATION)))

        (RULE SAFETY2
                (IF (STATE IS SYSTEM-STATUS-IS-NORMAL-OPERATION))
                (THEN (STATE IS RECOMMENDATION-IS-NO-ACTION-NEEDED)))

        (RULE SAFETY3
                (IF (STATE IS WATER-SOURCE-IS-UNDETERMINED))
                (THEN (STATE IS SYSTEM-STATUS-IS-UNDETERMINED)))

        (RULE SAFETY4
                (IF (STATE IS CHECK-VALVE-IS-UNDETERMINED))
                (THEN (STATE IS SYSTEM-STATUS-IS-UNDETERMINED)))

        (RULE SAFETY5
                (IF (STATE IS SYSTEM-STATUS-IS-UNDETERMINED))
                (THEN (STATE IS RECOMMENDATION-IS-SHUT-DOWN-REACTOR)))

        (RULE SAFETY6
                (IF (STATE IS WATER-SOURCE-IS-UNAVAILABLE))
                (THEN (STATE IS SYSTEM-STATUS-IS-ACCIDENT-STATUS)))

        (RULE SAFETY7
                (IF (STATE IS FLOW-IS-IMPOSSIBLE))
                (THEN (STATE IS SYSTEM-STATUS-IS-ACCIDENT-STATUS)))

        (RULE SAFETY8
                (IF (STATE IS SYSTEM-STATUS-IS-ACCIDENT-STATUS))
                (THEN (STATE IS RECOMMENDATION-IS-DECLARE-EMERGENCY)))

        (RULE SAFETY9
                (IF (STATE IS CHECK-VALVE-IS-OPEN)
                    (STATE IS PUMPING-IS-AVAILABLE))
                (THEN (STATE IS FLOW-IS-POSSIBLE)))

        (RULE SAFETY10
                (IF (STATE IS CHECK-VALVE-IS-CLOSED))
                (THEN (STATE IS FLOW-IS-IMPOSSIBLE)))

        (RULE SAFETY11
                (IF (STATE IS PUMPING-IS-UNAVAILABLE))
                (THEN (STATE IS FLOW-IS-IMPOSSIBLE)))

        (RULE SAFETY12
                (IF (STATE IS PUMP-A-IS-OPERATIVE))
                (THEN (STATE IS PUMPING-IS-AVAILABLE)))

        (RULE SAFETY13
                (IF (STATE IS PUMP-B-IS-OPERATIVE))
                (THEN (STATE IS PUMPING-IS-AVAILABLE)))

        (RULE SAFETY14
                (IF (STATE IS PUMP-A-IS-INOPERATIVE)
                    (STATE IS PUMP-B-IS-INOPERATIVE))
                (THEN (STATE IS PUMPING-IS-UNAVAILABLE)))))
```

Fig. 1 Knowledge-Base in the form of IF/THEN rules and the
associated hypotheses

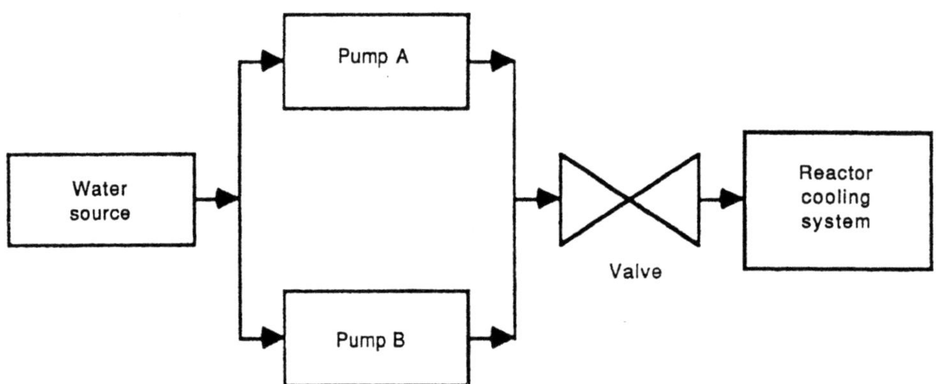

Fig. 2 Simplified model of a power plant coolant system

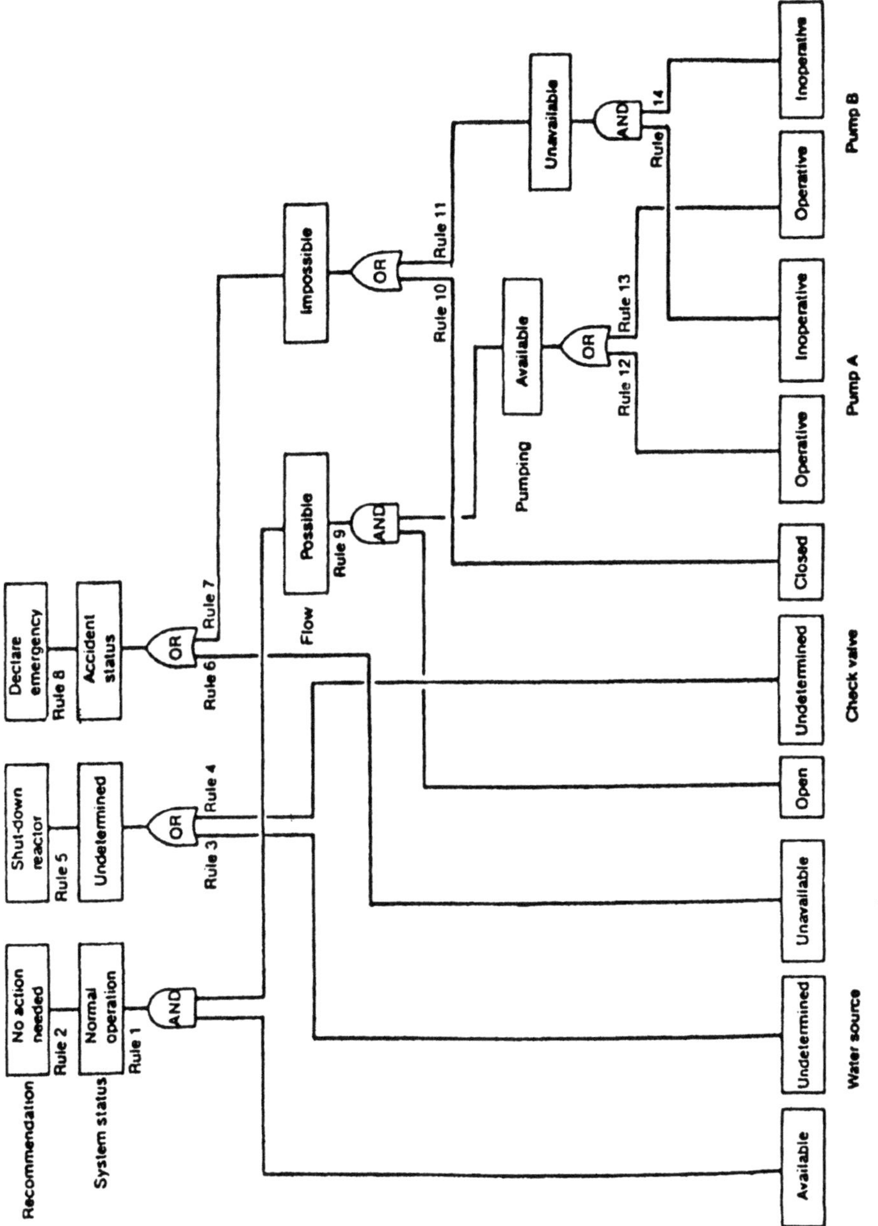

Fig. 3 Logical structure of Knowledge-Base describing physical system

```
**************************************************************
Hello.  I'm HAL, a general-purpose inference engine.

I can use either Forward-Chaining or Backward-chaining
techniques (your choice) to operate on the Knowledge Base
you give me in order to make conclusions.

Choose the Knowledge Base to use
     1. COOLANT
     2. OCONEE
Please enter the number of your response: 3 lisp> 1

Choose which technique you'd like HAL to use
     1. FORWARD-CHAIN
     2. BACKWARD-CHAIN
Please enter the number of your response: 3 lisp> 1

Choose the input data set that you'd like to be used
     1. COOLANT-SAMPLE-INPUT-DATA-1
     2. COOLANT-SAMPLE-INPUT-DATA-2
     3. COOLANT-SAMPLE-INPUT-DATA-3
     4. COOLANT-SAMPLE-INPUT-DATA-4
     5. COOLANT-SAMPLE-INPUT-DATA-5
     6. COOLANT-SAMPLE-INPUT-DATA-6
     7. COOLANT-SAMPLE-INPUT-DATA-7
Please enter the number of your response: 3 lisp> 1

********* HAL is now Forward-Chaining *****************
SIMPLIFIED NUCLEAR REACTOR COOLANT SYSTEM MODEL

 I have been able to induce that
RECOMMENDATION IS NO-ACTION-NEEDED

********* HAL has completed Forward-Chaining **********

Would you like to see the accumulated facts?
     1. YES
     2. NO
Please enter the number of your response: 3 lisp> 1

Here are the accumulated facts:
PUMP-B IS OPERATIVE
PUMP-A IS OPERATIVE
CHECK-VALVE IS OPEN
WATER-SOURCE IS AVAILABLE
PUMPING IS AVAILABLE
FLOW IS POSSIBLE
SYSTEM-STATUS IS NORMAL-OPERATION
RECOMMENDATION IS NO-ACTION-NEEDED
```

Fig. 4 Example of the Forward-chaining procedure in HAL-1987

```
********* HAL is now Backward-Chaining *****************
SIMPLIFIED NUCLEAR REACTOR COOLANT SYSTEM MODEL

WATER-SOURCE IS
        1. AVAILABLE
        2. UNAVAILABLE
        3. UNDETERMINED
Please enter the number of your response: 2 lisp) 1

CHECK-VALVE IS
        1. OPEN
        2. UNDETERMINED
        3. CLOSED
Please enter the number of your response: 2 lisp) 1

PUMP-A IS
        1. OPERATIVE
        2. INOPERATIVE
Please enter the number of your response: 2 lisp) 2

PUMP-B IS
        1. OPERATIVE
        2. INOPERATIVE
Please enter the number of your response: 2 lisp) 2

I HAVE DETERMINED THAT FLOW IS IMPOSSIBLE

I have been able to deduce that
RECOMMENDATION IS DECLARE-EMERGENCY

********* HAL has completed Backward-Chaining *********

Would you like to see the accumulated facts?
        1. YES
        2. NO
Please enter the number of your response: 2 lisp) 1

Here are the accumulated facts:
WATER-SOURCE IS AVAILABLE
CHECK-VALVE IS OPEN
PUMP-A IS INOPERATIVE
PUMP-B IS INOPERATIVE
PUMPING IS UNAVAILABLE
FLOW IS IMPOSSIBLE
SYSTEM-STATUS IS ACCIDENT-STATUS
RECOMMENDATION IS DECLARE-EMERGENCY
```

Fig. 5 Example of the Backward-chaining procedure in HAL-1987

SOPHIE: AN EXPERT SYSTEM FOR THE SELECTION OF HAZARD EVALUATION PROCEDURES

Stephen L. Nicolosi

Battelle
Columbus, Ohio

SOPHIE (Selection Of Procedures for Hazard Identification and Evaluation) was developed at Battelle-Columbus to assists users in the chemical industry select hazard evaluation procedures. The intent of SOPHIE is to provide managers with guidance on the reasonableness of selected hazard evaluation procedures for a given situation. Managers can use this guidance to assess whether an approach to hazard evaluation implemented by their staff will have high efficiency and effectiveness or whether a reevaluation of approach is desirable. In some cases, the methods selected by an expert hazard evaluation staff may not comply with those methods suggested by the expert system. In such cases, more efficient methods may have been overlooked, or there may be particular circumstances regarding the application at hand which require a different approach. In the latter case, the manager should be made aware of these special circumstances early in the analysis. Application of the expert system, thereby assists the manager and technical staff as a quality assurance mentor for methods selection.

SOPHIE is an expert system which is narrowly defined to perform a specialized function. It is not expected that SOPHIE will replace experts nor was it developed as a guide for experts. However, experts can use SOPHIE as a means to review their own approach for completeness and to help assure that they have not overlooked another approach which may provide specific advantages.

Applications of expert systems in the nuclear industry may also provide useful assistance as a quality assurance tool. In applications such as control rooms, personnel already are the experts, thereby not requiring the assistance of an expert system. Under selected conditions, however, when the plant is not behaving as expected, or other high stress situations, it may be desirable to have a knowledgeable person query an appropriate expert system. The purpose of this query is to provide a quality assurance measure. The expert system may report a situation assessment and recommended actions

which are identical to those chosen by the control room
operators. In other cases, the suggested situation or
actions may differ from those chosen by the actual human
experts. If the expert system recommendation is in
compliance with the assessment of the operators, it
provides them with a measure of assurance that their
assessment may be correct. If the recommendation of the
expert system is not in compliance with the assessments of
the operators, then the knowledgeable operator using the
expert system will assess the reasonableness of its
recommendations. If these recommendations are reasonable,
the operators have addition ideas to fold into their
evaluation of the situation and their response. In some
serious and high stress situations, it may be desirable
for a knowledgeable operator to perform this quality
assurance exercise, providing an appropriate expert system
is available.

SOPHIE is constructed as a rule based expert system.
It contains 124 rules which are implemented and processed
using the EXSYS expert system shell (EXSYS is a registered
trademark of EXSYS, Inc., Albuquerque, NM). To assure a
near minimum number of user directed queries, and to
reduce the likelihood of questionable responses, the
knowledge base was constructed using a decision tree
structure. This made the rules large through repetition
of several qualifiers throughout the rule base. This
approach can also make the expert system hard to modify to
account for new approaches or methods unless these can be
accomodated through extensions of the existing decision
tree structure. These disadvantages are offset however
through knowledge that operation of the expert system is
entirely predictable and that for criteria falling within
the knowledge base, the recommendations are reasonable.
Possibilities of errors on the part of the expert system
are thereby reduced. If the expert system is in error for
a given situation, it will most likely be due to the
situation evaluated lying outside the knowledge base.

Suggestions made by SOPHIE, are weighted on a 0 to 10
scale to provide a measure of their expected
applicability. If the method of choice by the user
receives a high recommendation by the expert system, then
compliance exists with the knowledge base. If there is a
disparity, then the user should assess whether the
suggested method is more efficient, whether there are
reasons out of convenience for the difference, or whether
the situation has not been adequately assessed by the
expert system.

The present version of SOPHIE addresses qualitative
hazard evaluation methods. Provision exists for extension
of this system to include quantitative methods in the
future. The current version provides recommendations of
the following methods: Preliminary Hazard Analysis,
Process/System Checklists, Safety Review, Relative Ranking
(such as Dow and Mond), What-If, Hazard and Operability
Study (HAZOP), Failure Modes and Effects Analysis (FMEA),
Failure Modes, Effects, and Criticality Analysis (FMECA),
Fault Tree Analysis, Event Tree Analysis, Cause-
Consequence Analysis, and Human Error Analysis.

ADVANCED TECHNOLOGY FOR NUCLEAR POWERPLANTS

Howard H. Rohm

Office of Nuclear Energy
U.S. Department of Energy
Washington, D.C.

INTRODUCTION

Advanced technologies have had a profound impact on our life and
work place. Advanced technologies include computers and related
software development, human factors engineering, systems analysis and
analytical techniques, and improved quantitative management processes.
New techniques, made possible by modern computer hardware and software
applications, are becoming indispensable for providing improved methods
to assure safe, reliable, and less costly means of accomplishing many
tasks. Advanced technologies are ideally suited for improving the
performance of new and existing nuclear powerplants.

During the early development of nuclear power, extensive use was
made of the then "current" advanced technology (i.e., large mainframe
computers) for activities such as core design and fuel management.
Conventional technology, such as analogue gauges, was used because of
the large experience base and operator acceptance. Most powerplants in
the U.S. were ordered prior to 1974, at a time when advanced technology
approaches had not been widely adopted. The long lead times required
to design and construct powerplants (and the very costly retrofitting
process) precluded using new technological advances that were developed
after the initial design was complete. Regulatory and financial
considerations also made it difficult to incorporate new technology and
plant modifications that were not required by the Nuclear Regulatory
Commission (NRC).

As a result of the accident at Three-Mile Island, several advanced
technology applications were imposed on the industry by NRC (1). These
included incorporating human factors engineering in control rooms,
utilizing plant process computer technology and color-graphic displays
for Safety Parameter Display Systems, and developing integrated
emergency operating procedures for transients and accidents (2). To
date, responses to these requirements have been implemented in most of
the operating nuclear power plants in the country. Some of the systems
are outstanding and have made a significant contribution to improving
plant safety and operating efficiencies. However, most of the TMI-2
requirements are still being refined and have the potential for
significant benefits. These systems represent a large investment that
the utilities have already made in utilizing advanced technology to

meet NRC requirements. The first steps have been taken; however, the
promise of improved plant performance has yet to be fulfilled.

The next section reviews objectives of advanced technology in
nuclear power plants. Section 3 highlights the scope of advanced
technology programs. Examples of DOE's and other organizations' and
countries' advanced technology programs are given in Section 4.
Finally, Section 5 summarizes benefits from utilizing advanced
technology in new nuclear plant design and construction.

OBJECTIVES OF ADVANCED TECHNOLOGY IN NUCLEAR POWER PLANTS

Advanced technologies hold the promise to reduce cost, increase
productivity, and enhance safety in nuclear power plants. Utilizing
improved methods can reduce costs associated with inefficient and
non-productive or non-critical operations. As better techniques are
incorporated in plant operations, existing requirements can be reduced.
Advanced technologies can provide the means of monitoring plant
performance and optimizing plant requirements.

Advanced technology can also provide methods to ensure better
availability/reliability of plant systems and components. This will
increase plant productivity and provide extended plant life through
increased quality control and assurance. In addition, advanced
technology can provide significant safety enhancements to enable
operators to understand plant conditions and respond efficiently and
effectively to plant upset conditions. Such understanding of safety
parameters can aid in adding much needed confidence to the regulatory
process by assuring safe operations. Finally, incorporating advanced
technology in nuclear powerplants improves the competitive ability of
the U.S. nuclear industry. This includes improving the general climate
and public acceptance for new nuclear power plants within the U.S. and
abroad.

AREAS OF ADVANCED TECHNOLOGY

Several advanced technology areas have been identified that could
enhance performance of current plant processes. These areas include:

o   Artificial Intelligence - Expert Systems
o   Robotics
o   Sensors
o   Modularization (Manufacturing and Maintenance)
o   Other Technologic Advances.

The following section briefly identifies these technologies and
indicates their relationship to nuclear power.

Artificial Intelligence (AI) - Expert Systems

AI encompasses the development of automated knowledge-based
systems to enhance decision-making capabilities for the design,
construction, operation, maintenance and management of nuclear
powerplants. AI provides the capability of capturing expert skills and
experience and making them available upon demand under any conditions.
This technology enhances and complements normal operation capabilities
by emulating the human reasoning process and interjecting expert
opinion into the analysis of problem situations.

Some examples of AI applications that have been proposed for
nuclear plants include: fuel-loading, construction management decision
support, module control, integrated control, automated reasoning
verification, automated intelligent diagnostics, maintenance
management, and an automated reasoning tool kit to provide a common
development environment for efficient expert system development (3).

In addition, AI has the potential to reduce nuclear life cycle
costs through efforts such as integrated knowledge base; management
decision support systems; designer verification, validation, and
integration methods; and integrated plant automation systems (3).

Robotics

Robotics includes systems designed to move materials, parts, tools
or specialized devices through a variety of motions to perform a
variety of tasks. The systems are generally programmable and use
flexible manipulators. In addition, they can be remotely operated and
controlled. Potential benefits of robotic systems include reduction in
plant downtime through maintenance and inspection systems, and
reductions in personnel exposures (e.g., radiation and hazardous
materials) through remotely operated devices. The techniques are
applicable throughout the powerplant life cycle, including:
construction, operations, maintenance, fuel fabrication and handling,
reprocessing, waste handling and decommissioning (4).

Sensors

Advanced technology can also significantly improve existing sensor
systems within nuclear power plants. Sensor technology is required to
measure the large variety of parameters (e.g., temperature, pressure,
flow and fluid level) needed to determine component, system and
functional performance. Sensors are directly coupled to the plant
process computer and its ability to quantify and compare acceptable
parametric values. Improved sensors are needed to meet the high level
of accuracy and speed required of the control systems that will be run
by the process computers. This information could be used by operators
or by AI systems and robotics systems to serve real-time operational
needs. Thus, costs can be reduced and performance enhanced for the
fabrication, operation, maintenance, and inspection processes. In
addition, improved sensors can be expected to extend plant life by
performing continuous on-line monitoring of conditions such as water
chemistry and radiation concentrations. Improvements in sensors can
also substantially improve leak detection and in-service inspection
capabilities (5).

Modularization (Manufacturing and Maintenance)

Future development of nuclear powerplants will depend up on the
ability of the industry to assure public health and safety and to
control costs of construction and operation. Modularization involves
segmenting a process into optimum blocks of activities to maximize
efficiencies and minimize costs. Modularization can provide the means
to meet both of these requirements by developing designs that are
optimized to prove their safety capability and that are small enough so
that large portions of the plant can be constructed in offsite
facilities where more predictable schedules and costs can be
maintained. Concepts such as readiness reviews that involve regulatory
approvals at early stages of construction are also effective methods to
assure that cost and schedule forecasts will be met. Modularization

should reduce costs in construction, maintenance, life-extension, and decommissioning of nuclear powerplants (6).

Reliability assurance programs can also significantly enhance safety and assure continued operation. These programs utilize plant risk models and provide continuous monitoring and corrective actions to assure the plant constantly maintains a maintenance standard and level of reliability/availability (7).

Other Technological Advances

Significant effort will be needed to assure industry acceptance of advanced technologies. Resources will be needed to provide testing, qualification, and computer aided training activities to accelerate technology transfer and to secure NRC approvals. This need could perhaps be the most critical in assuring that advanced technology will be utilized in future nuclear power activities.

Because of the extensive use of computers, software management will also become a major issue. Improved configuration management, validation and verification, and software quality assurance techniques will be needed to gain confidence in the software applications.

EXAMPLES OF ADVANCED TECHNOLOGY PROGRAMS

At present, DOE, EPRI and nuclear industry groups have established programs for developing relevant advanced technology applications. EPRI has sponsored programs in artificial intelligence and advanced computer concepts. Among these are a system to aid operators in recommending emergency action levels (8); an AI based alarm system that analyzes and presets reactor conditions (9); and an anticipated transient without scram (ATWS) problem analyzer for a boiling water reactor (10). An automated procedures tracking system is being developed and will be utilized in a simulator test next year. As reported during the 1985 conference at Pasco (11), a continuing DOE-Nuclear Energy (NE) activity is demonstration and evaluation of advanced control concepts and man-machine integration in conjunction with Experimental Breeder Reactor (EBR-II) and Fast Flux Test Facility (FFTF) operations. At FFTF, a master information and data acquisition system (MIDAS) has been implemented to track plant maintenance jobs. An automated fuel failure diagnostician has also been developed. At EBR-II, a trial application of DISYS ("Diagnostic System") and tests of advanced graphics, using on-line input signals, are being developed. A number of industry groups are also developing systems for analyzing pump failures, performing preventative maintenance, and evaluating reactor trips (12), (13), (14).

In the area of robotics, DOE has provided support to four universities, Florida, Michigan, Tennessee, and Texas, and to the Oak Ridge National Laboratory to pursue research leading to the development and deployment of an advanced robotic system. The system will be capable of performing tasks that are hazardous to humans (e.g., those that generate significant occupational radiation exposure), and/or tasks whose execution times can be reduced if performed by an automated system. The goal is to develop a generation of advanced robotic systems capable of performing surveillance, maintenance, and repair tasks in nuclear facilities, and other hazardous environments.

The strategy adopted in order to achieve the program goal in an efficient and timely manner consists of utilizing and advancing

state-of-the-art robotics technology through close interaction between
the universities and the manufacturers and operators of nuclear power
plants. There is a potentially broad range of applications for the
robotic systems developed in the course of this project. The approach
to achieving the program objective is a transition from teleoperation
to the capability of autonomous operation within three successive
generations of robotic systems.

Advanced programs are receiving more attention in developed
countries. France, Germany, and Japan have extensive research and
development activities underway to utilize advanced technologies in
their nuclear power plants. Government organizations and industry
groups are actively developing new applications. However, very few of
these projects have actually been incorporated into operating
facilities.

Perhaps the most extensive use of advanced technologies for
nuclear powerplants has been undertaken under the auspices of the
Organization for Economic Cooperation and Development (OECD) Nuclear
Energy Agency's Halden Reactor Project (15). In cooperation with the
U.S. (DOE, EPRI, and NRC) and the major western developed countries,
the Halden Reactor Project has adapted numerous advanced applications
to provide a test facility for computerized man-machine communication.
Within the framework of a simulator laboratory (NORS), more than 10
application packages have been or are being developed for the
following: integrated surveillance and control systems, early fault
detector and diagnosis capabilities, computerized procedures systems,
advising systems, emergency management, software reliability, and
computer structures and data communications. Table 1 lists the
specific packages that are in use or under development.

Another activity sponsored by DOE is the advanced control and
instrumentation program at Oak Ridge National Laboratory. The program
will support the design of automated control, monitoring, and
diagnostic systems for advanced reactor powerplants. A major thrust
of the program is to utilize the most advanced computer software
techniques to assure modular, reconfigurable, and transportable
real-time applications. This project and related work provides great
promise for the development of advanced technology applications to
nuclear power.

Table 1 - Artificial Intelligence Applications Packages
Developed for the Halden Reactor Project

Core Surveillance System (SCORPIO)
Success Path Monitoring System (SPMS)
Computer Aided Instruction (CAI)
Critical Function Monitoring System (HALO)
Integrated Surveillance and Control System (ISACS)
Fault diagnosis systems (DISKET-JAERI)
Computerized Procedures System (COPMA)
Project on Diverse Software (PODS)
Software Testing and Evaluation Methods (STEM)
Software Safety Tools (SOSAT)
Computer Safety Tools (COSAT) [to be developed]

The table lists advanced technology projects
underway at the Halden Reactor Project

SUMMARY

Advanced technology offers significant potential benefit to the nuclear industry. Improvements can be anticipated in plant performance, reliability, and overall plant safety as well as reduced life cycle costs. Utilizing artificial intelligence and expert systems, robotics, advanced instruments and controls, and modularization technologies can enhance plant operations and provide new insights and perspectives to plant risk and thus focus resources to areas of importance. Plant reliability, operability, availability, accident interdiction and limitation, and plant recovery are expected to improve.

However, utilizing these technologies is not an automatic process. In addition to the actual costs associated with developing and implementing the technologies, operator training and acceptance represents a potential significant problem. Traditional plant operators have little or no experience with computer technology. There has already been some difficulty getting nuclear plant operators to accept and use the new technologies that have been implemented thus far. The promise for significant nuclear powerplant improvement from advanced technology is at hand. The processes to develop and effectively implement these improvements are going to require effort by the technical and regulatory arms of government to work together with industry.

REFERENCES

1. NUREG-0737, "Clarification of TMI Action Plan Requirements," U.S. Nuclear Regulatory Commission, November 1980.
2. NUREG-0737, Supplement 1, "Requirements for Emergency Response Capability," Generic Letter No. 82-33, U.S. Nuclear Regulatory Commission, December 1982.
3. DOE/NE-0064, "Artificial Intelligence and Nuclear Power," June 1985.
4. DOE/NE-0065, "Robotics and Nuclear Power," June 1985.
5. DOE/NE-0066, "Sensors and Nuclear Power," June 1985.
6. DOE/NE-0067, "Modularization and Nuclear Power," June 1985.
7. M. A. Azam and E.V. Lofgren, "Effectiveness of Reliability Technology Applicable to LWR Operational Safety, NUREG/CR-4618, April 30, 1986.
8. R. Touchton, D. Genter, "Emergency Action Level Classification System," paper presented at EPRI workshop on "Artificial Intelligence Applications to Nuclear Power," Palo Alto, California, June 19 to 21, 1985.
9. MPR Associates, Power Plant Alarm Systems: A Survey and Recommended Approach for Evaluating Improvements, EPRI NP-4361, December 1985.
10. D.G. Cain, "BWR Shutdown Analyzer Using Certified Intelligence Techniques," EPRI NP-4139 Special Report, July 1985.
11. A. P. D'Zmura and S. E. Seeman, "The U.S. DOE Man-Machine Integration Program for Liquid Metal Reactors" in Proceedings of an International Topical Meeting on Computer Applications for Nuclear Power Plant Operation and Control, Pasco, WA, September 1985.
12. M. Rooney, "Expert Pump Diagnostic Systems," EPRI workshop on "Artificial Intelligence Applications to Nuclear Power."
13. R. Christopherson, "Maintenance Advisor," EPRI workshop on "Artificial Intelligence Applications to Nuclear Power."

14. M.W. Frank and A. Epstein, "Application of AI to Improve Plant
     Availability," Society of Computer Simulation, Simulation
     Series 17, pg. 92, January 1986.
15. OECD-NEA, Halden Reactor Project Programme, Institutt for
     Energiteknikk, Norway, November 1986.

EFFICIENT DIAGNOSTIC SEARCH VIA INFORMATION ENTROPY REDUCTION

Eugene L. Filshtein

Combustion Engineering, Inc.
1000 Prospect Hill Road
Windsor, CT. 06095
U.S.A.

ABSTRACT

This paper presents a methodology and algorithm that optimize a diagnostic
search in terms of the expected time required to reach the conclusion, i.e.,
to find a failed component. It is assumed that the fault tree describing the
relationship among the events is available. The procedure allows one to
locate the best candidate among the fault tree events for information
acquisition in the following sense. Once the event state is ascertained and
this knowledge is propagated through the fault tree, the uncertainty as to
which component is in a failed state is minimized. This procedure is based on
expected information entropy reduction and, when recurrently applied, should
optimize a diagnostic search. The entropy reduction algorithm proved to be
time-consuming for large scale systems. In order to make the algorithm
suitable for real time applications, the knowledge structure had to be
changed. A Transformed Fault Tree (TFT) knowledge structure was introduced.
It allows almost instantaneous assessment of fault tree events regarding
expected entropy reduction.

The above methodology is implemented in a man-machine interface module of
Combustion Engineering's Generic Diagnostic System.

INTRODUCTION

Suppose that the top event of the fault tree is in a failed state and our
objective is to locate a failed component that caused the top event to occur
by sequentially ascertaining states of the tree events using available
instrumentation. Once an event's state is determined, this knowledge can be
propagated through the fault tree, e.g., if the event state is determined as
"operational", then all the inputs to this event connected by an OR gate must
be operational.

Unless the diagnosis is reached, the process will continue by "interrogating"
another event, etc. It is intuitively clear that the duration of the
diagnostic search is affected by the sequence in which information is
acquired. As an illustration, compare the _expected_ duration to diagnose a car
problem by an experienced mechanic versus an educated novice.

Our objective is to find an optimal search strategy that would minimize the
time required to complete the diagnosis. Let us first assume that time
requirements for information acquisition regarding states of the events are
identical. Then an optimal strategy is the one which minimizes the expected
number of inquiries required to complete the search.

It is useful to look at the diagnosis as a stepwise search procedure with each
step being an information acquisition that increases the knowledge about the
state of the system under diagnosis or, which is the same, reduces the
uncertainty with regard to the system's state. Note that, when the states of
all the events comprising the system are known, the diagnosis is complete. An
optimal search strategy is such a sequence of inquiries that results in the
fastest reduction of uncertainty regarding the system's state.

<u>Entropy as a Measure of Uncertainty</u>

First of all, one has to quantify the term "uncertainty" for a given
application. To make the matter simpler, suppose that only one out of n basic
events (components) is in a failed state. Then one can use the Information
Entropy as a measure of uncertainty:

$$H = -\sum_{i=1}^{n} p_i \log_2 p_i , \quad \sum_{i=1}^{n} p_i = 1 \tag{1}$$

where p_i = component failure probabilities. Just because Equation 1 considers
only basic events (components), one should not conclude that the fault tree
structure has no impact on the entropy value. Component failure
probabilities, p_i, are <u>conditional</u> probabilities. They depend on failure
rates of basic events, the fault tree structure and the fact that the top
event is in a failed state. Note that entropy, H, assumes its largest value
when all components have equal failure probability, $1/n$; then $H = \log_2 n$.
Indeed, when all failures are equally likely, one cannot even have preferences
in a search for a failed component and, therefore, his uncertainty is the
largest. When the diagnosis is complete, then all failure probabilities but
one are equal to zero, and this makes $H = 0$, i.e., uncertainty is equal to
zero and our knowledge regarding the system's state is absolute.

It seems logical to define a criterion for an optimal diagnostic procedure as
the system's entropy reduction. At each step of a diagnostic search, one has
to make inquiries that result in maximum reduction of the system's entropy.

Suppose that a fault tree has a number of basic and intermediate identifiable
events, i.e., those events whose state can be determined using available
instrumentation. Then a straightforward procedure to optimize a diagnostic
search is described as follows:

1. Consider all identifiable events as candidates for information
    acquisition and calculate failure probability for each of those events to
    be in a failed state, $\Pr [E_s(+)]$, given that the top event takes place.

2. For a candidate, E_s, consider two hypothesis regarding its state:

        $E_s(+)$, which means that E_s is in a failed state, and $E_s(-)$, which
        means that E_s is operational

    Under each hypothesis, reduce the fault tree by propagating the status
    of E_s through the fault tree gates.

3. Under each hypothesis, calculate basic component failure probabilities and two conditional entropies, $H/E_s(+)$ and $H/E_s(-)$, using Equation 1.

4. Calculate the <u>expected</u> entropy given that inquiry S is made;

$$E(H_s) = \Pr\{E_s(+)\} \cdot H/E_s(+) + \{1 - \Pr(E_s(+)\} \cdot H/E_s(-) \quad \ldots \qquad (2)$$

5. Among all candidates select the one which corresponds to the minimal expected entropy value as the best candidate for information acquisition.

To evaluate failure probabilities for the fault tree events, one has to start with failure rates for the basic components. There are several conceivable sources of information for an assessment of basic component failure rates:

- manufacturer's specifications
- past observations of failures
- expert opinion
- assuming them all being equal (only ratios of the failure rates are important)

Any of the above assessments will suffice as long as it represents our knowledge of the hardware environment to the best extent possible.

We will evaluate failure rates of all intermediate events in an "upward" run (from the basic components to the top event) and then failure probabilities of the basic components for Equation 1 in a "downward" run (from the top event to basic components).

<u>Failure Rate Propagation - Upward Run</u>

The following relationships exist between the failure rate of a gate, α_g, and its input failure rates, α_i.

OR-gate: $$\alpha_g = \sum_{i=1}^{k} \alpha_i \qquad (3)$$

AND-gate:

A solution in a closed form for k inputs cannot be provided.

For k = 2
$$\alpha_g = \frac{1}{\dfrac{1}{\alpha_1} + \dfrac{1}{\alpha_2} - \dfrac{1}{\alpha_1 + \alpha_2}} \qquad (4)$$

For k = 3
$$\alpha_g = \frac{1}{\dfrac{1}{\alpha_1} + \dfrac{1}{\alpha_2} + \dfrac{1}{\alpha_3} - \dfrac{1}{\alpha_1 + \alpha_2} - \dfrac{1}{\alpha_2 + \alpha_3} - \dfrac{1}{\alpha_1 + \alpha_3} + \dfrac{1}{\alpha_1 + \alpha_2 + \alpha_3}} \qquad (5)$$

For k > 3 a close approximation is recommended: $\qquad (6)$

$$\alpha_g = \frac{\sum_{i=1}^{k} \alpha_i}{k(1 + \frac{1}{2} + \frac{1}{3} + \ldots + \frac{1}{k})}$$

By recurrently applying Equations 3-6, one can evaluate failure rates for all intermediate events of a fault tree in the "upward" run.

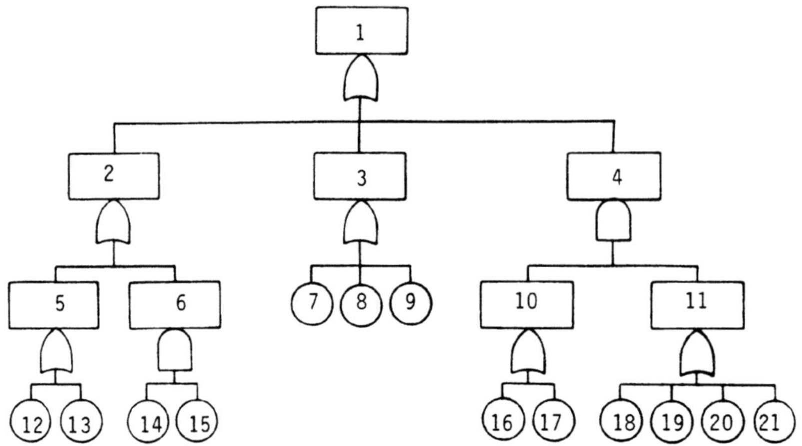

Figure 1A. Fault-Tree (Example)

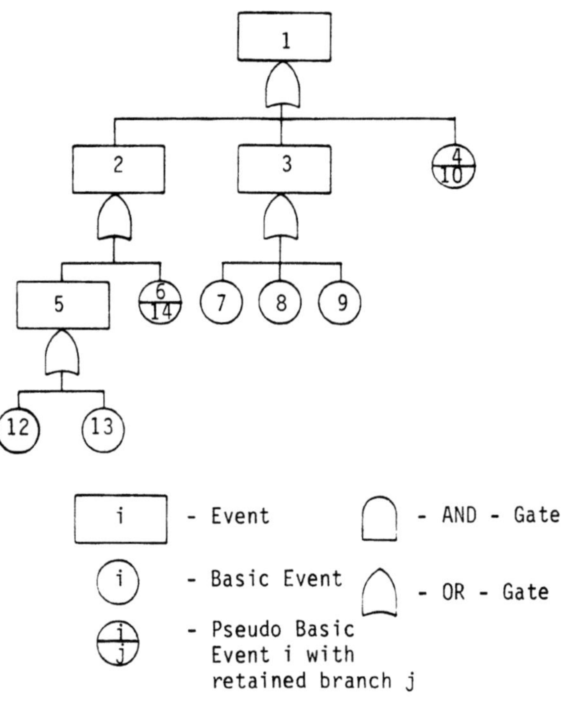

Figure 1B. Transformed Fault-Tree

Propagation of Failure Probabilities - Downward Run

To calculate failure probabilities for the fault tree events, the relationship between a single output event, E_m, and its inputs, E_j^m should be described. For AND-gates the relationship is trivial:

$$Pr\{E_j^m (+)\} = Pr\{Em(+)\},... \tag{7}$$

One can show that for OR-gates the following relationship exists:

$$Pr\{_j^m (+)\} = Pr\{Em(+)\} \frac{\alpha_j}{\sum\limits_{i=1}^{k} \alpha_i} \tag{8}$$

Since the top event, E_1, is known to be in a failed state, i.e., $Pr[E_1(+)]=1$, we can carry out a "downward" run by recurrently applying Equations 7 and 8 until failure probabilities for the all tree events, including basic, are evaluated. Note that all those probabilities are, in effect, conditional upon the top event taking place, so they should be understood as $Pr[E_i(+)/E_1(+)]$.

Transformed Fault Tree

An optimal search strategy based on a principle of maximum entropy reduction is described above. This strategy, when implemented, should result in a minimal number of inquiries required to carry out the diagnosis. It is anticipated, however, that a straightforward application of the search procedure would be too time consuming for a real life situation. This procedure must be repeated twice as many times as there are identifiable events on the tree for a single step in a diagnostic search. There is very little hope that by applying some kind of optimization it is possible to reduce significantly the number of identifiable events to be tested without running a risk to miss a winner. Thus, one has to substantially reduce the time required to evaluate each of the candidates. To accomplish this, we have to change the knowledge structure and to introduce a Transformed Fault Tree (TFT) that allows almost instanteous assessment of a tree's events regarding the expected entropy reduction.

First, let us consider a regular fault tree such as the one on Figure 1a. Suppose that the top event, E_1, is in a failed state, $E_1(+)$, and all the event failure rates and conditional failure probabilities, $Pr\{E_1(+)\}$, are evaluated using appropriate formulas. Note that the unconditional failure probabilities are negligibly small, since each of them is equal to the probability that a given event takes place in a given instant of time. For example, unconditional $Pr\{E_9(+)\}$ is equal to the probability that the plant will be in a state of emergency, say at 7:30 p.m., October 15, 1992, due to the failure of component 9. These unconditional probabilities can be evaluated, but for all practical purposes, we can assume that they are equal to zero.

Suppose that a state of one of the events, e.g., event 3 is ascertained. If no failure is detected, $E_3(-)$, then the branch where E_3 is a top event should be eliminated; thus, all conditional failure probabilities for the basic events E_7, E_8, and E_9 are equal to zero.

If E_3 is in a failed state, $E_3(+)$, then it can be shown that for all the events that are not included in this branch, their conditional failure probabilities are equal to zero. This is because the failure of the top event, $E_1(+)$, is already explained by $E_3(+)$.

In general, the above conclusions will hold for any fault tree that has only an OR-gate structure.

AND-gate structures constitute some problems regarding interpretation of conditional probabilities and evaluation of the entropy. The problem will manifest itself in the fact that the sum of basic event conditional failure probabilities will <u>not</u> add up to unity when the tree contains AND-gates, but will always exceed it. This is caused by the fact that some failure scenarios are inconsistent with the top event failed state, $E_1(+)$. For example, a scenario where basic event 20 is in a failed state and the rest are operational is impossible. Thus, one has to consider <u>joint</u> outcomes, e.g., $E_{20}(+) \wedge E_{16}(+)$ or $E_{19}(+) \wedge E_{17}(+)$, etc., which would significantly complicate entropy evaluation. To avoid a burden, we will introduce a psuedo-basic event, which substitutes as an AND-gate. A pseudo-event, E_p, has the following properties:

1. Its failure rate is evaluated as for the AND-gate using Equations 4, 5 or 6.

2. If it is determined that $E_p(-)$, then the event is eliminated as for an OR-gate structure.

3. If $E_p(+)$, then the events are replaced by a single input branch E_i^p, which has the smallest entropy value associated with it. Indeed, the knowledge that was obtained, $E_p(+)$, indicates that all the input branches of this event, contain failed components. But, <u>given this knowledge</u> we also have established the cause-effect relationship between E_p and a failure within <u>any single</u> input branch of E_p because, once this failure is found and eliminated, E_p will become operational. This implies that all but one input branch of an AND-gate can be eliminated from the search, and the question is which one of them should be retained. Probably, the one which is easier to diagnose, i.e., the one which has the smallest entropy value associated with it should be retained. Note that the entropies for those input branches can be evaluated just once, i.e., this is pre-processed information.

Now we can formulate the properties of a (TFT) - (see Figure 1b):

1. It is only an OR-gate tree.

2. Elements of the tree are events, basic events and pseudo-basic events.

3. The top event of this tree is always considered to be in a failed state, $E_1(+)$.

4. We will use the following symbols to describe the cause-effect relationship between the events that is useful for diagnostic purposes.

    $E_i \rightarrow E_j$ if a failure of E_j is explainable by the failure of E_i, and

    $E_i \nrightarrow E_j$ otherwise; For example, $E_{12} \rightarrow E_2$ and $E_{13} \nrightarrow E_3$ (see Fig. 1b).

It is important to add another condition for the relationship $E_i \rightarrow E_j$ to hold:

$Pr[E_i(+)] \neq 0$, i.e., failure of E_1 is not unconditional.

Suppose E_j is in a failed state, $E_j(+)$.

Then: $\quad e_i \rightarrow E_1 \quad$ if $e_i \rightarrow E_j$

$\qquad\quad e_s \nrightarrow E_1 \quad$ if $e_s \nrightarrow E_j$ $\hfill (9)$

Suppose E_j is operational, $E_j(-)$

Then: $\quad e_i \nrightarrow E_1 \quad$ if $e_s \rightarrow E_j$

$\qquad\quad e_s \rightarrow E_1 \quad$ if $e_s \nrightarrow E_j$ $\hfill (10)$

where e - basic and pseudo-basic events.

The relationships (9) and (10) allow effective pruning of the TFT once the state of a tree event is ascertained.

5. It follows from Equations 3 and 8 that, for any intermediate event, E_j, and a complete and exhaustive set of e_i with $e_i \rightarrow E_j$

$$\alpha_j = \Sigma \alpha_i 1$$

and $\quad Pr\{e_i(+)/E_j(+)\} = Pr\{e_i(+)\}/Pr\{E_j(+)\}$ $\hfill (11)$

Equations 11 allow one to reevaluate component failure probabilities in a diagnosistic process.

6. Now we will introduce a TFT property which is of major interest for our purposes.

Suppose one contemplates an inquiry as to the state of event E* which has a failure probability p*. It can be proven that the entropy reduction which is expected to result if an inquiry regarding the state of event E* is carried out is as follows:

$$\Delta^* = - P^* \log_2 P^* - (1-P^*) \log_2 (1-P^*) \qquad (12)$$

regardless of the number of components to be diagnosed.

This result is rather surprising. What it says is that the expected entropy reduction that would be attained by the inquiry regarding the state of a particular event in the TFT is <u>solely determined by its failure probability and does not depend on the tree structure.</u> This property would not hold for a regular fault tree and serves as a justification for utilization of the TFT.
Equation 12 allows one to develop a very effective search strategy. First, we will optimize Δ^* as a function of P^*.

$$\frac{d\Delta^*}{P^*} = \log_2 \frac{(1-P^*)}{P^*} = 0$$

(13)

Thus $\frac{1-P^*}{P^*} = 1$ and $P^*_{opt} = 0.5$

It means that the most effective inquiry would occur if the event failure
probability is equal to 0.5.

It is important that Δ^* as a function of P^* is symmetrical about $p^* = 0.5$
(see Figure 2) because this allows one to formulate the following optimal
search strategy. <u>An optimal search strategy is pursued when an inquiry is
made to the event which has a failure probability closest to 0.5 among all
other candidates.</u> This implies, for example, that an operator would get more
information regarding the system state by ascertaining the state of an event
with a 0.35 failure probability than for the one with a 0.9 failure
probability.

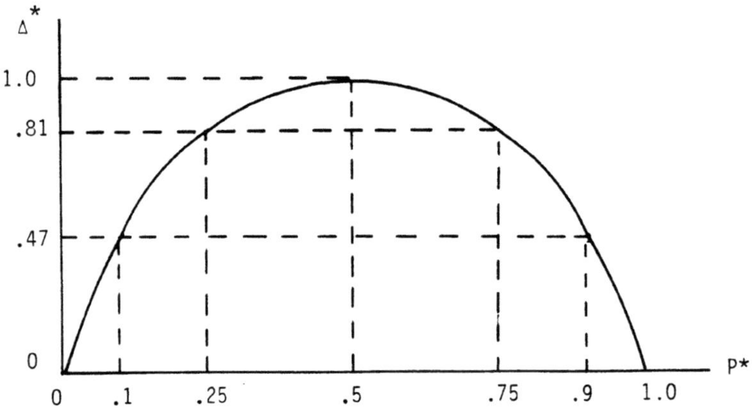

Figure 2 Entropy Reduction as a Function
of Failure Probability

An optimal search strategy described above is valid under the assumption that
all inquiries are equally time-consuming, which could be justifiable if
inquiries are made by interrogating the instrumentation panel. In some cases,
however, time requirements for information acquisition can differ
significantly.

Consider situations where some instruments are physically attached to components or where one has to run a test to place an inquiry. Suppose that each identifiable event, E_i, has its own time requirement, t_i.

To optimize a search strategy we will introduce the rate of information recovery, R, as a quantity that evaluates the entropy reduction per unit of time:

$$R = \Delta/t \tag{14}$$

where Δ is given by Equation 12

An optimal strategy corresponds to the maximum rate of information recovery:

$$\max_i \ R_i = \max_i \ \frac{\Delta_i}{t_i}$$

Rearranging Equation 14:

$$R = \log_2 \{p^{\frac{p}{t}} (1-p)^{\frac{1-p}{t}}\}^{-1}$$

Logarithm is a monotonic function of its argument, i.e.

if $\log x > \log y$ then $x > y$.

Therefore:

$$\max_i R = \max_i \ \{p^{\frac{p}{t}}(1-p)^{\frac{1-p}{t}}\}^{-1}$$

$$= \min_i \ \sqrt[t]{p^p (1-p)^{1-p}} \tag{15}$$

The quantity $C(p,t) = \sqrt[t]{p^p (1-p)^{1-p}}$ will serve as an optimization criterion. An optimal search strategy can be described as follows:

An optimal search strategy is pursued when an inquiry is made to the event which has the smallest value of the $C(p,t)$ criterion.

The methodology described in this paper is implemented in a man-machine interface module of Combustion Engineering's Generic Diagnostic system. This module is discussed in a paper "An Intelligent Man-Machinde Interface For Emergency Response" by E. Filshtein and P. Rzasa.

THE STRUCTURE OF AN EXPERT SYSTEM TO DIAGNOSE AND SUPPLY

A CORRECTIVE PROCEDURE FOR NUCLEAR POWER PLANT MALFUNCTIONS

B.K. Hajek[1], J.E. Stasenko[1], S. Hashemi[1],
R. Bhatnagar[1], W.F. Punch III[2]
and N. Yamada[3]

[1]Nuclear Engineering Program, Dept. of Mechanical Engineering
[2]The Laboratory for Artificial Intelligence Research (LAIR)
 The Ohio State University, Columbus, Ohio 43210
[3]Energy Research Laboratory, Hitachi Limited, Hitachi, Japan

ABSTRACT

During the past two years, two prototype knowledge based systems
have been developed at The Ohio State University. [1-3] These systems
were the result of collaboration between the Nuclear Engineering Program
and the Laboratory for Artificial Intelligence Research (LAIR). The
first system uses hierarchical classification to diagnose malfunctions of
the coolant system in a General Electric Boiling Water Reactor (BWR).
The second system provides a plan of action, through a process of dynamic
procedure management, to stabilize the plant once an abnormal transient
has occurred.

The objective of this paper is to discuss the structure that has
been designed to integrate the two systems. The combined system will be
capable of informing plant personnel about the nature of malfunctions,
and of supplying to the operator the most direct corrective procedure
available. Two important features of the integrated system are faulty
sensor detection, based on malfunction context and unlike sensor data,
and procedure management based on the initial state of the plant.

Since the two knowledge based systems were developed separately, the
integration has required a separate component currently under
development, the Plant Status Monitoring System (PSMS). The task of PSMS
is to monitor plant parameters in order to detect an abnormal condition
developing within the plant. Based on the nature of the event, PSMS is
capable of directing control to either the procedure management or
diagnosis component.

The integrated system plays only an advisory role, and any suggested
action would be executed by the plant personnel.

INTRODUCTION

Over the past two years, two knowledge based systems have been developed at The Ohio State University through the joint efforts of the Nuclear Engineering Program and the Laboratory For Artificial Intelligence Research (LAIR). These systems are known as the Dynamic Procedure Management System (DPMS) and the Diagnosis and Validation System (DVS).

The DPMS is capable of developing a corrective plan of action for any malfunction that may occur in the coolant system of a Boiling Water Reactor. [1]

The task of DVS [2,3] is to localize the source of a malfunction to a specific component or subsystem of the reactor. DVS also validates any of the data used to make the diagnosis.

An ideal system would be able to perform these functions as a single unit, so the scope of the current project work concerns itself with the task of integrating the two independently operating systems. To achieve the integration, it was necessary to develop two additional components: the Plant Status Monitoring System (PSMS), and an interactive System Database.

The function and structure of PSMS and the System Database are described in this paper. The combined system will be structured as shown in Figure 1.

The integrated system is being designed to interface with a specific existing plant system, namely the GEPAC Plus System designed by General Electric. An alternate data source would be the plant process computer plus a strictly SPDS system such as the General Electric Emergency Response and Information System (ERIS). The objective of using one of

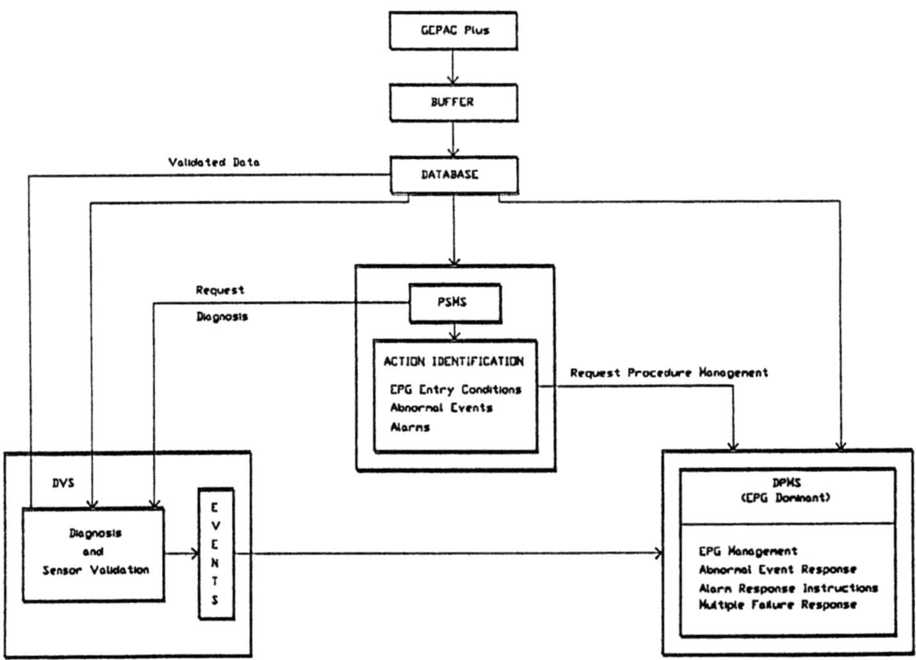

Figure 1. Integrated Operator Aid Expert System.

these two data sources is to provide some pre-processing of the data, to save processing time that otherwise would have to performed by the expert system, and to avoid having to add large numbers of sensors to an already heavily instrumented plant.

Testing of the expert system is being accomplished by using data from the Perry Nuclear Power Plant. Perry procedures, plant statuses, and transient analyses are being used for the knowledge base. The simulator is operated and hard copy of the simulator runs is analyzed. If necessary, the expert system knowledge base is modified to assure the expectations are the same as the simulator results.

DIAGNOSIS AND VALIDATION SYSTEM

The function of the Diagnosis and Validation System (DVS) can be divided into two parts. The diagnosis subcomponent performs the task of identifying the primary cause(s) of an abnormal plant state. The sensor validation subcomponent validates/rejects any suspicious data encountered during diagnosis. Ideally, the identification process will be able to localize the malfunction source down to a component level, where component therapy will be executed.

If suspicious data is encountered, DVS performs a context analysis of this data, determines what the correct data should be, changes the values in the System Database, and runs another analysis to determine the specific malfunction that has occurred. [4]

DVS normally will be run whenever a fault is detected by PSMS.

DVS has been written using a tool known as the Conceptual Structures and Design Language (CSRL). [5] CSRL performs the hierarchical classification process using an establish and refine strategy specific to diagnostic reasoning. It is one of the generic task tool kits developed at The Ohio State University. [6] In the generic tasks approach, the representation and organization of the knowledge follows the function of the task, and the control or inference strategy is specific to it.

DYNAMIC PROCEDURE MANAGEMENT SYSTEM

The Dynamic Procedure Management System contains a library of plans that provide procedural guidance to the reactor operator to recover the plant from a malfunction or to maintain the safety goals within the plant. This component is triggered at the request of PSMS and is able to synthesize a procedure based on the information provided by PSMS. (This component of the expert system was referred to as the Procedure Synthesis System in previous reports.)

Since DPMS was initially designed to operate independently of a diagnosis system, it was necessary for it to perform a limited fault diagnosis. Thus certain abnormal events could be detected by DPMS. This function has been removed from the present version of the program since all detection and initial diagnosis will be done by PSMS in the combined system. However, DPMS will continue to monitor the results of actions taken by the operator to determine if they have been successful in their intent. This will be performed automatically if the data are available from the plant computer. Otherwise, DPMS will wait for the operator to input the data.

DPMS has been reorganized based on the generic tasks approach and coded using a subset of the generic task tool kit known as the Design Specialists and Plans Language (DSPL). [7]

Sometimes, due to insufficient available data, the identification process may be able only to localize the failure source to a plant subsystem. In this case, DPMS maintains one (or more) of the safety goals that would be threatened by the malfunction identified by PSMS or DVS. That is, DPMS will display a procedure that maintains a set of parameters that assure the plant is in a safe condition. If no safety goals are being threatened, then DPMS will wait for additional analysis to be completed by DVS. The assurance that these parameters are in their normal range will assure maintenance of the plant safety goals. [8]

In all cases, if an entry condition to the Emergency Operating Procedures (EOPs) as defined by the Emergency Procedure Guidelines (EPGs) is exceeded, DPMS will default to the appropriate EOP(s) to provide actions to the operator based on the symptoms detected rather than any perceived specific malfunction or event.

PLANT STATUS MONITORING SYSTEM

The main component involved in the integration of the two previously developed components is the Plant Status Monitoring System. PSMS continually monitors a plant database, and detects abnormal plant transients. All of the sensors are monitored at some time, according to a predetermined schedule. Certain critical parameters, however, may be monitored more often. The critical parameter set includes the entry conditions for the Emergency Procedure Guidelines which establish the safety goals, the events identified in the Final Safety Analysis Report, and the events defined in DPMS.

Once an abnormal state is detected, PSMS performs the first level of decision making by instructing which component should be executed first. This decision making is based on the abnormal events classification of the plant. For any malfunction, PSMS must be able to operate in one of three modes:

1. Response to direct indications of plant changes that can be resolved by simple actions.

2. Response to changes in plant status where parameters are outside of normal ranges, but alarms or conditions do not satisfy entry conditions to one of the alarm response instructions or an event procedure.

3. Response to multiple alarms/sensors indicating that a safety goal is being threatened.

The first mode of operation is where an alarm(s) and/or sensor(s) gives a direct indication of a malfunction. That is, if the alarms and sensors satisfy all the entry conditions of one of the abnormal events defined in DPMS, then the corresponding corrective procedure from DPMS is suggested to the operator. In this case, diagnosis initially is performed in PSMS by using the multiple alarms and sensors. DVS will also be executed concurrently to confirm the preliminary diagnosis and to sense the possibility of multiple failures.

Another directly indicated event is when an alarm has been activated, but no sensors are out of their "normal" range (e.g. a component cooling pump trip alarm occurs, and the temperature of all cooled components remain normal for a few minutes). In these situations, no diagnosis can be performed by DVS until the effects of the malfunction detected by the alarm are sensed by parameters measured elsewhere in the plant, which could be minutes or hours later. To avoid eventually

entering one of the more complicated abnormal event procedures, PSMS
should directly invoke the Alarm Response Instructions (ARI) which will
be stored within PSMS.

A second mode of operation occurs when no alarms are initiated, but
there are sensors indicating outside their normal range of operation. In
this case, the state of the plant does not satisfy the entry conditions
of any PSMS abnormal event, alarm response procedure, or emergency
operating procedure. Thus, diagnosis is required.

If DVS is able to identify a single malfunction, then it will
provide a direct instruction to DPMS to execute. If DVS determines that
multiple failures have occurred, control will return to PSMS which will
direct DPMS to display corrective procedures for the diagnosed failures.
If DVS is unable to determine the malfunction source, then PSMS will
direct DPMS to maintain the highest level safety goal that is threatened.

An example of this latter type of response is that PSMS detects a
low feedwater flow rate for the existing plant steady state operating
conditions. Because of an earlier controller failure, the feed pumps are
being operated in manual. A valve that recirculates water to the hot
surge tank (or to the condenser in plants other than Perry) drifts open
and its limit switch detecting full closed sticks, thus precluding a
signal that the valve has moved off its seat being provided to the
computer.

PSMS will direct DVS to diagnose the problem of low feed flow.
Without sufficient data to determine the cause of the loss of feedwater
flow, control will return to PSMS which will direct DPMS to display a
procedure to the operator for manually increasing feedwater flow.

If the limit switch breaks loose, or if there is a flow sensor in
the recirculation line, then DVS will be able to identify the event and
provide this information to DPMS. DPMS may then direct the operator to
stabilize feed flow while closing the recirculation valve.

The final mode of operation is when there are multiple
alarms/sensors indicating that a safety goal is being threatened. (Note
that the EPGs are procedures designed to maintain the safety goals of the
plant, and safety goals are different from the abnormal events described
earlier.)

At this point, the procedures stored in DPMS (EOPs) are displayed.
For this malfunction type, PSMS must freeze the set of data just prior to
initiation of the EOPs. Diagnosis is then initiated on these data. The
data must be frozen because the DVS is not able to detect whether the new
state of the plant is due to a malfunction or to a corrective action.
Work is continuing to resolve this difficulty.

In this case, if the malfunction can be identified by DVS, at the
component level, then DVS will inform PSMS and DPMS so that direct action
can be determined to correct the initiating malfunction. The data can
then be unfrozen, and PSMS can return to its normal monitoring routine.

In the mean time, DPMS will have begun execution of the procedure(s)
to protect the threatened safety goal. Both the corrective procedure and
the EOP(s) will then be executed in parallel, with the EOP having
precedence when conflicting actions require. If the component procedure
is executed successfully, and it is determined interactively by PSMS and
DPMS that the conditions which had threatened the safety goal have
cleared and the safety goal is no longer threatened, execution of the
EOP(s) can be discontinued. A recommendation for this discontinuance
will be provided to the operator for his decision while tracking of EOP
performance continues.

In all of the above cases, the DVS can do the tasks of confirming the partial diagnosis performed by PSMS and continue to find a component fault. In the event that DVS detects a faulty sensor, then PSMS is informed of the validation results. This is done so that if the actions of DPMS are based on a faulty sensor, then DPMS should stop and the effects of the executed procedure steps should be re-evaluated.

For PSMS to operate in the above modes, it is necessary for it to have several properties.

The PSMS should be able to detect normal and abnormal plant states. It also must be able to detect the status of the components and systems required to keep the plant operating in a safe state for the given operating mode.

Another property of the PSMS is that it should be able to distinguish between abnormal and normal transients. That is, it should recognize if changes in a parameter are due to a normal operational change or an abnormal event. This requires a knowledge of the normal operating parameters as a function of power and operating mode.

Finally, PSMS should be able to output messages to DPMS, DVS, and plant personnel. The messages to DPMS and DVS would primarily instruct each system to start/stop.

EXPERT SYSTEM DATABASE

All the data required for PSMS to perform its decision tasks are contained in a periodically updated database.

This database receives data through a buffer from the plant computers. It is separate from the plant computers to assure that it doesn't interfere with any of their functions.

Some of the data contained in this database includes analog and digital sensor data, alarm states (On/Off), component states (On/Off, Open/Closed, Intermediate), equipment availability, Operating Mode, normal condition values, and limit and range values.

The database also has an inference capability that enables it to provide composed data. This function will be run only when such data is required.

The database provides data directly to PSMS, DVS, and DPMS. It operates independently of PSMS and DPMS. However, it is interactive with DVS. If DVS detects suspicious data, it has the power to halt the database updating. DVS will then operate to determine corrected values for suspected faulty sensors, and then will alter the values in the database. The altered data will be marked as such. DVS then will run again using this new dataset to complete the diagnosis process. When this process is complete, appropriate information will be provided to PSMS and DPMS, and the database will be unfrozen.

The buffer system is necessary to enable future updates of the faulty data to be handled in a way that conserves run time.

SUMMARY

System integration has been accomplished through a new component, the Plant Status Monitoring System along with an expert system database.

The integration of the Dynamic Procedure Management System and the
Diagnosis and Validation System allows DPMS to use the comprehensive
diagnosis from DVS, while it also enables DVS to confirm any of the
diagnoses initially performed by PSMS. The task of PSMS is to monitor
the critical plant parameters and upon detection of an abnormal
transient, direct control to either of the two sub-systems.

PSMS operates in one of the three modes described earlier. Once the
displayed procedures are executed, PSMS continues to monitor the plant
data to assure the plant has recovered from a malfunction, and the
condition has been stabilized.

ACKNOWLEDGMENTS

The authors acknowledge the encouragement and support of B.
Chandrasekaran, D. W. Miller, and D. D. Sharma.

The development of the two prototype systems was partially supported
through a grant from the National Science Foundation (Grant No. 8400840).
The research described in this paper is supported by The Department of
Energy (Grant No. DE-AC02-86NE37965).

We also would like to express our appreciation to the staff of the
Perry Nuclear Power Plant Nuclear Training Department for their extensive
assistance in obtaining plant data and for working with us to run plant
transients on the Perry simulator. This collaboration has provided a
reference plant for the work we are performing.

The collaboration of GE Nuclear Energy, Electronic and Computer
Services also is appreciated. GE's collaboration is making it possible
to assure that our expert system will properly interface with the plant
computer systems.

Finally, the authors wish to acknowledge the support provided by The
Ohio State University.

REFERENCES

1.   Sharma D. D., D. W. Miller, B. K. Hajek, B. Chandrasekaran,
     "Intelligent Process Control Operator Aid - An Artificial
     Intelligence Approach." Presented and published at the Sixth Power
     Plant Dynamics, Control and Testing Symposium, Knoxville, Tennessee,
     April 1986.

2.   Hashemi, S., D. W. Miller, B. Chandrasekaran, B. K. Hajek, "Expert
     Systems Application to Plant Diagnosis and Sensor Data Validation."
     Presented and published at the Sixth Power Plant Dynamics, Control
     and Testing Symposium, April 1986.

3.   Hajek, B. K., S. Hashemi, B. Chandrasekaran, and D. W. Miller,
     "Artificial Intelligence Enhancement to Safety Parameter Display
     Systems." Presented and published in the Proceedings of the
     International Topical Meetings on Advances in Human Factors in
     Nuclear Power Plants, Knoxville, Tennessee, April 1986.

4.   Hashemi, S., Hajek, B. K., Miller, D. W., Chandrasekaran, B., Punch
     III, W. F., "An Expert System for Sensor Data Validation and
     Malfunction Detection." Presented and published in the Proceedings
     of the ANS Conference on Artificial Intelligence and other
     Innovative Computer Applications in the Nuclear Industry, Snowbird,
     Utah, 1987.

5.   Bylander, T. and Mittal, S., "CSRL: A Language for Classificatory
      Problem Solving and Uncertainty Handling." _AI Magazine_ 7(2):66-77,
      1986.

6.   Chandrasekaran,   B.,   "Towards   a   Functional   Architecture   for
      Intelligence   Based   on   Generic   Information   Processing   Tasks."
      Presented and published in the Proceedings of the International
      Joint Conference on Artificial Intelligence, Italy, August, 1987.

7.   Brown, D. C., and Chandrasekaran, B., "Knowledge and Control for
      Mechanical Design Expert System." _IEEE Computer_ 19(7):92-100, 1986.

8.   Curcoran, W. R., J. F. Church, M. T. Cross, and W. N. Guinn, "The
      Critical Safety Functions and Plant Operation." Nuclear Technology,
      Vol.55, December, 1981.

CAA - AN EXPERT SYSTEM FOR ANALYSIS MANAGEMENT IN CHEMICAL PROCESSES

Andreas Jaeschke, Werner Konrad[*], Helmut Orth,
Ono Tjandra[*], and Gerd Zilly

Kernforschungszentrum Karlsruhe GmbH
Postfach 3640, D-7500 Karlsruhe
Ferderal Republic of Germany

[*]SYNERGTECH
Sendlinger Str. 57, D-8000 München 2
Federal Republic of Germany

INTRODUCTION

Due to the lack of sufficient in-line instrumentation, process control
and process monitoring of a nuclear fuel reprocessing plant require an
efficient analytical laboratory for the physical and chemical assay of
the various process constituents and intermediate or final products.
The demand for efficiency, reliability and short response time calls
for the use of computer support in the laboratory, which is usually
realized by laboratory information systems covering both levels:
analytical instruments automation and laboratory organization.

Within the frame of laboratory organization the selection of the
appropriate analytical method, the definition of the analytical steps
(including sample preparation, measurement procedure, evaluation etc.)
and scheduling the sequences in laboratory work is an inherently
complex task. It depends on a variety of input informations and
implies multiple decision steps, which often must be revised
repeatedly during the analytical treatment because of unexpected
analytical results.

THE LABORATORY INFORMATION SYSTEM KALAU

For the analytical laboratories of the nuclear fuel reprocessing pilot
plant at Karlsruhe (WAK) the information system KALAU was installed in
1984 and since then expanded in several steps /1/2/3/. The major
functions (and with this the major components) of this system are
shown in fig. 1. The orders for chemical analyses are input by the
plant operator in the main control room. From the items of the
requested analyses the system derives the amount of samples to be
extracted from the process vessels by the central sampling station.
Then the disposition module of the system splits the analytical order
into sub-orders - one for each variable to be measured -, defines the
steps of sample preparation and analysis and distributes the sub-

orders optimally on the appropriate analytical workstations and instruments. Finally, the system picks up, checks, evaluates and combines the individual results and reports them immediately to the main control room. Besides this, the system comprises components for analytical quality control, long-term data storage and evaluation etc.

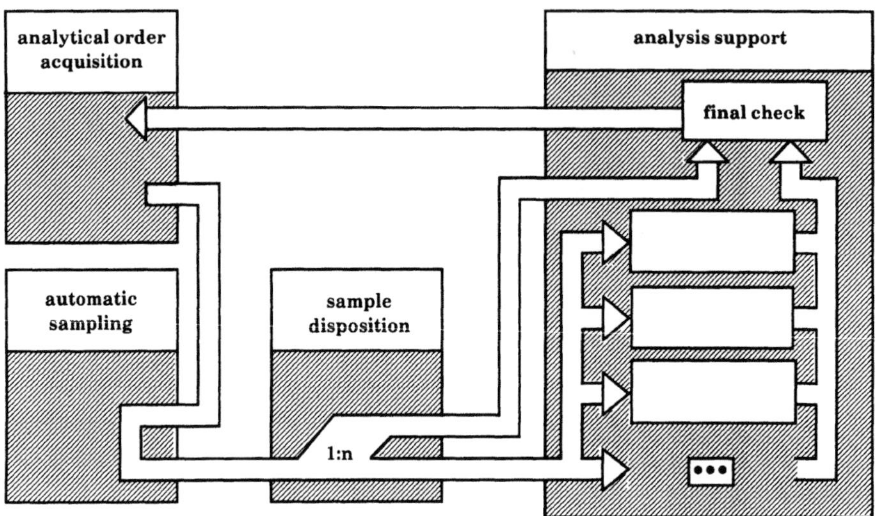

Fig. 1. Functions and information flow of the KALAU system

Since its integration in the WAK-laboratories, the system has proved to be an excellent tool for operating and supervising the laboratories.

However, concerning conception and realization, there is still one problem inherent probably in all systems of this type. In order to provide high reliability and high user friendliness, the concept must be based on predefined and fixed operation steps (i.e. predefined dialogs, evaluation procedures, data structures etc.). This entails an unwelcome *inflexibility* of the system against unforeseen analytical investigations outside the laboratory routine (e.g. in case of a process failure) and it entails an unwelcome *inadaptability* of the system against structural changes in the laboratory organization. These changes, however, are very frequent in the present case and may be caused by insertion or modification of analytical instruments, by improvements of the analytical methods or by new internal or external regulations to be obeyed.

In order to preserve the functionality and the user friendliness of the lab-information system, it must dynamicly be adapted to these changes and that ends in ceaseless reprogramming, compiling and testing. This, however, involves that the application software loses its original structures step by step and becomes increasingly cluttered up with patches and thus gets unreliable and unmaintainable. A high degree of modularisation and an ample use of parameters have proved to facilitate the reprogramming task, but are no final remedy for the problem in discussion.

The basic reason for this type of problems is the fact that the whole knowledge on the process, on laboratory organization and instrumen-

tation, which is necessary to perform the various disposition
functions, is represented implicitly in the conventionally realized
programs of the application system. So each change of this knowledge
results inevitably in a modification of the programs. The intention
therefore must be to separate the knowledge from its interpretation
and to represent it explicitly and externally to its executing
programs. This suggests to apply the methods and structures of expert
systems (xps) for this application.

THE CONCEPT OF AN XPS-BASED LABORATORY INFORMATION SYSTEM

The basic logical concept using xps techniques is illustrated in
fig. 2. The process and laboratory knowledge hitherto distributed on
the various conventional programs is now extracted and concentrated in
the knowlegde base (kb) of an xps which - within a hybrid system -
controls the activities of the remaining conventional components as
there are user dialogs, report generation modules, db-updating modules
etc. The easy maintenance of the kb - not affecting its interpreting
program modules - essentially improves the *adaptability* as demanded
above.

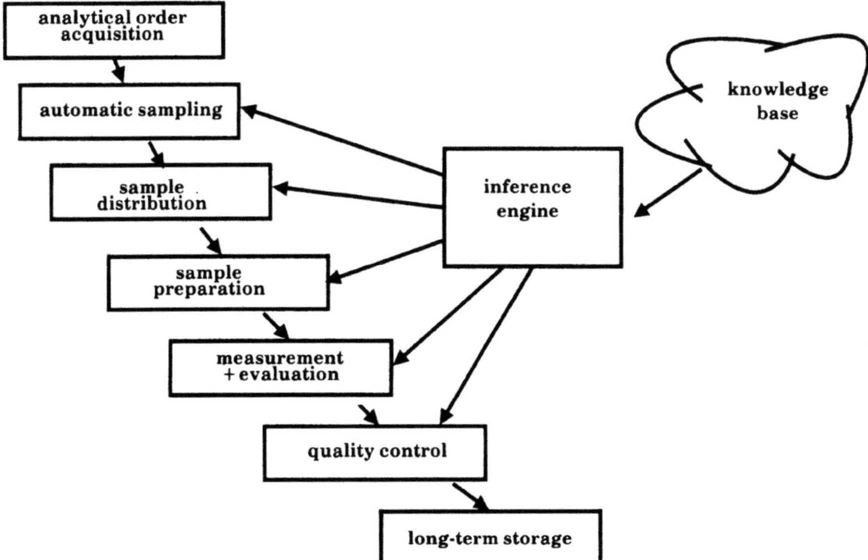

Fig. 2. Xps-controlled laboratory information system

The integration of an xps as supervising nucleus is also apt to
increase the *flexibility* of the system against non-routine analytical
requirements. To illustrate this, three different operation modes
feasible in a laboratory shall be contrasted here:
   1) not-computer-supported lab,
   2) lab supported by a conventional information system and
   3) lab supported by an xps-based information system.

1) In this case the organization in the laboratory is controlled by a
standard analytical scheme (documented in the lab-manual), which was
deduced and defined by the laboratory chemist on the base of his
knowledge of analytical chemistry and process-/lab-facts. Standardization implies simplification and a strong loss of flexibility and

effectivity. The scheme primarily covers the routine work in the
laboratory and all exceptional situations require the intervention of
the lab-chemist and a treatment outside the standard scheme.

2) The concept of most lab-information systems implemented up to now
integrates the standard analytical scheme as an internal model of the
lab. These systems therefore offer no increase of flexibility. On the
contrary, experience shows, that in most realizations the compromise
between system complexity and flexibility was made in disfavour of the
system flexibility.

3) Xps-based information systems, however, permit to forget the
standard scheme with its inherent constraints. Instead of the scheme
the basic knowledge originally used to derive the scheme can be
integrated in the system representing the model of laboratory and
process. This step from 'derived (surface) knowledge' to 'primary
(basic) knowledge' enables the system to perform all disposition
functions on the same level and with the same flexibility as the lab-
chemist.

The basic knowledge required for this comprehends the domains:

- general and analytical chemistry
  (analytical methods, sample preparation ...),

- analytical instruments and laboratory equipment
  (characteristics of instruments and tools, measurement ranges
  and accuracy of instruments, throughput ...),

- process and laboratory
  (topography of sample stations and lab-instruments, operational
  mode of the process (start up, failure etc.), estimated values,
  actual state of the lab (queues, availability of instruments and
  personnel ...).

THE CAA LAB-INFORMATION SYSTEM

A laboratory information system CAA (Chemical Analysis Assistant)
/4/5/ corresponding to the ideas outlined above is being developed in
our institute. The realization is done in two steps. The first step
includes the implementation of a pilot system (CAA-P) to prove and
demonstrate the feasibility and efficiency of an xps-based structure.
The concept of CAA-P concentrates to the xps-nucleus including:

- the knowledge base (kb),
- the inference engine,
- an explanation component and
- an user interface (specific for the pilot system).

The conventional components of the entire hybrid system are realized
only fragmentarily in the pilot version. As to the functional volume
CAA-P covers all relevant laboratory activities, but is restricted to
a sub-area of the laboratory sufficient for demonstration purposes.

The knowledge of CAA-P is represented in the kb as an attributed
semantical net. The net consists of entities connected by directed
relations. Each entity represents a semantic unit ('knowledge
schema'), which is supplied with declarative and procedural knowledge.
The former is used by the explanation component of the system. The
latter is executed, when the respective entity is traversed. To all
relations premises are associated which formulate the preconditions to
be met before passing them.

The functional scope of the CAA-P system can roughly be split into two main functions:

1.) Main function: *Chemical analysis*

1a) *Laboratory operation*

This comprises the handling of an analytical task from order acquisition via all disposition steps (including redisposition in case of unexpected results) to evaluation and output of the final results. Within this function the underlying explanation facility is a dominant factor. During the kb-construction it is used to verify the correctness, completeness and consistency of the kb and during operation it helps the lab-operator to understand the reasoning of the system. The acceptability of the system heavily depends on this capability to explain its inferences.

1b) *Laboratory information*

This function serves to inform the lab-personnel on the current state of laboratory and lab-operation, e.g. by listing the queue of pending analytical orders or the present state of an instrument (active, stand-by, overburdened, defect). These informations are generated on the basis of the kb as well as the system data base.

2.) Main function: *Maintenance*

This main function provides the tools to maintain the system (i.e. primary the kb) in an up-to-date state. Two levels of maintenance activities are to distinguish here.

The first level comprises all kb-modifications which do not alter the structure of the kb, i.e. do not add or erase kb-entities or relations. Such modifications are e.g. altering of parameters or switching of flags (instrument intact/defect) within existing relations etc. The implementation of maintenance functions of this level rises no problems and the handling of this function by lab-personnel normally represents no risk for the system integrity and reliability.

The second level, however, comprehends all kb-modifications that impact the structure of the kb. Here two problems are coming up:
  - preservation of the kb-consistency and
  - preservation of the compatibility with the conventional compo-
    nents.

The former is a general problem of xps and not generally solved. For the latter solutions are conceivable only for some special cases e.g.
  - insertion or erasure of an entity the associated procedural
    knowledge of which does not contain interfaces to any
    conventional component or
  - insertion of a mono-linked sub-net (i.e. linked only by one
    relation to the original semantic net of the kb).

It is doubtful wether a general solution can be found here. Certainly, provisions and constraints in the structure of the conventional frame are necessary. In our future work we intend to investigate these problems in more detail. For the present the CAA-P system admits modifications of the type in question, but does not offer any support as to preserving consistency and compatibility. (Some research related to maintain the kb by means of machine learning methods can be found in /7/).

The user interface of the CAA-P system reflects the functional structure of the application. It extensively uses menu techniques and graphical representations (pictograms, diagrams and figures taking pattern from lab-reality). The user selects the activities he wants to perform by clicking in the nodes of a decision tree that reflects the operations of the analytical process (fig. 3). Since these nodes are layed out as pictograms, strong support is provided for selecting the right operation and to stay informed about the operation context. The decision tree remains available permanently in one window of the display and indicates the current status and the history of the dialog. Activities outside the user's authorization are issued in the tree, but are marked as not selectable. Selecting a non-terminal node of the tree causes the output of the sub-tree belonging to this node (fig. 3; selection of node 3 causes output of node 8-11). Selecting a terminal node results in the execution of the activity associated to this node (fig. 3; selection of node 8 initiates the output of the form 'order-acquisition' (fig. 4) and starts the respective dialog).

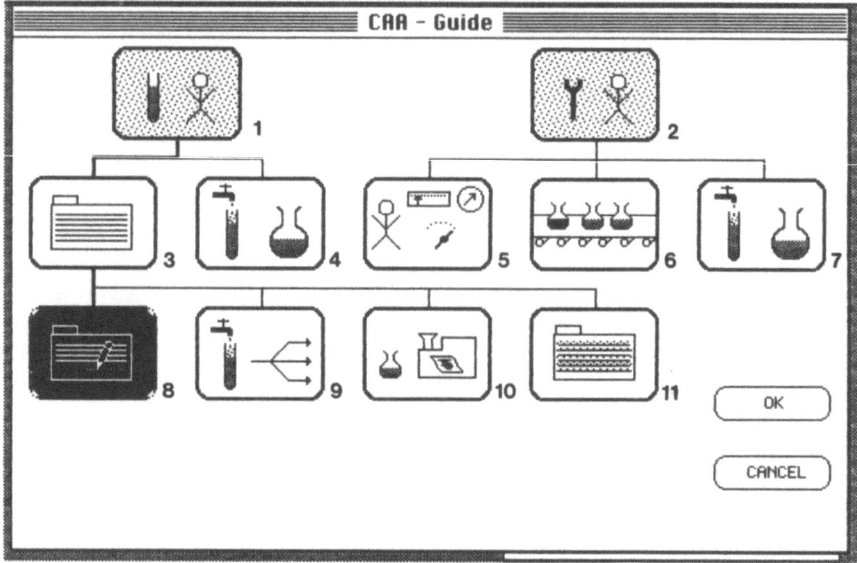

Fig. 3.   Pictogram decision tree of the CAA dialog guide
          (1 = chemical analaysis, 2 = maintenance,
          3 = laboratory operation, 4 = laboratory information,...
          8 = analytical order acquisition, 9 = sampling and sample
          distribution etc.)

The CAA-P system is implemented on a TI-Explorer and is written in PROLOG. It uses parts of the XCAT system /6/. The realization of the pilot system is completed. Some refinements will be added in the next few month.

The second step in the realization of the CAA system comprises the extension of the knowledge base to all lab-areas and the link of the CAA-P system with appropriate and adapted modules of the KALAU system. The concept aims to integrate the existing components with the least possible modifications. It therefore provides a weak coupling between both components preserving the independency of both. This is reflected by the hardware configuration - a hybrid system - consisting of the TI-Explorer dedicated to the CAA-P modules and a VAX computer dedicated to the KALAU modules (linked by DECNET).

Though the discussed hardware and software configuration of the hybrid system emerged from our specific development circumstances, we feel that - considering the current state of hardware and software tools -

the conceived structure is optimally suited for this type of
application. The experiences with the CAA-P system encourage us to
continue the concept outlined above.

Fig. 4. Display form 'analytical order acquisition'

REFERENCES

/1/   J. Lausch, R. Weh;
      Proceedings of the Conference on Safeguards Technology:
      The Process Safeguards Interface, Hilton Head Island, Nov. '83,
      pp. 356-359

/2/   A. Jaeschke, H. Orth;
      Proceedings of MECO-82 (Measurement and Control), Athinai, GR,
      August '83

/3/   R. Friehmelt et al;
      Proceedings of the 23rd IEEE Conference; Productivity an Urgent
      Priority, COMPCON, Fall '81, Washington, D.C., Sept. '81,
      pp. 226-231

/4/   A. Jaeschke et.al;
      Proceedings of the 7th International Workshop; Expert Systems and
      Their Applications, Avignon France, May '87, pp. 353-369

/5/   H. Orth et.al;
      Proceedings of the Fachtagung Prozeßrechensysteme '88, Stuttgart,
      March '88, to be published

/6/   W.Konrad et al;
      Proceedings of 'Economics and AI',
      Aix-en-Provence, France, Sept.'86, pp. 99-104

/7/   W. Konrad et al;
      Proceedings of the 11. Fachtagung GWAI '87, Geseke, W.-Germany,
      Oct. '87, to be published

DIAGNOSTICS AID FOR MASS SPECTROMETER TROUBLE-SHOOTING

E. E. Filby,   R. A. Rankin, and G. W. Webb

Westinghouse Idaho Nuclear Company, Inc.
P.O. Box 4000, Idaho Falls, Idaho 83403

ABSTRACT

The "MS Expert" system provides problem diagnostics for instruments used in
the Mass Spectrometry Laboratory (MSL). The most critical results generated
on these mass spectrometers are the uranium concentration and isotopic
content data used for process control and materials accountability at the
Idaho Chemical Processing Plant. The two purposes of the system are: (1) to
minimize instrument downtime and thereby provide the best possible support
to the Plant, and (2) to improve long-term data quality. This system
combines the knowledge of several experts on mass spectrometry to provide a
diagnostic tool, and can make these skills available on a more timely basis.
It integrates code written in the Pascal language with a knowledge base
entered into a commercial expert system "shell." The user performs some
preliminary status checks, and then selects from among several broad diagnostic categories. These initial steps provide input to the rule base. The
overall analysis provides the user with a set of possible solutions to the
observed problems, graded as to their probabilities. Besides the trouble-
shooting benefits expected from this system, it will also provide structured
diagnostic training for lab personnel. In addition, development of the
system knowledge base has already produced a better understanding of instrument behavior. Two key findings are that a good user interface is necessary
for full acceptance of the tool, and a development system should include
standard programming capabilities as well as the expert system shell.

INTRODUCTION

Diagnostic "knowledge based" or "expert" systems are one of the more
highly developed areas within the Artificial Intelligence (AI) field, with
many applications[1]. Some typical manufacturing uses include the following: CORA was produced by Westinghouse Electric engineers using a Carnegie-
Mellon University tool to design relay protection systems[2]. University of
Delaware researchers produced FALCON to identify causes for upsets in a
chemical process[3]. The Carnegie-Mellon and Westinghouse team also
developed PDS, which aids in the diagnoses of process machinery malfunctions. A more recent application combines rule-based methods with a video-
disc display system to troubleshoot an automated assembly line[4]. There
have also been several systems related to the nuclear field. GAMMA was
developed by Schlumber-Doll Research to help interpret gamma-ray activation

spectra[5]. The "Nuclear Power Plant Consultant" (NPPC) from the Georgia
Institute of Technology helps operators diagnose abnormal events[6].
REACTOR was developed at EG&G Idaho for much the same purpose using a different approach[7,8]. Workers at Battelle Columbus Labs devised WELDEX to
identify weld faults by interpreting data from radiographic inspection[9].

Numerous systems have been developed to help diagnosis problems with
electronic instrumentation, including FOREST, developed by researchers at
the University of Pennsylvania and RCA Corporation. FOREST isolates and
diagnoses problems with complex electronics circuits[10]. Developers at the
Lockheed Palo Alto Research Laboratory have the "Baseband Distribution
System" (BDS) to locate defective components in signal-switching networks[11]. Mass spectrometry was involved in one of the earliest developments of expert systems. The Stanford University DENDRAL system was devised
in the late 1960s to interpret mass spectral and other data to elucidate
chemical structures for an extensive range of organic molecules[12,13]. And
the TQMSTUNE program from the Lawrence Livermore National Laboratory uses an
expert system to fine tune a triple quadrapole mass spectrometer[14,15].

The "MS Expert" system is a diagnostic program developed by the MSL to
help Lab personnel minimize instrument downtime and improve long-term data
quality and consistency. This paper will first discuss the facility background and related MSL functions that established the system requirements.
After an overview of the system development and operation, some of the
benefits and key points learned will be presented. Finally, a number of
possible improvements and changes to the system will be proposed.

BACKGROUND

Process and Project Environment

The Idaho Chemical Processing Plant (ICPP) is located at the Idaho
National Engineering Laboratory (INEL), near Idaho Falls, Idaho. It is
operated for the U.S. Department of Energy by Westinghouse Idaho Nuclear
Company, Inc. (WINCO). The production facilities at the ICPP are designed
to reprocess spent nuclear fuel owned by the U.S. Government[16]. The
operation provides a continuous materials flow to reprocess nuclear fuels so
the uranium can be returned to the fuel cycle. All of the chemical processes must be monitored for proper operation, and safeguards and accountability measurements must be performed for the recovered uranium. The MSL
provides uranium isotopic and stream content results for these functions.

In addition, the MSL provides support for several key nuclear fuel
development, safety, and environmental programs. Over the years, the MSL
has measured actinides, fissions products, and other nuclear fuel materials
for such projects. A few of the programs supported by mass spectrometry at
the ICPP include General Atomic fuel studies[17], fuel behavior in the Power
Burst Facility [18], fission yield measurements[19], and measurements of
iodine in the environment[20]. All of these program require a high degree
of precision, accuracy, and sensitivity in the mass spectrometric data.

MSL Functions

As noted above, the MSL measures actinides, fission products, and many
other elements for the various programs supported. However, the major MSL
function is to provide uranium concentration and isotopic content data for
production at the ICPP. One of the two Plant-related uses of these data is
process monitoring and control. Generally the Plant processes operate continuously, and run for two to ten months at a time. During these periods,
the MSL must cover all work shifts, seven days a week. Because some of the

analyses are on the critical path, a rapid turnaround time must be maintained. Peak workload can also be quite high, and backlogs may exceed 100-150 samples. Thus, although various strategies have been devised to minimize the possible problems, instrument downtime can sometimes have a severe impact on the overall ICPP operation.

In addition to these process support functions, MSL results are used along with other Plant information to meet safeguards and accountability requirements. This means the Lab must provide highly precise and accurate data while also maintaining a good turnaround time on a large load of samples. Generally, the development programs served by the Lab have a similar need for very high quality results. Finally, the environmental and safety programs served by the Lab add yet another facet to this data-quality requirement. These samples require the best possible sensitivity, as well as data that is as accurate and as precise as possible.

The various functions performed by the MSL define the technical requirements of the methods and equipment used. The MSL must provide highly precise and accurate results, in a timely manner for a large workload. To meet these throughput, turnaround time, and data quality standards, MSL instruments must be maintained in peak operating condition.

<u>Instrument Troubleshooting</u>

There are two keys to timely and effective instrument maintenance: (1) competent and rapid diagnosis of problems, and (2) proficient execution of repairs. Problem definition on a mass spectrometer can be very difficult because these instruments use a highly complex array of components. This includes an extensive ultra high vacuum (UHV) system, composed of stainless steel housings, precision-fit flanges and gaskets, high performance valves, and a variety of UHV pumps and gauges. Each of the key mass spectrometers is a magnetic-sector instrument, so each requires one or more high-quality magnets with very precise controls. The solids units use thermal ionization sources, requiring multiple power supplies and controls to produce sample ions and focus them into a path through the magnetic field. Specialized components collect and measure the ions. This requires additional precision equipment, including systems that can accurately and reproducibly measure electrical currents on the order of 10^{-14} amp. The UHV, magnet, source, and collector units all require monitoring and control electronics.

Also, an electronic interface is required between the analog hardware of the instrument and the dedicated computer. The computer adds yet another set of hardware problems to confuse the situation: Power supplies, memory, central processor, as well as the interfaces with the computer peripherals such as the printer, console, and mass memory. The system software and applications programs can also become corrupted or contain unsuspected bugs.

All of the instrument components must be working properly to provide timely, high quality data. But because of the wide range of integrated modules, root causes for mass spectrometer problems can be very difficult to diagnose. Locating the weak link can be an intricate and time-consuming effort. Moreover, a single problem may impact several individual subsystems. Also, minor weaknesses in several subunits that would not cause a failure individually could combine synergistically to create a problem.

The most difficult problems to diagnose are those that appear only in the data output. Again, there are many potential sources for such anomalies. For example, the chemistry of the sample itself can impact how well the material runs. Thus, if a problem appears in the quality of the data generated, it is extremely difficult to determine whether the cause is the instrument or the sample preparation.

Normally, the Lab Supervisor or other specialist is called upon to help
with diagnosis and repair. But even an experienced person may find it
difficult to diagnose problems that occur infrequently, while complex
problems may require consultation among several specialists. If the expert
for a given area is not available, the time lost can be significant.
Technical problems may be aggravated by the shift schedule. On the off-
shift or weekend, the operator usually makes a first attempt to diagnose the
problem without help. If this fails, the Lab Supervisor or other specialist
is called at home and these two try to "walk through" a diagnosis over the
telephone. If that fails, the specialist must travel to the ICPP, which
obviously takes much more time. There is clearly a significant improvement
to be gained by having an expert system available at the job site.

SYSTEM DEVELOPMENT

Project Approach and Goals

The development program for the "MS Expert" system used the same basic
approach used for most expert system projects. The system combines the
knowledge of the mass spectrometry experts to provide a general diagnostic
tool. This tool, in turn, makes the expertise available to the Shift
Scientist on a more timely basis.

The goals for the project address the two most critical requirements
imposed on functions of the MSL. These requirements are (1) maintenance of
timely turnaround and continuous coverage, and (2) assurance of high quality
data. Even at the prototype stage, industry experience has shown that an
expert system will reduce the time needed for troubleshooting and diagnosis.
Use of the system will not solve every problem, but it should shorten the
troubleshooting cycle for many cases. On the other hand, diagnosis of the
causes for a degradation in data quality requires subtle reasoning; this
goal will demand much more development and experience to attain.

Initial Development

Initial development of the system was constrained by budget limitations.
The system was implemented on an existing IBM PC/AT. The development was
carried out on a commercially available expert system shell, "Exsys"[21].
At the time this development work began, this package was one of the least
expensive shells that offered a probabilistic, knowledge based expert system
environment. It provides the necessary inference engine as well as an
editor for entering and testing the knowledge base. The first development
stage focused on the printer and the link to the computer. The results
confirmed that the overall approach was sound, but a lot of work would be
required to obtain a truly useful system.

Several problems were also identified based on the limitations of the
inexpensive system shell. First, the user interface provided by the system
shell was inflexible. It was difficult to tailor the text-based system for
complex questions and more involved input from the user. Moreover, the
rigid question and answer sequence imposed by the package can be cumbersome
and time-consuming to use. The time spent getting basic information into
the system was considered unacceptable. The pre-set displays also created
problems because they are all built purely on text listings and make no use
of more advanced visual cues. Also, the system makes only minor use of the
color capabilities of the IBM PC, and the color that is used cannot be
modified by the developer. It should be noted that examination of several
shells showed that all of the less costly systems shared these problems.

<u>The "MS Expert" System</u>

The second phase of development used an external program to alleviate these deficiencies. The system has the ability to accept input from external programs, and to feed output back to such programs. To take advantage of this capability, a custom program was written in Pascal as a pre-processor for the expert system rule base and inference engine. The custom program has most of the features lacking in the shell: use of colors allows key information to be highlighted, presentation on the full screen organizes the information, and overlay windows make it possible to move rapidly through several complex input sequences.

The main framework for this dual system uses DOS batch and text files to execute the required programs. Thus, the user actually invokes the batch file, MSHELP.BAT, to start his session. The first command in this file executes MSEXPERT.COM, which is the compiled program generated in Turbo Pascal[22]. Procedures written specifically for this program provide the windowing environment as well as the explanatory text required to lead the user through what might be called the problem definition phase. As will be seen, it is entirely possible that the consultation session will not go beyond the use of this pre-processor.

But if the rule base must be queried, the program will write the appropriate command line to a second batch file, called NEXTSTEP.BAT. The other command in the first batch file executes NEXTSTEP.BAT when the user exits the Pascal program. NEXTSTEP.BAT calls the expert system with the command "EXSYS <symptom>" where <symptom> refers to the rule base pertinent to the subsystem selected in the pre-processor session. The Pascal program also writes the major symptom selected within the subsystem area to a DOS textfile, RETURN.DAT. This latter file is the default used by the EXSYS system to obtain input from the external program. (This default file name can be changed.) The information in RETURN.DAT is called by a command built into the rule base search by the expert system.

The sequence is basically transparent to the user. Once the user has moved into the rule based system, he is coached through a conventional question-answer sequence to help diagnose the problem. At some points, the expert system may call another external program, and pass information to it via the default file PASS.DAT. After execution of the external program, the system returns to where it left the expert system within the rule base. Like the pre-processor, the other external programs are used to streamline the entry of a information into the system.

The approach described above has three advantages. First, as noted above, it provides a more flexible and interesting user interface to quickly handle larger amounts of information. Second, the restricted knowledge base selected during pre-processing is much faster to search during the question-answer period. Third, this approach makes it easier to develop the knowledge base in modules. This method does have two disadvantages. First, the development team must have the appropriate programming skills to complement knowledge of the expert system shell. Second, some redundancy in the separate rule bases is required to handle situations where a given major symptom can point to a root problem in any one of several areas.

DEVELOPMENT RESULTS

<u>System Operation</u>

As noted above, the user starts by typing "MSHELP" to invoke the batch file. After a pause, the lead screen shown in Figure 1 appears.

MASS SPECTROMETRY TROUBLE-SHOOTING

Welcome to the MS Laboratory "Expert System" for problem diagnostics. This system will help you solve some more common instrument problems.

Initially, you will be asked to highlight the general area where you see something wrong: Vacuum system, electronics, computer, etc. Or, you may select a more general class of problem area: Beam, peak, or just data. You will be asked to check the status of various items related to that area. Once the correct status is verified, you then select from a list of problem symptoms. That will allow the system to begin a detailed analysis. If necessary, you can switch to some other problem area, or abort the program early.

During the analysis, you must respond to the questions asked by the "Expert." You will then receive a set of possible solutions, rated as to how well the results are likely to address the problem(s).

■ *** Press SPACE to begin *** ■

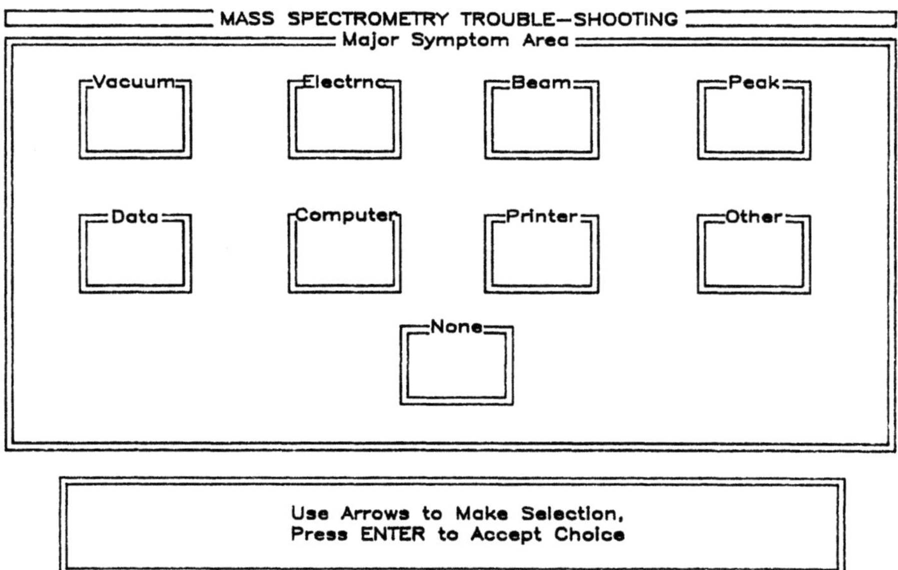

Figure 1. Text Description of MS Expert.

Figure 2. Major Symptom Areas.

The initial display briefly describes the MS Expert, telling the user how the system will proceed and what it will produce. Experienced users will immediately hit the space bar to go on. Figure 2 shows the display asking the user for the main subsystem where symptoms are observed: Vacuum, electronics, beam or peak features, data, computer, printer, or other.

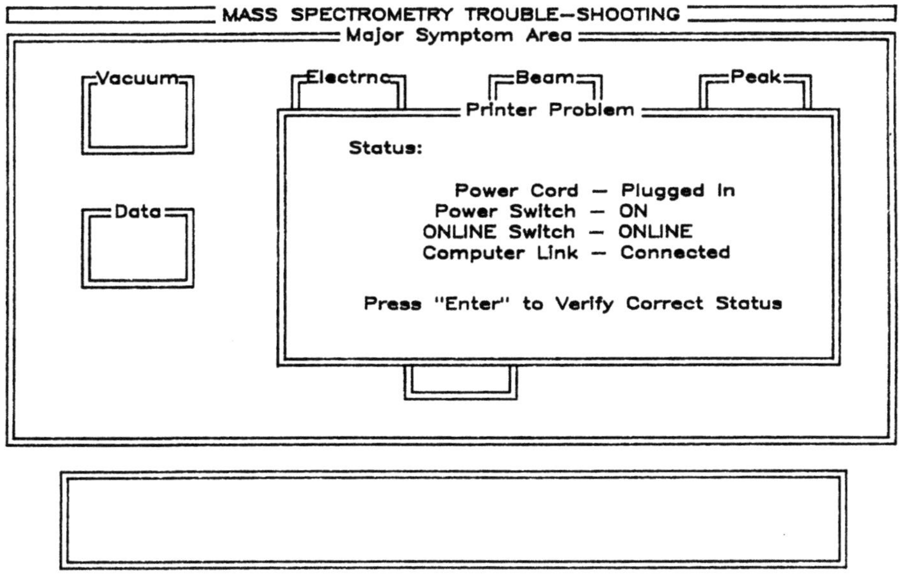

Figure 3. SubSystem Status Review.

The actual display uses multiple colors to heighten interest and stress specific items. The sub-windows have different colors, and the instruction window at the bottom uses a distinctive color set. To further guide user actions, the currently selected item is shown in a brighter color and the box blinks.

The user selects an area by hitting the Enter key. A window quickly expands from the chosen box to show material specific to that subsystem. Figure 3 shows one such display. This screen requires the user to verify the status of critical settings. Typical items include the power status, gauge readings, and other "obvious" -- but sometimes overlooked -- settings. As before, multiple colors are used to highlight key items. Because only a single keystroke response is needed, directions are given within the window, and the instruction window across the bottom is clear. Once the status has been validated, a window with a list of common symptoms for that area overlays the screen, as illustrated in Figure 4.

The symptoms display reiterates the assertion that the "correct status [has been] verified." The user selects one of the symptoms by moving to it with the arrow keys and hitting Enter. (Instruction for this screen have re-appeared in the bottom window.) If a common symptom is selected, the program exits to the DOS command level, where the batch file execution calls the expert system with the relevant subsystem rule base. Recall that the general information about the major symptom area specifies the knowledge base used in the expert system. Thus, identification of a vacuum problem means that a rule base pertaining only to that area will be used by the package.

The lowest line in the major symptom window presents three options: The user can note that his problem is not on the list, indicate that the problem has been corrected, or change to another area. If this selection is made, the continuation window reproduced in Figure 5 appears.

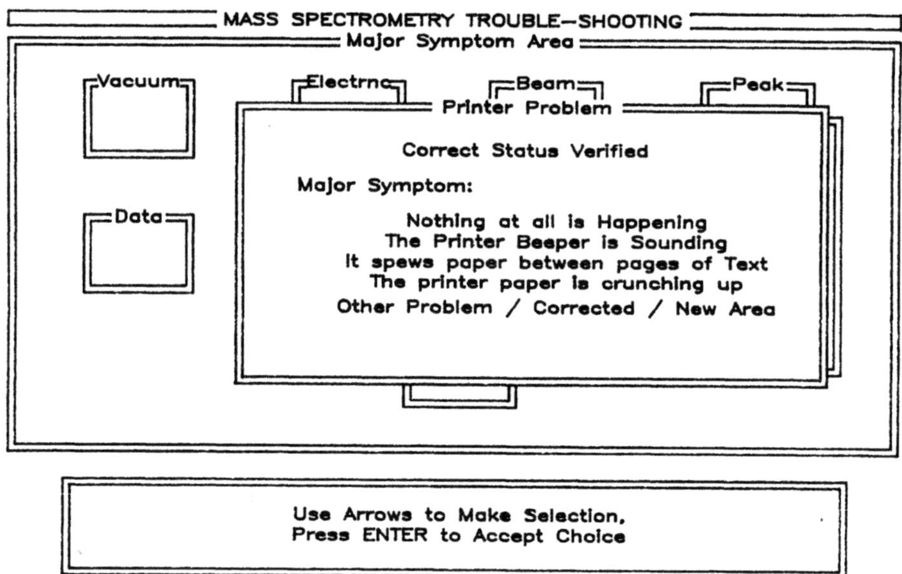

Figure 4. Commonly Observed Symptoms.

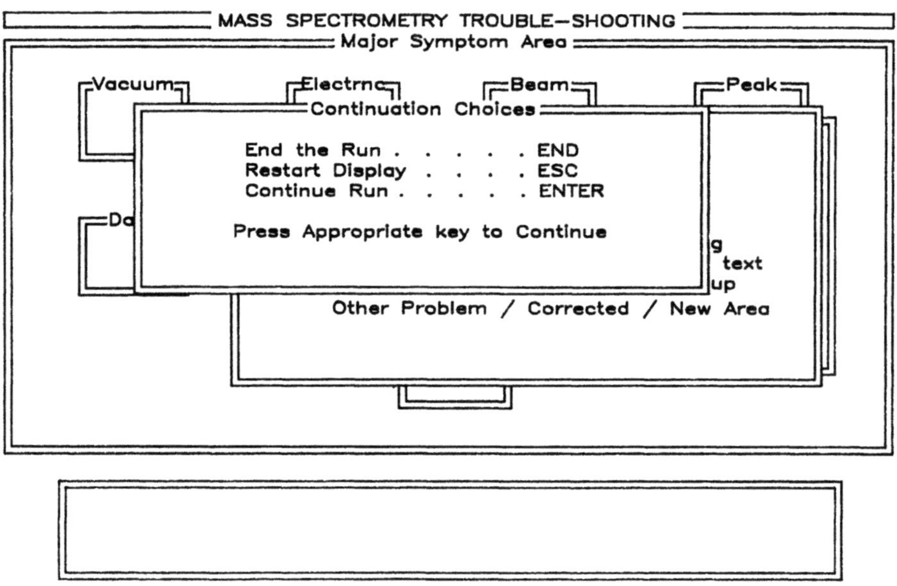

Figure 5. Continuation Menu for MS Expert.

Three allowed entries are shown. The "End" key aborts the run, perhaps
because the user fixed the problem during the status check. Alternately,
the user can restart the display so another subsystem can be selected and
reviewed. Lastly, the user can "continue run", in which case the program
exits to DOS and then proceeds into the expert system. Again, the calling
statement includes the name of the rule base for the selected problem area.
The transition to the expert system requires no further input from the user.

When the batch file calls the rule-based package, the user first views
some lead-in information. This phase requires only a continuation response
from the user. One of the first actions by the expert system after this
introductory phase is to call the RETURN.DAT file to obtain the key symptom
information. This step is also transparent to the user.

From here, the expert system branches into the set of rules and prompts.
But there is vital advantage attained here from the use of the head-end
program. The system "knows" that the status of the system has been checked
and verified to be correct, streamlining the structure of the individual
rules as well as the total rule base.

The expert shell can also call external programs during the rule review.
This means a Pascal program can be used for a response beyond the scope of
the packaged system. The session output is a list of likely causes, ordered
by their probabilities on a scale of 1 to 10. In some cases, the list might
contain specific advice on what needs to be done next. This could guide the
user through further troubleshooting. Of course, the list may include the
comment, "I have no idea what's going on. Call your Supervisor."

<u>Benefits and Findings</u>

This work confirmed the positive potential of the expert system. As the
system becomes "smarter," there should be a tangible reduction in trouble-
shooting time. Future inclusion of more subtle rules for data degradation
problems should help maintain a more consistent level of data quality. The
expert system will become an invaluable training tool when it is broadened
beyond levels attainable by conventional methods. In the short term, the
development effort provided some insight into the diagnostic process by
imposing a systematization of knowledge.

There are two other significant findings. First, the user interface
should employ every possible visual cue to provide a logical course through
the system. If available, color coding should be used to enhance interest
and stress key information. Textual materials must be carefully reviewed to
insure that it is presented clearly and concisely. The focus must be on
solving problems, with minimal distractions. A system that is difficult or
confusing will frustrate the users and they will avoid using it.

Second, the "hybrid" approach was extremely powerful. The conventional
program provided the best vehicle for moving quickly through a great deal of
simple, but vital, information. It also gave a good user interface to start
the session with attention-getting windows and color presentations. Then
the expert system shell provided the inference engine and rule base organization for the less structured part of the sequence. The more versatile
packages provide this hybrid capability. The inexpensive package used for
this work has a limited ability to link with external programs, but even
that narrow method added considerable capability. The process to select
packages for this kind of work should weight this feature heavily.

FUTURE PLANS

System development will continue for several years, as new problems
arise, are solved, and then are added to the rule base. It is possible that
a different expert system vehicle will be adopted in the near future. The
rule-based session needs improvement, with more flexible presentation of
textual information and better use of color. A system that uses more of the
language-based approach (perhaps Prolog, Pascal, or Lisp) might also be
selected. Several "toolkits" that provide a core of data structures for the
rule base and a pre-coded inference engine are now available.

REFERENCES

1. D. A. Waterman, _A Guide to Expert Systems_, Addison-Wesley Publishing Company; Reading, Mass. (1986)

2. D. B. Davis, "Artificial Intelligence Goes to Work", _High Technology_, 7 (4), p. 16 (April 1987).

3. D. Chester, D. Lamb, P. Dhurjati, "Rule-Based Computer Alarm Analysis in Chemical Process Plants," _Proc. Seventh Annual Conf. on Computer Technology_, IEEE Computer Soc. (March 1984) p. 22.

4. J.D. Folley, R. J. Hritz, "Embedded AI Expert System Troubleshoots Automated Assembly," _Industrial Engineering_, (April 1987) p. 32.

5. D. R. Barstow, "Knowledge Engineering in Nuclear Physics," _Proc. Inter. Joint Conf. on AI_, (1979) p. 34.

6. W. E. Underwood, "A CSA Model-Based Nuclear Power Plant Consultant," _Proc. Natl. Conf. AAAI_, (1982) p. 302.

7. W. R. Nelson, "REACTOR: An Expert System for Diagnosis and Treatment of Nuclear Reactor Accidents," _Proc. Natl. Conf. AAAI_ (1982) p. 296.

8. W. R. Nelson, _Response Trees and Expert Systems for Nuclear Reactor Operations_, NUREG/CR-3631, Idaho Natl. Engineering Lab. (Febr. 1984).

9. S. Mahalingam, D. D. Sharma, "WELDEX -- An Expert System for Non-Destructive Testing of Welds," _Proc. Second Conf. on Artificial Intelligence Applications_, IEEE Computer Soc. (1985) p. 572.

10. T. Finin, J. McAdams, P. Kleinosky, "FOREST: An Expert System for Automatic Test Equipment," _Proc. First Conf. on Artificial Intelligence Applications_, IEEE Computer Soc. (December 1984).

11. T. J. Laffey, W. A. Perkins, O. Firschein, "LES: A Model-Based Expert System for Electronic Maintenance," _Proc. Joint Workshop on AI in Maintenance_, (October 1984) p. 1.

12. E. A. Feigenbaum, B. G. Buchanan, J. Lederberg, "On Generality and Problem Solving: A Case Study Using the DENDRAL Program," _Machine Intelligence_, (B. Meltzer, D. Michie, Ed.), Vol. 6, Edinburgh University Press (1971) p. 165.

13. R. Lindsay, B. G. Buchanan, E. A. Feigenbaum, J. Lederberg, _Applications of Artificial Intelligence of Chemical Inference: The DENDRAL Project_, McGraw-Hill Book Company, New York (1980).

14. C. M. Wong, R. W. Crawford, J. C. Kunz, T. P. Kehler, "_Application of Artificial Intelligence to Triple Quadrapole Mass Spectrometry_," IEEE Transactions on Nuclear Science, Vol. NS-31, (1), 804 (Feb 1983).

15. C. M. Wong, H. R. Brand, _Development of an AI (Artificial Intelligence)-Based System for Tandem Mass Spectrometer_, UCRL-95378, Lawrence Livermore National Lab (1986).

16. Staff publication, _Idaho Chemical Processing Plant_, Exxon Nuclear Idaho Company, Inc. (January 1981).

17. D. E. Adams, et al, "GA Fuel Research - Charcoal Trap Studies," _Tech. Quarterly Progress Report, July - September 1979_, D. L. Plung (Ed.), ENICO 1033, Exxon Nuclear Idaho Company, Inc. (March 1980) p. 104.

18. T. F. Cook, _An Evaluation of Fuel Rod Behavior during Test LOC-11_, NUREG/CR-0590, EG&G Idaho, Inc. (1979).

19. W. J. Maeck, R. L. Tromp, _Revised EBR-II Fast Reaction Fission Yields for ^{233}U, ^{235}U, and ^{238}U_, ENICO-1091, Exxon Nuclear Idaho Company, Idaho Falls, ID (1981).

20. R. A. Rankin, et al, "Determination of Iodine-129 in the Environment Using a Two-Stage Mass Spectrometer," _Thirty-First Annual Conference on Mass Spectrometry and Allied Topics_, Boston, MA (1983).

21. Staff publication, _EXSYS Expert System Development Package_, Exsys, Inc.; Albuquerque, NM (1985).

22. Staff publication, _Turbo Pascal Reference Manual_, Version 3.0, Borland International, Scotts Valley, California (1985).

UNCERTAINTY PROPAGATION IN PRODUCTION-RULE

SYSTEMS USING A PROBABILITY-POSSIBILITY APPROACH

Lefteri Tsoukalas and Magdi Ragheb

Department of Nuclear Engineering
University of Illinois at Urbana-Champaign
103 S. Goodwin Ave., Urbana, Illinois 61801

ABSTRACT

A methodology is developed to provide a framework for the quantification and propagation of uncertainty in Production-Rule Systems, when such uncertainty consists of both random - or probabilistic - and fuzzy - or possibilistic - components. Randomness in this context refers to the uncertainty of occurrence of an event, and fuzziness describes the imprecision of the meaning of the event. In the statement, "Power will increase about 10% with a probability of 80%", the random component is given by "a probability of 80%", and the fuzzy component by "about 10%". The concepts of information granularity and of the probability of a fuzzy event are used in the development of this methodology. The approach is demonstrated through an application to a model Knowledge-Base constructed in the form of qualified if/then rules. Quantification of the "performance level" and the relationship to the concept of "reliability" is discussed.

THEORY AND APPLICATIONS

A typical rule in Rule-Based Production system[1] looks like:

$$\text{If A then B}$$

where A is a condition (antecedent) and B is the action (consequent). This structure is applicable in situations where there is usually one correct decision. Inexact reasoning on the other hand, requires a structure that can take into account phenomenological data in which one may have both probabilistic and possibilistic knowledge[2]. In order to account for such a combination we use the generalized description of an event as a fuzzy granule[3].

Let X be a variable taking values in U, and let G be a fuzzy subset of U. Typically, $U=R^n$ and G is a convex fuzzy subset of U characterized by a membership function μ_G. A fuzzy granule g is a proposition of the form:

$$g = X \text{ is } G \text{ is } \Delta \tag{1}$$

where Δ is a fuzzy probability characterized by a possibility distribution over the unit interval.

"X is G" is considered to be a fuzzy event, admitting both a component of randomness and a component of fuzziness, thus capturing in a more general way the "knowledge" that we possess about the functioning of a device. For example, if U=R1 we have:

g = (device-A-output) is (adequate) is (likely)

Here X is "device-A-output", G is the fuzzy subset "adequate" with membership function $\mu_{adequate}$, and Δ denotes the certainty (likelihood) of g, with membership function μ_{likely}.

If the proposition "g=X is G" is interpreted as a fuzzy event, then Eqn. (1) is interpreted as the proposition:

Prob $\{X$ is $G\}$ is Δ

which translates into:

$$\pi Prob \{X \text{ is } G\} = \Delta \tag{3}$$

Now, we use the probability of a fuzzy event[4] as:

$$\phi(pX) = Prob\{X \text{ is } G\} = \int_U pX(u) \ \mu_G(u) \ du \tag{4}$$

where $p_X(u)$ is the probability density associated with X. Thus, the translation of Eqn. (1) may be:

$$g = X \text{ is } G \text{ is } \Delta \rightarrow \pi(p_X) = \mu\phi(\int_U pX(u) \ \mu_G(u) \ du) \tag{5}$$

where -> stands for "translates into". The desired possibility distribution of $\phi(p_X)$ may be expressed in a symbolic form as a fuzzy set:

$$\Delta = \int_{[0,1]} \pi(pX)/\phi(pX) \tag{6}$$

in which the integral signs denotes the union of the singletons $\pi(p_X)/\phi(p_X)$. This signifies that g induces a possiblity distribution of the probability distribution of X, with the possiblity of the probability density p_X given by the right-hand member if Eqn. (5). For example, in the case of Eqn. (2), we get:

(device-A-output) is (adequate) is (likely) ->

$$\pi(pX) = \mu_{likely} (\int_U px(u) \ \mu_{adequate}(u) \ du \tag{7}$$
$$= \mu_{likely}(\phi(pX))$$

If u is a device's power level and u_0 is the nominal power level, the probability distribution describing the observed data on the device power about a mean value may be a normal distribution as shown in Fig. 1. The membership function $\mu_{adequate}(u)$ and $\phi(pX)$ are also shown. Fig. 2 shows the membership function μ_Δ as a function of the probabilities $\phi(p_X)$ and the process of inversion, once $\phi(pX)$ has been determined, to deduce the corresponding $\pi(p_X)$.

Consider the system shown in Fig. 3 Componenets A,B,C,H,I and J are connected in a series-parallel configuration[5]. If we want to generate a Model-Based system to describe the functioning of the device, a goal-tree is constructed as shown in Fig. 3, where D and E appear as intermediate inferences, and F as the top event. The quantities $(1-\beta_1)$, $(1-\beta_2)$ and $(1-\beta_3)$ are suggestive evidences describing the confidence in the rules[6]. An event occurrence (rule) "X is G" is uncertain in the

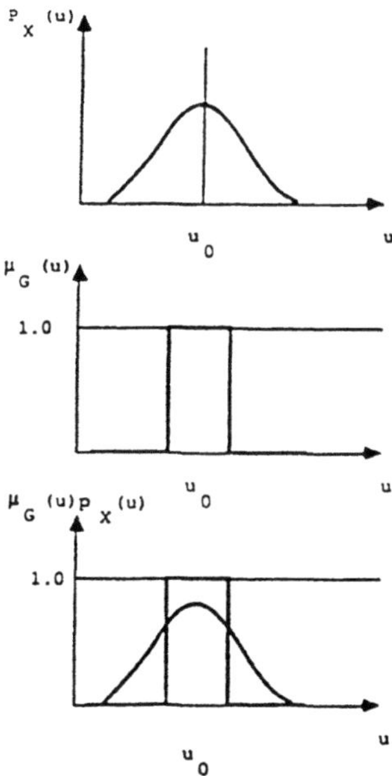

Fig. 1 Probability density function
$p_X(u)$ and membership function
$\mu_G(u) = \mu_{adequate}(u)$.

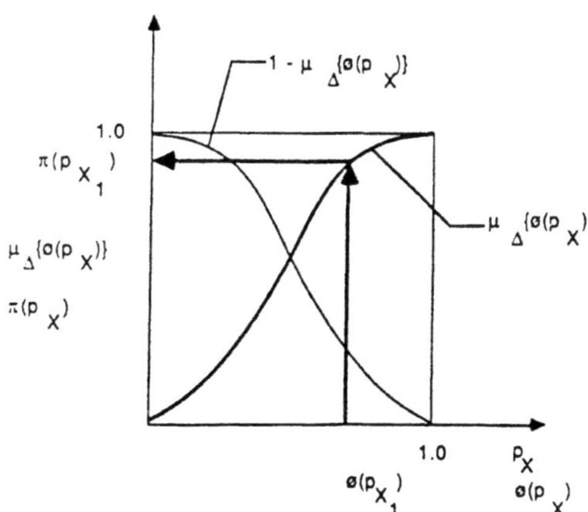

Fig. 2 Membership function $\mu_\Delta\{\emptyset(p_X)\}$
and possibility distribution $\pi(p_X)$.

sense that we are not sure the value of X belongs to the values which
are more or less compatible with G. Our confidence in the fact "X is G"
is only equal to $(1-\beta)$ where total confidence would correspond to 1.
This can be represented by:

$$u \ \epsilon \ U_X, \pi_X(u)=\max\{\mu_G(u),\beta\}$$

as shown in Fig. 4. To describe the functioning of the individual components, we can write the following rules for the Model-Based system.

Rule 1: If (1) A is G_A is Δ_A, AND $\qquad\qquad\qquad\qquad\qquad$ (8)
$\qquad\qquad$(2) B is G_B is Δ_B, AND
$\qquad\qquad$(3) C is G_C is Δ_C,
$\qquad$Then D with degree of certainty: $(1-\beta_1)$

Rule 2: If (1) H is G_H is Δ_H, AND $\qquad\qquad\qquad\qquad\qquad$ (9)
$\qquad\qquad$(2) I is G_I is Δ_I, AND
$\qquad\qquad$(3) J is G_J is Δ_J
$\qquad$Then E with degree of certainty: $(1-\beta_2)$

Rule 3: If (1) D, OR $\qquad\qquad\qquad\qquad\qquad\qquad\qquad\qquad$ (10)
$\qquad\qquad$(2) E
$\qquad$Then F with degree of certainty: $(1-\beta_3)$

The rules incorporate the probabilistic knowledge, obtained through
measurements and sensor data, in the probability density functions
(pdf's): $p_A(u),p_B(u),p_C(u),p_I(u)$, and $p_J(u)$ appearing in Eqn. (7). The
knowledge base includes the definitions of the membership functions,
$\mu_{GA}(u),\mu_{GB}(u),\mu_{GC}(u)$, $\mu_{GH}(u),\mu_{GI}(u),\mu_{GJ}(u)$, which can be modified by the
user. The system then can compute, using Eqn. (4), $\phi(p_A),\phi(p_C)$,
$\phi(p_H),\phi(p_I)$ and $\phi(p_J)$. Upon querying the user, a membership function is
adopted for each device, e.g., μ_{likely}, $\mu_{not-likely}$, $\mu_{very-likely}$,
etc. and the corresponding: Δ_A, Δ_B, Δ_C, Δ_H, Δ_I,Δ_J. The uncertainty can
then be propagated in the goal tree using the following relationships:

Rule 1: $\pi_D=\text{Max}\{\beta_1,\text{Min}[\pi(p_B),\pi(p_C)]\}$ $\qquad\qquad\qquad\qquad\qquad$ (11)

Rule 2: $\pi_E=\text{Max}\{\beta_2,\text{Min}[\pi(p_H),\pi(p_I),\pi(p_J)]\}$ $\qquad\qquad\qquad\quad$ (12)

Rule 3: $\pi_F=\text{Max}\{\beta_3,\text{Max}[\pi_D,\pi_E]\}.$ $\qquad\qquad\qquad\qquad\qquad\quad$ (13)

The methodology presented here is applied to an example modelling a
reactor subsystem in order to demonstrate its usefulness. Both random
(probabilistic) and fuzzy (possibilistic) types of uncertainty in the
knowledge base of a Production-Rule Analysis System that monitor its
performance are used. This approach makes a synthesis of "reliability"
and "performance level" in a quantitative manner. By accounting for
both the probabilistic and possibilistic aspects of the uncertainty
associated in decision-making, it provides the capability of supporting
applications involving man-machine interfaces.

APPLICATION TO RESIDUAL HEAT REMOVAL SYSTEM

A model of a Residual Heat Removal System (RHR)[7] consists of a
booster pump, which provides suction for high-pressure and low-pressure
coolant injection subsystems designated HPCI and LPCI respectively. In
the event of a minor leak the high-pressure pump (HP-Pump), is actuated

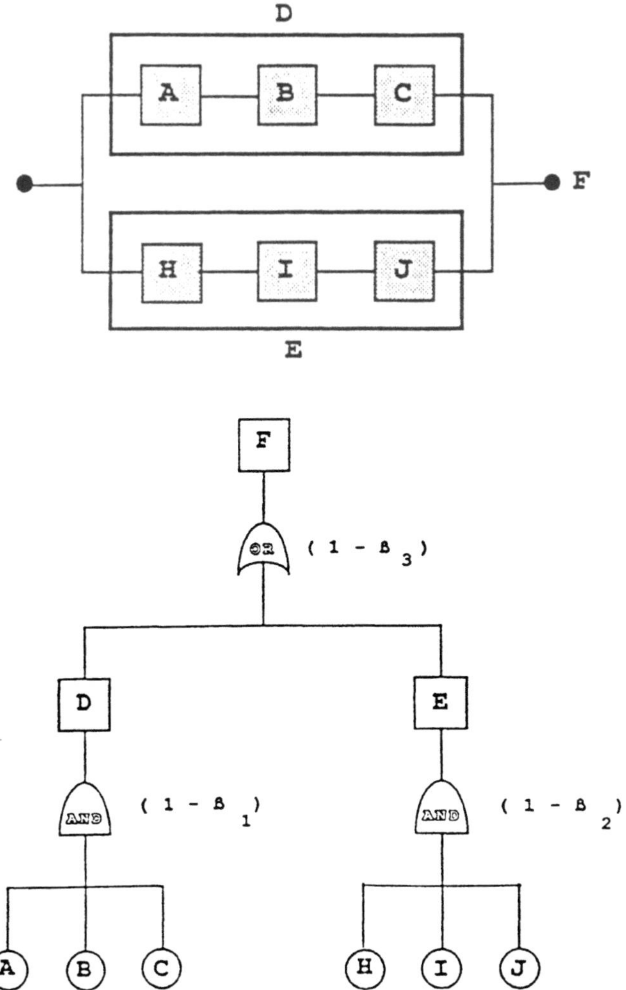

Fig. 3 Physical configuration and associated goal-tree for parallel-series system of engineering components.

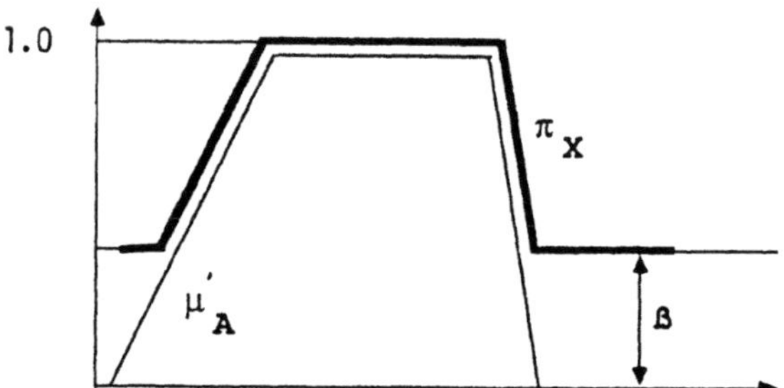

Fig. 4 Possibility distribution π_X associated with a rule having a membership function μ_A and degree of certainty (1-ß).

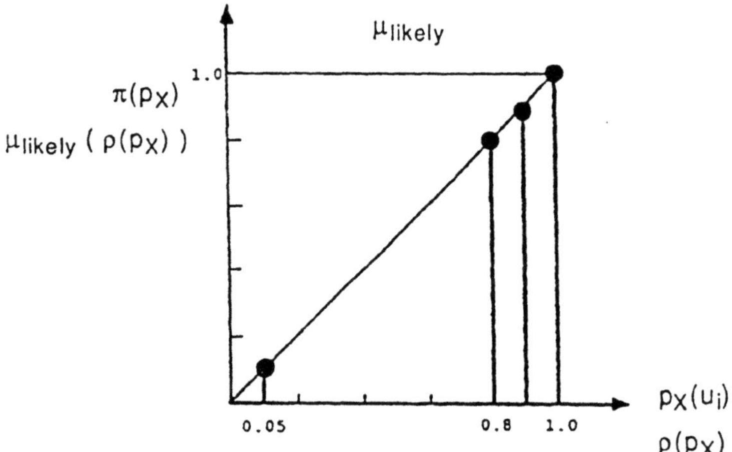

Fig. 5 Membership function corresponding to μ_{likely}.

in order to inject coolant into the core. If there is a major leak or
the HP-Pump fails to start, the low pressure pump (LP-Pump) is activated
(provided that the reactor pressure has been reduced below a certain
threshold). The knowledge of the functionally significant activities of
the system are represented in the goal-tree of Fig. 6.

For each component a salient operational parameter (such as
pressure, temperature, flow-rate, etc.) is chosen as the main indicator
of its performance. To each of these parameters a fuzzy variable is
associated. The fuzzy variables presently used are given in Table I.

TABLE I

Fuzzy Variables Used in the Goal-Tree Modelling the System

Fuzzy Variable	Description
X_1	containment penetration shell pressure
X_2	flow through valve 2
X_3	flow through valve 3
X_4	booster pump discharge
X_5	flow through check valve 5
X_6	high pressure pump discharge
X_7	flow through check valve 7
X_8	flow through high pressure motor valve 8
X_9	flow through outboard containment isolation valve 9
X_{10}	flow through inboard containment isolation valve 10
X_{11}	flow through low pressure motor valve 13
X_{12}	flow through low pressure motor valve 14
X_{13}	low pressure pump dischrage
X_{14}	flow through check valve 16
X_{15}	flow through motor valve 17
X_{16}	flow through outboard containment isolation valve 18
X_{17}	flow through inboard containment isolation valve 19
X_{18}	flow through motor valve 20
X_{19}	flow through motor valve 21
X_{20}	ΔT ($T_{primary}-T_{secondary}$) of heat exchanger I
X_{21}	ΔT of heat exchanger II
X_{22}	flow through motor valve 24
X_{23}	flow through motor valve 25
X_{24}	emergency cooling water pump discharge

Using the goal-tree of Fig. 6, we can write the following relations
to describe the uncertainty of the inferred assertions, taking into
account the imprecision in the rules.

$$\pi_A = \text{Max}[\beta_1, \text{Min}(\pi(px_1), \pi(px_2), \pi(px_3), \pi(px_4), \pi(px_5))] \tag{14}$$

$$\pi_B = \text{Max}[\beta_2, \text{Min}(\pi(px_6), \pi(px_7), \pi(px_8), \pi(px_9), \pi(px_{10}))] \tag{15}$$

$$\pi_C = \text{Max}[\beta_3, \text{Max}(\pi(px_{11}), \pi(px_{12}))] \tag{16}$$

$$\pi_D = \text{Max}[\beta_4, \text{Min}(\pi(px_{13}), \pi(px_{14}), \pi(px_{15}), \pi(px_{16}), \pi(px_{17}) \tag{17}$$

$$\pi_E = \text{Max}[\beta_5, \text{Min}(\pi(px_{18}), \pi(px_{19}), \pi(px_{20}), \pi(px_{21}), \pi(px_{22}), \pi(px_{23}), \pi(px_{24}))] \tag{18}$$

$$\pi_F = \text{Max}[\beta_6, \text{Min}(\pi_C, \pi_D, \pi_E)] \tag{19}$$

$$\pi_G = \text{Max}[\beta_7, \text{Max}(\pi_B, \pi_F)] \tag{20}$$

$$\pi_H = \text{Max}[\beta_8, \text{Min}(\pi_A, \pi_G)] \tag{21}$$

In the knowledge base of the system there exist propositions of the form of Eqn. (1) which describe the basic facts. For example the performance of the booster pump is encoded in the propostion g_4:

g_4=(booster-pump operability) is (adequate) is (likely)

The probability density function and the selected membership function for g_4, $\mu_{adequate}$, are shown in Fig. 7 and μ_{likely} is shown in Fig. 5 Using the discretized version of Eqn. (4):

$$\phi(p_X) = \sum_{i=1}^{n} p_X(u_i) \, \mu_X(u_i) \tag{22}$$

we obtain the probability of the fuzzy event: $\phi(p_{X4}) = 1.0 \, (0.80 + 0.10) = 0.9$, and thus the possibility value of the assertion g_4 is $\pi(g_4) = 0.9$ in accordance with the definition of μ_{likely} given in Fig. 6. This value represents a quantified version of the level of performance of the containment penetration shell.

In the case where the containment isolation valve is leaking the system has the probability distribution function p_{X5} which is based on sensor information and the membership function $\mu_{adequate}$ in its knwoledge base. Both distributions are shown in Fig. 7. For computational expediency the distributions are analyzed using triplex arithmetic. The adopted confidences in the rules are: $(1-\beta_1) = (1-\beta_2) = \ldots = (1-\beta_5) = 1.0$ and $(1-\beta_6) = (1-\beta_7) = (1-\beta_8) = 0.9$. The probability of the fuzzy event "X_{18} is adequate" is given by Eqn. (22), $\phi(X_{18}) = 1.0 \, (0.1) = 0.1$, and the uncertainty in the proposition:

g_{18} = (flow through outboard containment isolation valve 18) is (adequate) is (likely),

is given by the possiblity value: $\pi(g_{18}) = 0.1$. Thus the performance of the low pressure section is $\pi_D = 0.1$, and on the basis of this low performance level the production-rule system is able to identify this troubled section. The overall performance of the RHR remains satisfactory however, as long as the performance of the other branches remains satisfactory. For example when $\pi_A = 1.0$, $\pi_B = 1.0$, $\pi_C = 1.0$ and $\pi_D = 0.1$ (based on the fact about the containment isolation valve 18 not performing adequately) using Eqns 15-21 the system calculates the overall performance of the system to be $\pi_H = 1.0$, that is, satifactory. When however the performance of the booster section in Fig. 6 is low, $\pi_A = 0.1$, the system concludes easily that the overall performance is low, since it calculates $\pi_H = 0.1$, and provides a diagnostic function by reporting that the global performance is low because of the low performance of the booster section.

DISCUSSION AND CONCLUSIONS

The enunciated methodology makes a distinction between the performance level and the reliability of a system. If we consider for example a system made of two components, connected in parallel, with reliabilities r_1 and r_2, the overall reliability of the system is:

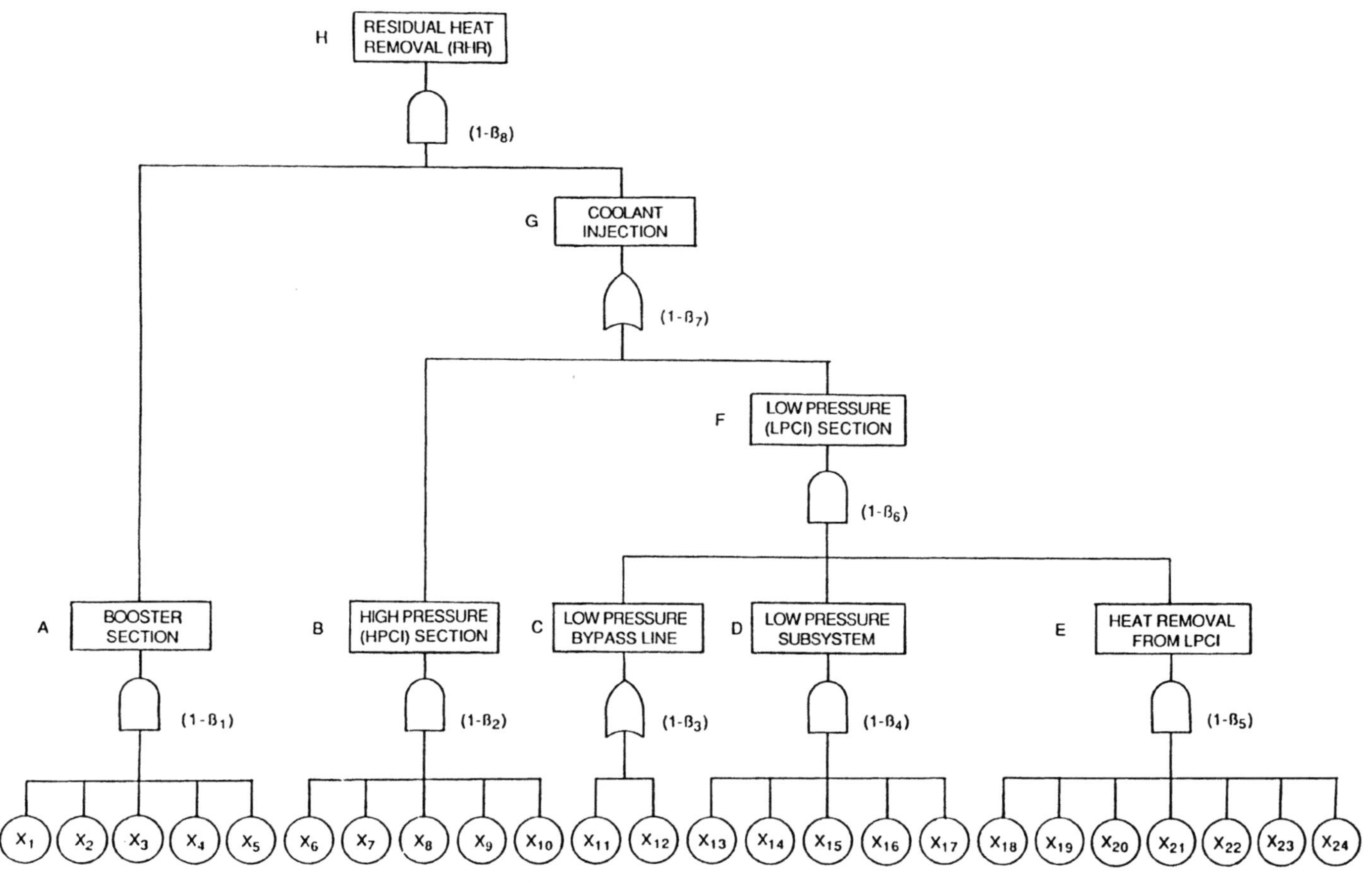

Fig. 6 Goal-tree for the Residual Heat Removal system.

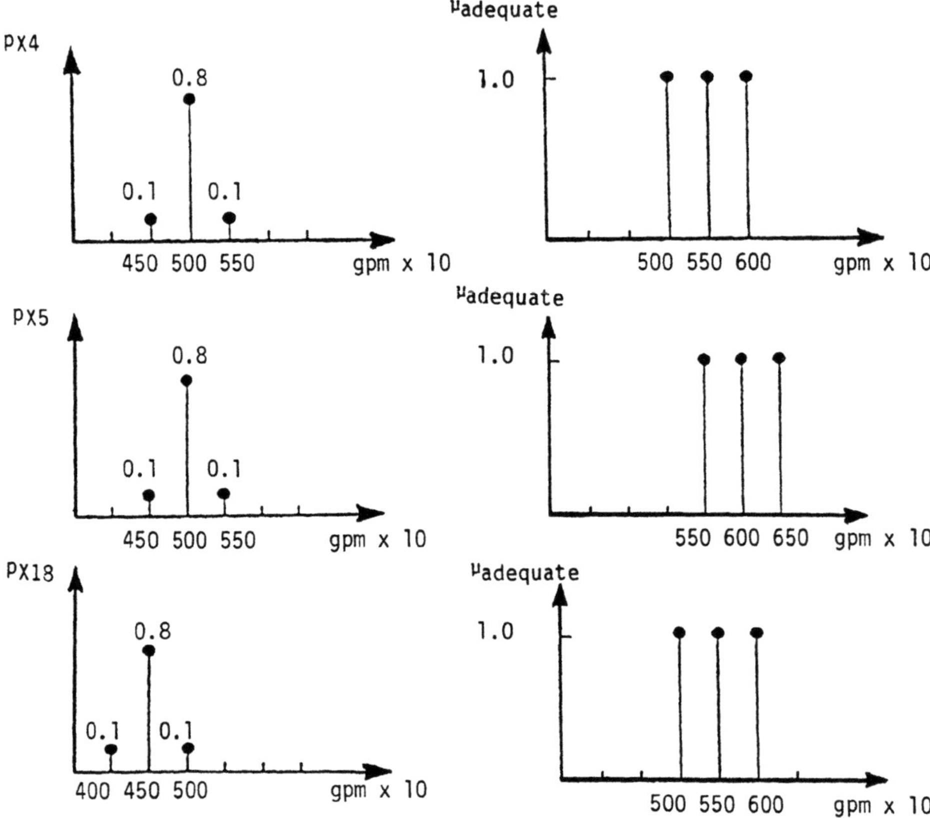

Fig. 7   Probability distributions and membership
functions for the Residual Heat Removal system.

$$R = 1 - \prod_{i=1}^{2} (1-r_i) = r_1 + r_2 - r_1\,r_2$$

If $r_1 = r_2$, then $R = 2r_1 - r_1^2 > r_1$, and $R > r_1$ if $r_1 \neq 0$ or $r_1 \neq 1$, which implies that redundancy improves the reliability of the system. If we consider the performance level of the system, P, however, we get: $P = r_1 \vee r_2 = \mathrm{Max}\,(r_1, r_2)$, in accordance with the laws of fuzzy algebra. If again $r_1 = r_2$, then $P = r_1$, implying the redundancy does not modify the performance level. Redundancy improves the reliability of the system but not its performance level. If each of two components on operation is operating moderately well, their parallel combination results in moderate operation, and in increased reliability.

The proposed methodology provides a framework for the quantification of both the performance level and the reliability of an engineering system. It can account for both the probabilistic, random contribution to uncertainty, which arises in the process of acquiring sensor data, and the possibilistic, fuzzy contribution arising from the processing and interpretation of its information and knowledge content. It is based on the concepts of information granularity and the probability of fuzzy event. Such an approach is useful for both diagnostic and control functions performed by rule-based systems. It is currently under implementation on IZE, an expert system written in OPS5 and running on a VAX-8500 in conjunction with a knowledge-base for fault-identification in nuclear power plants.

REFERENCES

1. M. Ragheb and D. Gvillo, "Development of Model-Based Fault-Identification   systems on Microcomputers," Applications of Artifical Intelligence III,   J. F. Gilmore, Ed., Proc. of SPIE, 635, pp. 268-275, (1986).
2. L. A. Zadeh, "Fuzzy Sets as a Basis for Theory of Possiblity, "Fuzzy Sets and Systems, pp 3-28, (1978).
3. L. A. Zadeh, "Fuzzy Sets and Information Granularity, "Advances in Fuzzy Set Theory and Applications, M. M. Gupta, R. K. Ragade, and R. R. Yager, Eds., North-Holland Publishing Company, pp. 3-18, (1979).
4. L. A. Zadeh, "Probability Measure of Fuzzy Events, "Journal of Mathematical Analysis and Applications," 23, pp. 421-427, (1968).
5. M. Ragheb, D. Gvillo and H. Makowitz, "Symbolic Simulation of Engineering Systems on a Supercomputer, "Applications of Artifical Intelligence III," J. F. Gilmore, Ed., Proc. of SPIE, 635, pp. 368-374, (1986).
6. R. Martin-Clouaire and H. Prade, "SPII-1: A Simple Inference Engine Capable of Accomodating Both Imprecision and Uncertainty," Computer Assisted Decision Making, G. Mitra, Ed., North-Holland Publishing Company, (1986).
7. A. Kuhlmann, "Introduction to Safety Science," Springer-Verlag, New York, p. 65, (1986).

SIA: A SAFEGUARD INSPECTOR ASSISTANT TO PLUTONIUM ISOTOPIC COMPOSITION MEASUREMENTS BY MEANS OF GAMMA RAY SPECTROMETRY

Biancamaria Carniel[1], Gianfranco De Grandi[2]
and Alberto Stefanini[1]

[1]CISE s.p.a. P.O. Box 12081, 20134 Milano, Italy

[2]Commission of European Communities — JRC Ispra
21020 Ispra (Varese) Italy

ABSTRACT

In this paper we describe the design of a knowledge-based supporting system for Non Destructive Analysis of Plutonium isotopic composition, which fits into an integrated data evaluation scheme for safeguards. This system is intended to be an intelligent assistant to the inspection procedure, able to generate as much as possible correct measurements, to guide the inspector throughout the procedure and to report descriptive data to the data evaluation system. The problem of improving the measurement quality is tackled using expert system approach, which should convey into this tool some of the physicist knowledge.

The system has been designed at CISE laboratories, in the frame of the cooperation among the JRC of Ispra, the International Atomic Energy Agency (IAEA) and the Safeguards Directorate (DCS) of Luxembourg for the fissile material control and management program.

Introduction

Safeguards are technical means for verifying the fulfilment of political obligations undertaken by States under international agreements related to the peaceful uses of nuclear energy. Safeguard practices are based upon verification activities, aimed to establish the truth of statements regarding amounts, presence and use of nuclear materials. Among the different activities involved in the safeguard inspection job (inspection planning, sampling, measurement planning and execution, data collection and interpretation, etc.) it has been decided to focus on measurement execution. As a matter of fact, this activity (involving instrumentation set-up and preliminary evaluation of the measurement consistency and accuracy), turns out to be particularly critical for many reasons:

- correct execution of the measurement procedure requires a large amount of skilled expertise about the measurement process and the instrumentation employed; this kind of knowledge subsumes good competence about the physical principles which the measurement is based upon;

- on the other hand, this expert knowledge is not always fully available to the safeguard inspectors, whose pre-professional background is various and whose training time is usually rather short;

- different kinds and models of instruments are currently in use; thus quite often safeguard inspectors are required to deal with equipments they are not fully acquainted with;

-    safeguard inspectors have to deal with rather rigid procedures to be carefully followed during the
     measurement process; these procedures are generally too simple and schematic, thus turning out
     to be useless in complex cases.

The above features are especially true for such complex measurement activities as the Plutonium isotopic composition measurement by means of gamma-ray spectrometry. As a matter of fact, this activity
turns out to be a very complex and knowledge intensive one; it requires a large amount of expert and
ill-defined knowledge not easily available in user manuals and equipment guides.

Because of the aforementioned reasons, this problem seems almost paradigmatic for the application of a
knowledge-based approach to the design of a computer-based assistant devoted to support the safeguard
inspectors in their measurement activities. The main fields where expert systems have been applied, are
those where a large amount of expert knowledge is needed in order to make decisions; in other words
fields where skilled specialized staff has to be present, as in this case of measurement activities. Therefore, it would be highly desirable to support the experts, often not easily available, and their cognitive
processes with intelligent and powerful computer-based assistants.

Requirements analysis

We can roughly subdivide a typical Pu isotopic composition measurement procedure, as currently
performed by inspectors, into the following main steps:

[a]   detection chain and measurement equipment set-up.

[b]   preliminary collection and analysis of a spectrum from the analized sample.

[c]   actual gathering of the sample spectrum and recording of related data for subsequent examina-
       tion.

A careful analysis of the problem has outlined how the first and the second step of the measurement
process are the most critical ones, in particular the preliminary spectrum analysis involves a large
amount of skilled and knowledge in order to judge the spectrum quality.

In the following we will describe when and how SIA turns out to be useful during the measurement procedure.

Supporting equipment set-up

Let us first stress the great importance of this phase for the whole measurement procedure. Often
safeguard inspectors have attended training courses too short and shallow to turn them into experts of
measurement activities. Furthermore, time shortage often compels the inspectors to perform their
activities with less care and attention than needed. An on-line support system which can provide in a
effective way all information relevant for the good fulfilment of their work, avoiding as much as possible human errors, would be much helpful.

SIA should guide, coordinate and verify the sequence of actions and tests performed by the operators.
It should coordinate standardized actions (as suggested by the IAEA agency and instrumentation
guide), and also check the correctness of the independent choices of the operators. Therefore SIA
should:

-     for each action performed by the inspector show all the parameters to be set within their correct
      values ranges;

-     provide all the information about the particular action the inspector has to perform;

-     allow the inspector to retrieve information about previously performed actions or about the next
      ones to be performed;

-     suggest optional tests to do in order to make the whole set-up procedure safer;

-     provide in every moment understandable and complete explanations about the whole set-up phase
      and its implications on the following measurement procedure phases.

Supporting spectra evaluation

During phases [b] and [c], the most important activity to be performed by the inspector, consists
of the evaluation of the growing spectrum in order to detect possible malfunctions or incorrect set-up of
the detection chain.

The spectra evaluation phase is the most delicate of the whole measurement process. This activity, in order to be quickly and well done, needs a lot of expertise to recognize at a glance the good ar bad quality of the resulting spectrum. It is not very easy to describe all the graphical features of a spectrum, and there is no coded method that can enable an unexperienced inspector to judge the spectrum quality.

Evaluating the spectrum quality means to analyze carefully the spectrum shape during acquisition, in order to decide if its quality is good enough to carry on the measurement procedure. Bad quality means that the spectrum shape is affected by some distortions due to erroneous adjustments of the detection chain or to sudden troubles arisen in the meanwhile.

Spectra shape distortions are detectable by the employment of some specific dedicated statistical software packages for Gaussian curve fitting. These programs return numerical data relevant to the curve shape i.e. resolution factor FWHM, FWTM and FWFM, their ratios, skewness and kurtosis factors to evaluate the asymmetry of the curve, etc.. In order to support this evaluation activity, SIA should:

- provide an automatic analysis based upon the above mentioned mathematical statistical computations

- give skilled interpretations of the numerical values returned by the mathematical routines, in order to

    . establish if the spectrum is good enough so that the measurement procedure may go on;

    . provide a list of symptoms whenever the preliminary spectrum analysis has detected some anomaly.

Supporting diagnostic activity

The previous function, as we have seen, is aimed to detecting faults, i.e. at evaluating the spectrum quality. The next task supported by SIA consists of the search of the particular events that may have caused those symptoms.
Many symptoms may be caused by more than one trouble.
The search of the correct cause has to be carried on according to the context i.e., chain set-up, the environment, etc.).

The skilled expert has particular criteria to make out the right cause, often based upon the probability that a particular event may or may not occur. Often, to find out the correct cause, additional tests over the detection chain may be required. Any contact with the detection chain (i.e., glance over it, adjustments, etc.) may provide more information about the state of the whole measurement environment, without deteriorating it.

Once the possible cause has been pointed out, the inspector has to decide whether or not to carry on the measurement procedure. As a matter of fact if the acknowledged cause is a malfunctioning electronic component, the inspector is not able, in most cases, to mend it or to change it. On the other hand, if the detected cause concerns a bad adjustment or some other action misleaded by the inspector himself, he should be able to correct them, and to continue the measurement procedure.

Therefore SIA should suggest to the inspector possible causes relevant to the detected symptoms.

Suggesting recovery actions

When a recoverable malfunction has been detected the inspector has to follow carefully a specific procedure in order to readjust the whole measurement process. Sometimes small adjustments are sufficient to correct it, but often the inspector has to make substantial changes in the parameter values so that the whole detection chain must be readjusted, as in the set-up phase.

SIA should:

- provide the best therapy to be carefully followed by the inspector;

- look after the execution of the suggested therapy;

- recognize during therapy administration, if some other causes may be present and suggest the most appropriate corrective actions.

The functional architecture of SIA

In order to meet the requirements analyzed in the previous section, the functional architecture, as shown in Fig.1, has been designed.

The safeguard inspector can interact with SIA through a terminal equipped with a keyboard and a mouse. Through the terminal screen, the inspector receives all the suggestions and he performs all the prescribed actions directly on the detection chain.

SIA receives its input data from the detection chain through an appropriate connection interface, and from the user when information is not available on-line (for example the initialization of SIA may take place through a dialogue with the user: employing a simple menu-driven language it is possible to provide SIA with all the relevant information to the employed equipment). It is very important that SIA is connected directly to the detection chain in order to minimize errors due to human factors; parameters values that can be collected directly by SIA are for example: input counting rate, high voltage value, gain factors, shaping time constant, pile-up rejector on/off, etc.. SIA is able, handling directly these information, to control in every moment the detection chain state. Also, the spectrum shape can be visually analyzed on the terminal screen.

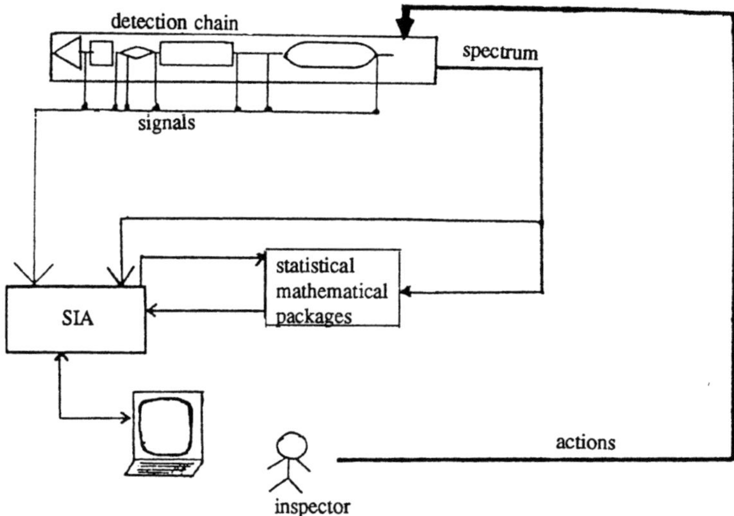

Fig. 1 - Functional architecture of SIA.

Actually in nuclear measurement by means of gamma-ray spectrometry, software packages are employed in order to compute all the relevant mathematical values needed for the determination of isotopic composition such as resolution factors, peaks area, background, etc .. These packages can discover peak locations and compute all the relevant numerical information. Once appropriate acceptability ranges for statistical errors are given, this software can provide quantitative information about the spectrum quality. Besides the resolution parameters, the package can provide for example, the skewness of certain peaks (this value "describes" the good or bad symmetry of the examined peak), the kurtosis of the peaks (how those peaks look like pure gaussian curves), etc..

These statistical tests may be fired on user's request, or in an automatic way, i.e., SIA should have a full control over those programs. It is possible to set a starting time (e.g., 50% of the measurement time), and to point out some particular ROI (Region of Interest, i.e. a spectrum region requiring a careful analysis) before the measurement procedure starts SIA should then automatically, when it reaches the half of the measurement time, activate the statistical packages over the predefined ROI and analyze an instantaneous picture of the growing spectrum, i.e. without stopping the spectrum acquisition process.

The statistical analysis may be fired also run-time on user's request employing the keyboard and the mouse facility.

The logical architecture of SIA

According to the analysis developed in the previous section, we can now propose a logical architecture for SIA, as shown in Fig.2. For each component of SIA a brief description is presented.

* knowledge base

Our aim is to integrate effective and natural representations of pieces of procedural knowledge to describe set-up procedures, and execution of corrective actions, with declarative knowledge like rules adopted for data interpretation, fault detection, diagnosis, and remedy identification. This approach offers the great advantages of the traditional rules-based systems such as the neat separation between knowledge representation and its use, the possibility of describing ill-defined and uncertain knowledge, the possibility of incrementally creating, updating, validating large knowledge bases; on the other hand, we can adopt the efficiency and the strength of a powerful procedural formalism (e.g. Event Graphs) see [5] and [6] for further details about this new kind of knowledge based system architecture. The two formalisms are efficiently connected, i.e. handled by an unitary inference mechanism.

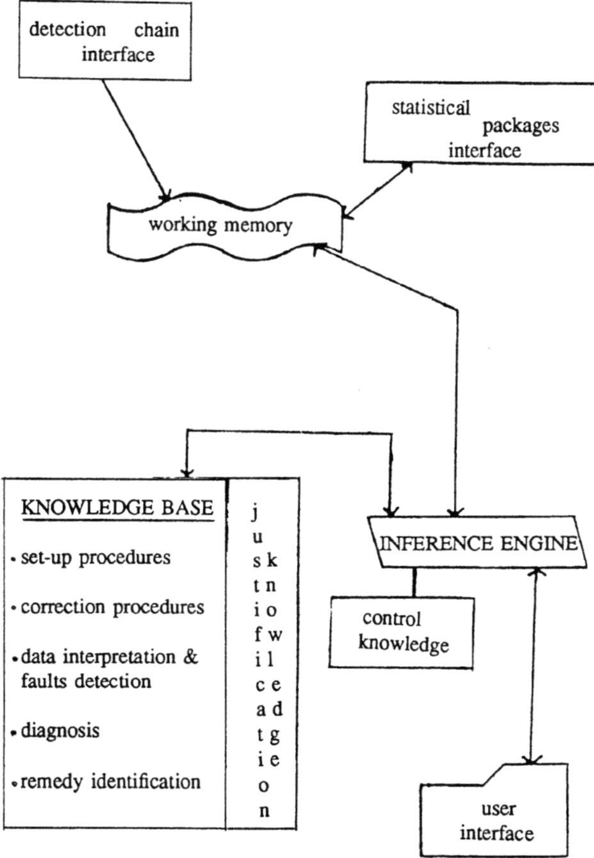

Fig.2 - Logical architecture of SIA.

Justification knowledge should always be available during the whole measurement session. In every moment SIA should be able to give the user explanations of its own reasoning. This means that SIA must keep a trace of the all deductions performed. It will keep the facts, the conditions that have matched a particular rule or procedural area (the state of the detection chain, the environmental conditions, etc.), the reasons (i.e. the detected symptoms) that have leaded the diagnosis task to get to certain conclusions, etc..

* inference engine

in order to handle the above mentioned knowledge, a specialized problem solving program must be designed. As the knowledge base may contain several kinds of representation, similarly the inference engine should support several reasoning paradigms (pattern matching, forward and backward chaining, etc.). The inference engine works following the general procedure employed by the safeguard inspectors. SIA should be initialized by the user, receiving the characteristics of the employed equipment; it stores all information relevant to the global context (i.e. the detection chain conditions, the environment conditions, etc.), then a procedure of set-up is fired, if the tests of the set-up procedure are satisfied, the process goes on with the energy calibration spectrum acquisition, if everything is correct the final measurement may be carried on otherwise we detect the possible symptoms of malfunctions and a diagnostic task is fired. It should be stressed that from a procedural area(i.e. an Event Graph) it is possible to fire rules and viceversa; thus the two formalisms are strictly connected.

* control knowledge

some words must be spent separately about another kind of knowledge: the control knowledge. This is an highly declarative representation of the strategies employed to handle the domain knowledge. It reflects the heuristics owned by the experts, it is generally not well defined, it is unstructured, so that it is often implemented with a rule-based formalism.

* working memory

this is a dynamic area needed in order to keep information about the computational state of the system. It is the buffer for the communications with the external interfaces.

* interfaces

Great care must be taken in the design of the interfaces with the external world:

- towards the detection chain: it must transmit on-line to SIA all the relevant information about the detection chain state i.e. the parameters values, the adjustments, etc..

- towards the statistical packages: it must provide to SIA all the information about the spectra shape (numerical data); SIA has the full control over these routines, that could be activated or in ana automatic way or on user's request;

- towards the user: the interaction with the opertor should be very easy: a multiwindows environment where the user may follow the computation, have a dialogue with SIA for further explanations, fire statistical packages, require more tests and so on, pop-up menus selectable with the mouse, fault tolerance ....

- towards the knowledge engineer and the expert: SIA should provide a module specifically devoted for the creation, debugging, validation, extension, and updating of the knowledge base. This implies for example a syntax-directed editor designed following the knowledge base organization.

Both these latter interfaces must be very easy to use cause often domain experts and operators are not well acquainted with computers.

Implementation remarks

Actually we have started with the development of SIA implementation. Let us sum up the main features offered by SIA:

- SIA should be an easy-to-use on-line assistant to the safeguard inspectors;

- it should integrate procedural and declarative knowledge representation formalisms into an unitary architecture.

Although on the current software market available a large number of knowledge-based system shells is available, we do not see any of them suitable to satisfy our requirements; as a matter of fact we should find a rule-based system with forward and backward chaining, gracefully interfaced with some traditional programming environment, suitable to medium-size machines.

On the other hand many researches in the Artificial Intelligence field have been successfully proposed several expert systems based upon a novel kind of mixed architecture. [5],[6]. In particular the example described in [5], shows how event graphs and rules may be treated together employing a general purpose symbolic language as LISP, at least for the first phase of prototyping.

As far as to be concerned with hardware, we are adopting for

prototyping a kind of machine of small/medium-size e.g. SUN-3 workstation.

for the target application , a portable PC should be tailored ad hoc.

Acknowledgements

We would like to thank Donato D'Adamo of JRC Ispra for his remarkable contribution to the definition of the knowledge base of SIA.

Professor Giovanni Guida for his valuable suggestions to the design of SIA architecture.

References

1)   Adams F. and Dams R., **Applied Gamma-ray Spectrometry.** Pergamon Press, 1970.

2)   Benoit R., Coppo N., de Grandi G., and Haurie Y., **The Ispra approach to a standardization of NDA measuring procedures and data treatment by use of automated instruments.** Rep. EUR 7049e , 1980.

3)   Carniel B., **Non-destructive determination of the isotopic composition of plutonium samples: a preliminary proposal for a knowledge-based software system.** CISE report 3001, Milan, Italy 1986

4)   D'Adamo D., **High Resolution Gamma-ray Spectroscopy. General Aspects and Particular Applications in the Safeguards Field.**
Rep. JRC-Ispra , october 1985

5)   Gallanti M., Guida G., Spampinato L., and Stefanini A., **Representing procedural knowledge in expert systems: an application to process control.** Proceedings of the 9th International Joint Conference on Artificial Intelligence, Los Angeles, CA 1985, pp. 345-352

6)   Georgeff M.P., and Bonollo U., **Procedural expert systems.** Proceedings of the 8th International Joint Conference on Artificial Intelligence, Karlsruhe (FRG) 1983, pp. 151-157

7)   Waterman D.A., **A Guide to Expert Systems.** Addison-Wesley, Reading, MA, 1986.

OVERVIEW OF DIVA:

AN EXPERT SYSTEM FOR TURBINE-GENERATOR DIAGNOSIS

Benoit Ricard[1], Bernard Monnier[1], Jacques Morel[2],
Jean-Marc David[3], Jean-Paul Krivine[3], and Jean-Pierre Tiarri[4]

[1]Electricité de France, Direction des Etudes et Recherches
6, quai Watier, F-78400 Chatou, France

[2]Electricité de France, Direction des Etudes et Recherches
1, av. du Général de Gaulle, F-92140 Clamart, France

[3]Laboratoires de Marcoussis, Division Informatique
Route de Nozay, F-91460 Marcoussis, France

[4]Alsthom, Service PEM/SE
3, av. des Trois Chênes, F-90018 Belfort Cedex, France

PROBLEM DESCRIPTION

DIVA* is an expert system project for failure diagnosis of turbine generators in electrical power plants. Using available data -esp. vibratory measurements- signaling an abnormal situation, the expert system should, as experts do, interpret them and identify the event sequences which initiated this situation.

This joint research project is developed by Alsthom (French turbine generator manufacturer), Direction des Etudes et Recherches of Electricité de France (Research and Development Division of the French Electricity Board) and Laboratoires de Marcoussis (Compagnie Générale d'Electricité Research Center). This article describes the fundamentals of this system and the advantages expected from the choices that have been made.

The main symptom of an incident on a turbine generator usually is an abnormal behavior of vibration measurements along the shaft. Any abnormal deviation of this vibratory state must therefore be analyzed and its causes must be identified.

The primary role of the expert is to identify the causes of such vibratory abnormalities.

---

* acronym for "DIagnostic de Vibrations d'Arbres" - shaft vibration diagnosis

The expert must then estimate the risk and, if necessary, propose emendations. The diagnosis must include what operations are necessary, what the consequences on the availability of the power plant will be and sometimes whether it is possible and safe to wait before undertaking a necessary maintenance operation.

The diagnosis made by an expert is not the sole elicitation of a fault from a list of symptoms. The expert tries to come up with a satisfactory description of the encountered phenomenon. The consistency of this description can serve as a confirmation of a proposed diagnosis. Reciprocally, a diagnosis could be reconsidered or estimated to be insufficient if the description cannot account for every observed symptom, or if some symptoms that are expected given the proposed description are absent.

The interest of an expert system for this application is justified by:
- the complexity of the problem, induced by the large number of possible failure causes and by the imprecise elements of the reasoning process. It should be noted that the main problem is not an information acquisition problem, as sensors and automatic surveillance systems do exist: it lies in the difficulty of a correct interpretation of these informations (exhibit fundamental elements of complex signals), and of their correct use (derive the right conclusions from these informations).
- the limited number of domain experts. The complexity of the problem implies that an expert should have a large experience of possible failures. Turbine generator problems nevertheless are only a limited part of all possible problems in a power plant and many of them hardly occur in a given plant. Therefore, it is not possible to have a domain expert in every plant. On the other hand, it seems important to make diagnosis knowledge on turbine-generator available in every plant;
- the fact that theoretical knowledge (that could give way to an algorithmic solution) is of little use in expert diagnosis; this implies that the only possibility for an automatic system seems to be to imitate the reasoning process of experts, or, more accurately, to elaborate knowledge representation and manipulation methods that will be strongly inspired by the way experts perform a diagnosis;
- the importance and the high cost of some incidents on a turbine generator.

The main steps in expert knowledge elicitation for this project has been described in previous publications[1]. This process is long and iterative. It would have been vain to gather knowledge in a first stage by extensive interrogation of experts and then to develop the expert system without letting them intervene. On the contrary, they are closely associated to the development process. Especially, knowledge representation elements constantly evolved from the reaction of experts to what had already been implemented.

FUNDAMENTAL ELEMENTS OF EXPERTISE

Analysis of expert reasoning exhibited some key points which we tried to keep when building the expert system:

1) the great <u>variety of processed informations</u> (vibratory measurements along the shaft, plant operation parameters, technical characteristics of the turbine generator, maintenance operations, historical data on the group or on groups of the same type), and the great diversity in the useful aspect of these informations (value, classes of values, existence or absence of a property, correlations between informations...). As an example, Fig. 1 displays some typical shapes of possible evolution of a vibratory level (in a simplified way). The fact that a vibration follows one of these patterns is one element experts use for diagnosis. This is one piece of information among many others: vibratory level is only one among measurements or data used for diagnosis and its shape is only one aspect of it (others might include maximum level, slope, recurrence of a situation, correlation with other parameters...). A fundamental point is that, when such a piece of information is studied by an expert, this is

done with some purpose in mind. Such a piece of information will usually be meaningful and possibly useful only in given circumstances. Furthermore, an expert will only ask for it when knowing <u>how</u> it can help for diagnosis. It can have two main uses: (a) at an early stage of a diagnosis, it will be used to build up hypotheses (*"a vibration of this shape makes me think of..."*), and (b) later to enforce or reject an hypothesis (*"if I suppose it was a rubbing problem, what should the vibration look like, and is it the case?"*).

Moreover, it should be noted that the information is usually not used for diagnosis as available in a control room or on monitoring systems. Therefore, data abstraction can be necessary to extract the important element from "raw" data. For instance, if the useful information is the existence of thermal variations on the turbine, this information will have to be established from a study of the existence of a possible cause for such a thermal variation (modification in operation conditions, evolution in the environment of the turbine-generator...).

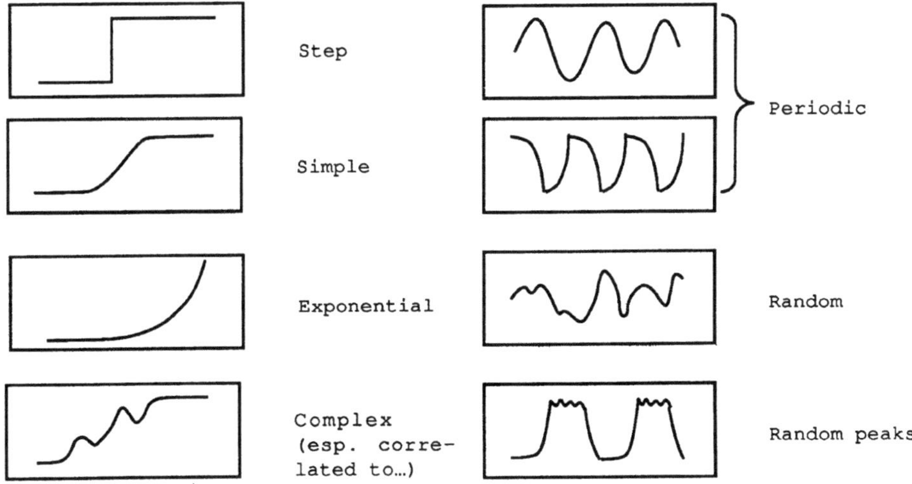

Fig. 1. Simplified examples of typical vibratory level behavior

2) the fact that the reasoning process is based upon a <u>recognition</u> mechanism: the expert knows a set of possible situations and tries to see how a given case matches these typical situations. Of course -and that is probably the most significant difference between an expert and an automatic system-, the expert can "imagine" new situations when experience is not sufficient.

Reasoning on a hypothesized situation must address two points:
   a) how well does this situation match reality?
   b) can I give a more precise (or better adapted) description of the situation?

3) the fact that most "deductive" knowledge (relations between noticeable symptoms and diagnoses) is only meaningful under <u>assumptions</u> on the encountered situation: experts do not usually handle global knowledge such as *"if the vibration behaves in such a way, then I can conclude that the failure is..."*, but knowledge related to a situation. Such knowledge can serve two goals: what can be expected under a given hypothesis *(when I think of a crack in the shaft, I expect the evolution to be exponential)* and how should reasoning be pursued after an hypothesis has been studied.

4) the encompassment of some kind of <u>imprecision</u> and <u>uncertainty</u> attached to knowledge elements: uncertainty refers to the fact that some data cannot be fully established *(for instance, one can sometimes only "think that" a thermal variation is "likely" to have occurred, but this cannot be proven nor measured)*; imprecision refers to the fact that some knowledge is only used in a "global" form *(the turbine-generator has been dismounted "recently"* is an example of such knowledge: how can we represent this "recently" with the whole meaning it bears to an expert ?).

The key point in knowledge representation is to find out a way to describe such situations and to arrange these descriptions in a structure that will allow the implementation of an efficient recognition mechanism. The main point in automated diagnosis will be to design an efficient way of matching these pre-defined typical situations with reality.

KNOWLEDGE REPRESENTATION

<u>Prototypes</u>

In order to match the requirements induced by expertise analysis, the main concept used for knowledge representation is the notion of <u>prototype</u>: a prototype is the representation of a typical situation studied by an expert during a diagnosis; such representations can evolve from a very general frame *(an incident in steady operation conditions)* to the recognition of a precise kind of failures *(a rubbing problem)* or even up to an hypothesis on the presence of an elementary failure *(a blade loss in HP cylinder)*.

Basically, DIVA considers a prototype at every stage in the diagnosis process, tries to check whether it is well adapted to the real encountered situation, using already available data and, when necessary, complementary information which must then be acquired. A new and better adapted prototype will then be selected. It can be "better" by being the description of a more precisely identified situation if the previous prototype itself was "realistic" enough or by being a different situation, for instance if the previous prototype did not seem very well adapted or as an alternative possibility.

 Part of the knowledge is included in the <u>description</u> of these prototypes and of their attached informations. An other part of knowledge lies in the <u>relations</u> between these prototypes. The basic relation between prototypes is the hierarchical link "...is a prototype which precises...". It follows the "natural" refinement diagnosis method (Fig. 2 presents an example of prototypes structured through that link).

A prototype can be seen seen as a schematic description of the mental representation of a typical situation by an expert. Therefore it bears two complementary meanings:

- a prototype is an intermediate or final statement issued by the system on the current situation: *"a vibratory step happening under steady operation conditions"* is a possible intermediate statement (of course, if it were to be the only final conclusion, the system would be of very poor help), *"a broken blade in HP cylinder"* could be a final statement.

- a prototype is a unit where knowledge on how to reason further is stored: how to be sure that the prototype is relevant in a given situation, how it can be refined, what data should be acquired... From that viewpoint, a prototype can be seen as a local mini expert system, collaborating with other mini expert systems. These other "expert systems" are described by prototypes corresponding to related more and less precise situations.

<u>Knowledge elements</u>

Knowledge elements used in DIVA can be separated in two categories:

a) <u>basic information</u>: an important part of knowledge used for diagnosis indeed consists in knowing how one can formalize descriptive information on the machine, its current situation and its past. This includes technical knowledge on machines, phenomena physics, knowledge on plant operation and on maintenance, etc.
Such knowledge will be stored in data bases. This is designed to address:
- the encompassment of a large variety of data structures and relations (machine structure data and operation description cannot be treated in the same way),
- the need for a great flexibility in the way such data will be addressed (some requests might require a processing of relations between data).

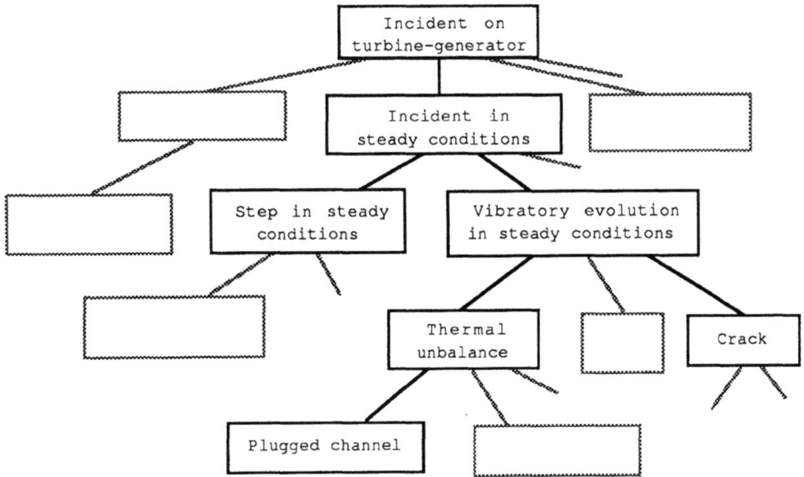

Fig. 2. Excerpts from a prototype hierarchy
(with refinement links)

b) <u>expert knowledge</u>: this designates knowledge elements which are directly related to the diagnosis process itself, as opposed to the basic knowledge which can have a wider use. This knowledge can be classified in three categories:

1) data abstraction knowledge: to obtain a piece of information useful for diagnosis, processing of other data (such as basic information elements) is often necessary. Such inferences might be done by classical computation (*derive an operation duration from startup and*

shutdown dates) but can also require goal-driven deduction through rule chaining (*if there has been an incident on the boiler, then a thermal shock has probably occurred*).

2) knowledge for prototype confirmation: when a prototype is hypothesized, the expert system must check whether this hypothesis is consistent with reality. This checking includes two kinds of operations: data acquisition and confirmation itself. As it is the description of a situation, a prototype holds knowledge on relevant parameters and their expected or prohibited values (*"when studying a coupling slide assumption, look for the vibration evolution magnitude: it is expected to be less than 30µm and it can't be greater than 60µm"*). It also bears knowledge on how to confirm that this prototype is a good description of reality. This is done by rules attached to the prototype. The role of these rules is to yield to a global interpretation of available data to assess or retract the prototype (with a weighing factor).

3) knowledge for prototype refinement: when a prototype has been confirmed, what the expert knows will help him/her to propose a new hypothesis: instead of trying all possible prototypes one after an other, it is usually possible to guide the refinement mechanism. This kind of knowledge is stored in control rules such as *"when studying a step phenomenon in steady operation conditions, if the magnitude is greater than 45 µm, then evoke a blade loss before evoking a coupling slide and don't bother studying a generator rod sliding problem"*.

It is important to note that these mechanisms for refinement and for diagnosis and therefore the associated rules are only meaningful in a given prototype. This is why rules are associated to the task they contribute to (confirmation or refinement) and located in the prototype in which they are applicable.

Another point of interest in separating the way data is acquired from the reasoning of the expert system, which includes the way it will be used, is that both might evolve independently.

REASONING PROCESS

The global reasoning mechanism follows the following pattern (a more detailed description can be found in [2]):

At every stage in reasoning, DIVA considers a given prototype as corresponding to a possible description of the problem being studied.

It first acquires a certain number of data which characterize this prototype; this is the data acquisition stage. It should be reminded that this acquisition may not be a straightforward process: to obtain the useful information, a complex process including deductions, checks, transformations is sometimes necessary. Characteristic information on a prototype is described through components which put together a parameter (e.g. vibratory measurement on a given sensor), an attribute of this parameter (e.g. its level), expected and rejection values for this attribute in the situation described by the prototype. Components can be seen as elements holding the basic typical description of the situation represented by a given prototype.

If there is no obvious reason to reject the prototype (i.e. if no component matches its rejection value), the system then seeks to confirm that this prototype is consistent with the available elements of description of the problem (or to reject the prototype as irrelevant); this is the confirmation stage. This confirmation can imply additional information acquisition (some might require special manipulations in the plant) in order apply relevant rules.

Consequently, a prototype can be recognized with a certain confidence level (from rejection *"I am sure the situation described in this prototype did not occur"* to acceptance *"I am sure this situation did happen"*) or DIVA can only estimate that it does not have enough information to conclude on its existence.

From the considered prototype, when it has been confirmed, the system tries to set up a new and more precise prototype using available data (either newly or previously acquired); for this new prototype, the process will be continued until a satisfying conclusion is reached. This is the prototype <u>refinement</u> stage.

If the considered prototype is rejected, the system tries alternative prototypes that have been left aside in the first place (e.g. siblings in the refinement hierarchy). The same research of alternative solutions is applied when no conclusion on a prototype is possible or when there is no other refinement possibility for a prototype (terminal node in the hierarchy). It is important to note that this search for alternative solution has a very different meaning in terms of diagnosis when it is done because of a rejection or when it is only a try of another possible situation.

A complete diagnosis session of DIVA brings out a set of prototypes with various levels of confidence. The system must then help the user to interpret these results. This interpretation can rely on the way retained prototypes are consistent with the real situation. The confidence level is an indicator of such adequacy but a more detailed analysis of how this was obtained is necessary. As diagnosis knowledge and relevant information on the turbine generator are associated to a prototype, this knowledge is useful to keep track of the reason why a prototype was rejected or accepted. Another element to interpret is the refinement chaining of prototypes that has been followed (e.g. how did the adequacy between prototypes and reality evolve when a more precise prototype was selected ?).

CONCLUSIONS AND FURTHER DEVELOPMENTS

The development of this expert system for turbine generator diagnostics is an ambitious project because of the complexity of the problem especially when knowledge representation is concerned and because of the expected precision of the answer.

The complexity of the system implies that special care must be devoted to developer and user interfaces. This includes the way questions are asked to refer to actually available data (where is this datum to be found and under what form) and, for user queries, the best way to avoid ambiguity as much as possible. It also means text and graphic interface to help understand the meaning of a question, to display what the expert system is doing, how it reached a conclusion, why it is trying to perform a certain task on a certain prototype...

The notion of prototypical situations (prototypes) is an interesting approach as :
    - it follows the approach of human experts progressively refining their understanding of a given situation,
    - it might lead to a partial answer if the system does not have sufficient informations, as an "intermediate" prototype can be seen as a partial answer;
    - a delegation of knowledge and of control is therefore made possible: informations to acquire, confirmation or rejection operations are associated to a given prototype in order to remain adapted to the solution of the problem it represents.

This allows a modular description of the problem; prototypes can be detailed one at a time with experts (corresponding knowledge to elicit will be an answer to the question "how would you describe the situation represented by this prototype and what do you do then?"). Partial knowledge bases can be built using such partial knowledge. This is especially useful as this project includes knowledge from different experts.

Furthermore, it can take into account the fact that diagnosis is not an homogeneous process in every cases: differences in data to examine, in ways to make and validate assumptions vary according to the generality level of the prototype, to the specificity of hypotheses, the existence (or absence) of a phenomenon...

These ideas for knowledge representation and processing have been tested through mock-ups and are now implemented in "real-size" prototype development. This prototype will address the whole problem of turbine generator diagnosis in a realistic manner. Development and testing of this prototype run a two-and-a-half year period (ending 1st term of 1988). The working effort is estimated to 10 man-years.

ACKNOWLEDGEMENTS

Every participant to this joint project has contributed to the results presented here.
Expertise on turbine generators:
Alsthom: B.Michau, B.Godfrin, E.Veith
EDF: J. Morel, R.Chevalier, D.Le Révérend, P.Neau

Expert system development:
Alsthom: P. Vaudrey, J.P.Tiarri, H.Roessel
EDF: B. Monnier and B.Ricard
LDM: J.M. David and J.P.Krivine

REFERENCES

1.   B.Ricard, B.Monnier, J.Morel, "Système Expert d'Aide au Diagnostic des Défauts d'un groupe turbo-alternateur", in: "Proceedings of the 2nd International Conference on Artificial Intelligence (CIIAM 86)" , Marseille (France), 1986

2.   J.M.David and J.P.Krivine, "Utilisation de prototypes dans un système expert de diagnostic: le projet Diva", in: "Proceedings of the 7th International Workshop "Expert Systems and their Applications" , Avignon (France), 1987

A COMPUTERIZED EXPERIMENT FOR COMPARISON BETWEEN

METHODS OF UNCERTAINTY PROPAGATION

Shimshon Arueti

Mechanical, Aerospace, and Nuclear Engineering Department
5532 Boelter Hall, UCLA
Los Angeles, CA 90024

1. INTRODUCTION

1.1 An Overview

One of the fields in artificial intelligence (AI) which has already
been applied to productive programs is the expert systems field. Among
other areas, expert systems are applied in engineering (e.g. diagnosis).
This paper summarizes an experiment, the results of which can be used as a
tool to help evaluate methods for propagating uncertain information to be
used for diagnosis and planning for maintenance in a nuclear power plant,
as well as in other industrial facilities.

Very often the data that are used in an expert system contain uncertain
parameters which are dealt with in various statistical and probabilistic
methods. Some of those methods have already come to maturity in fields such
as proababilistic safety analysis (PSA), while others are new methods which
have been introduced through AI related research (e.g. Certainty Factors).

1.2 The Role of Uncertainty in Expert Systems

Since most decisions are based on uncertain data, some expert systems
try to express that uncertainty both for the control of the expert system
(such as for ordering rules by their probabilities of being true in order
to choose the next rule for firing) and for output to the user for his
decision making process. One of the problems that has to be solved for
using those methods is how to propagate uncertainty along a decision tree
such that information about uncertainty of the components' behavior will
play its role in the overall calculated uncertainty of the system's behavior.

Several methods for uncertainty propagation are used, or proposed to be
used, in different expert systems, as well as in other fields:

a.   The Bayesian propagation method (BM), which is widely used in PSA.
b.   The "PROSPECTOR"'s method[1].
c.   The certainty factor (CF) method[2] as used in MYCIN, as well as in some
     other systems.
d.   Several expert systems use fuzzy sets logic[3].
e.   Dempster-Shafer theory[4].

1.3 Scope

An experiment has been performed, which simulates a situation similar
to that of a large industrial facility, such as a nuclear or a chemical
plant, which is subject to occasional failures in some of its equipment.
The failures are followed by specific symptoms which are detected by the
plant operators and used in order to identify the root cause for those
symptoms. In this study a computer program is used, which can be denoted
as 'symptom producer and cause identifier', with a given set of parameters
which may be changed in order to choose both the relations cause-symptom
and the number of causes, number of symptoms, and the accuracy of cause
identification. The impact of those parameters on the accuracy of cause
identification has been studied, and the applicability of two of the above
methods, BM and CF, has been compared for use in diagnostics expert systems
in engineering.

Two aspects have been chosen for comparison in the present paper:

a. The accuracy of the results.
b. The time complexity of the computation.

As a secondary task we chose to look for an appropriate heuristic function to reduce the time complexity.

2. THE EXPERIMENT

This experiment is totally carried out by a computer program. The
program deals with a finite set of NC causes C_j and a finite set of NS
symptoms, s_i. Each of the symptoms can be created by any of the causes with
a conditional probability $p(S_i/C_j)$. The program does the following:

a. It creates a cause C_j, one at a time. The causes are chosen by a random number generator (RNG) according to their relative frequency.
b. Once a cause has been created, the RNG generates a set of symptoms, S_i,
   again relative to the conditional probabilities $P(S_i,C_j)$ for each
   symptom-cause combination.
c. From the given symptoms and their conditional probabilities, the system
   "guesses" which may have been the causes that generated those symptoms,
   and with what probability this is the "right" guess. A "success" is
   declared if the correct cause is guessed with high probability. The
   results have been compared for the two methods.

3. SYMPTOM-CAUSE RELATIONS AND UNCERTAINTY PROPAGATION

In the most general case, the probability of finding that a given cause
occurred depends both on the set of symptoms that is detected an on the
existence of other causes. Furthermore, this probability is calculated
from the knowledge of a set of conditional probabilities $P(S_i/C_j)$ and
$P(S_i/\bar{C}_j)$ which are not necessarily independent from each other.

Considering all those relations in our computation makes the whole process prohibitive, even for a not very complex system (for any of the uncertainty propagation methods), since the amount of storage that is required
is of the order of $N_s*2^{NC+NS-1}$. In order to reduce the storage requirement
to the order of NC x NS, a set of assumptons is used. In most cases those
assumptions will be practically valid.

a. The causes are mutually exclusive and exhaustive. In most cases we can assume a single event at a time, or at least a single event which is the original parent for all the other occurrences.

b. The symptoms are independent form each other. This assumption is more difficult to implement in reality, and it requires one to choose a set of independent symptoms, and describe other possible symptoms such as function of that independent set. One way in which such an independent set is accomplished is by having several independent readings on the same parameter (e.g. multiple temperature sensors that are fed from different power sources) or even by calculating a paraemter by various different methods (e.g. reactor power is calculated separately by counting neutron flux and from the combination of inlet and outlet temperature and coolant flow rate).

c. For completeness a cause called "SUCCESS" has been added to the set of causes. This cause does not create any symptoms other than spurious symptoms.

Based on the actual symptoms that have been produced, the conditional probabilities $P(S_i/C_j)$ are propagated upward to the cause level in order to calculate the degree of belief that each of the causes is the real cause of the set of given symptoms. Those computations are made with the two different methods (BM and CF) and their results are compared to each other. In both cases the relative frequencies of occurrence of the causes are used as prior measures of belief, and the belief changes to a posterior belief, given the set of detected symptoms.

The set of data that each of the two methods experiences is:

a. A set of symptoms, as produced by the RNG.

b. A set of conditional probabilities which connect between the symptoms and the expected cause according to each expert.

4. PROPAGATING UNCERTAINTY WITH BM

Bayes' theorem is named after the English reverand T. Bayes[5] who did early work in probability theory. A broad discussion of application of the theorem can be found in various statistics and probility text such as Winkler[6]. The current application can be summarized as follows:

Given a prior probability $P(C_j)$ of cause C_j to occur, and the conditional probability $P(S_i/C_j)$ of symptom S_i to be detected, given cause C_j did occur, then

1. If the evidence shows that S_i was detected, the posterior probability for cause C_j being the "right" one is:

$$P(C_j/S_i) = \frac{P(S_i/C_j) \cdot P(C_j)}{P(S_i/C_j)\,P(C_j) + P(S_i/\bar{C}_j)\,P(\bar{C}_j)} \cdot \qquad (1)$$

2. If the evidence shows that S_i was not detected, then

$$P(C_j/\bar{S}_i) = \frac{P(\bar{S}_i/C_j) \cdot P(C_j)}{P(\bar{S}_i/C_j)\,P(C_j) + P(\bar{S}_i/\bar{C}_j)\,P(\bar{C}_j)} \cdot \qquad (2)$$

3. While repeating the computations for another symptom $P(S_{i+1})$, the posterior probability from the previous computation is used as a prior for the new one, instead of $P(C_j)$.

Usually following each occurrence of a cause C_j and a set of symptoms S_i we want to update our data base due to our new information on those events. Interesting research in this area has been done by Professor J. Pearl of UCLA (references 7 and 8) on the development of a "dynamic tree" and "dynamic graph" for continuous updating of uncertainties. This methodology, however, is not used in this study, since we want to keep our input data stationary for making parametric studies and comparisons with the CF method.

5. THE CERTAINTY FACTOR (CF) MODEL

In reference 9 the author says that "the [CF] model is, in effect, an approximation to conditional probability" and "We have introduced a new quantification scheme, which although it makes many assumptions similar to those made by subjective Bayesian analysis, permits us to use criteria as rules and to manipulate them to the advantages described earlier. In particular, the quantification scheme allows us to consider confirmation separately from probability and thus to overcome some of the inherent problems that accompany an attempt to put judgmental knowledge into a probabilistic format." As a conclusion to those quotations we can say that the authors prefer to have an approximation to Bayes theorem, which reflects better the way in which experts express their knowledge. This approximation has though, the same limitations that Bayes theorem has, including the large error in dealing with non-independent symptoms, or alternatively - the necessity for a storage of a large matrix of conditional probabilities for the separate combinations of symptoms and causes.

The rules for propagating uncertainty with the CF model are as follows:

a. The belief in a hypothesis (C_j) given some evidence (S_i) is measured by the value $CF[C_j,S_i]$, which may range between -1 and 1. Positive values of CF mean belief in the hypothesis while negative values mean disbelief.

b. $$CF[C_j,S_i] = MB[C_j,S_i] - MD[C_j,S_i] \tag{3}$$

    where $MB[C_j,S_i]$ is the measure of increase belief in the hypothesis C_j based on the evidence S_i and $MD[C_j,S_i]$ is the measure of increased disbelief inthe hypothesis C_j based on the evidence S_i.

c. $$MB[C_j,S_i] = \begin{cases} 1 & \text{if } P(C_j) = 1 \\[2mm] \dfrac{\max\{P(C_j/S_i),\, P(C_j)\} - P(C_j)}{\max\{1,0\} - P(C_j)} & \text{otherwise} \end{cases}$$

$$MD[C_j,S_i] = \begin{cases} 1 & \text{if } P(C_j) = 0 \\[2mm] \dfrac{\min[P(C_j/S_i),\, P(C_j)] - P(C_j)}{\min[1,0] - P(C_j)} & \text{otherwise} \end{cases} \tag{4}$$

    where $P(C_j)$ is the prior expert's belief in the hypothesis C_j and $P(C_j/S_i)$ is his posterior belief, given the evidence in S_i.

d. In each step the values of MB and MD are updated by the combining function:

$$
MB[C,S_1 \ \& \ S_2] = \begin{cases} 0 & \text{if } MD[C,S_1 \ \& \ S_2] = 1 \\ MB[C,S] + MB(C,S_2](1 - MB[C,S_1]) & \text{otherwise} \end{cases}
$$

$$
\tag{5}
$$

$$
MB[C,S_1 \ \& \ S_2] = \begin{cases} 0 & \text{if } MD[C,S_1 \ \& \ S_2] = 1 \\ MB[C,S] + MB(C,S_2](1 - MB[C,S_1]) & \text{otherwise} \end{cases}
$$

An important characteristic of this model is that both MB and MD can only increase. They are very vulnerable, therefore, to spurious symptoms, which are not recoverable by appearance of competitive symptoms. A spurious symptom can either increase MD for the "right" cause, thus limit the value of CF from above; and/or increase MB for the "wrong" causes, thus limiting their CF from below. This effect is especially noticed in cases of multiple symptoms per cause, especially if the symptoms are not reliable.

6. A HEURISTIC FUNCTION OF BM

As an alternative way of approximating the Bayesian model, a heuristic function has been applied to the basic equations.

Equation (1) above can be modified in the following way:

$$
P(C_j/S_i) = \frac{P(C_j)}{P(C_j) + \dfrac{P(S_i/C_j)}{P(S_i/\bar{C}_j)} \cdot P(\bar{C}_j)} = \frac{P(C_j)}{P(C_j) + \lambda P(\bar{C}_j)} \tag{6}
$$

λ can be any number $0 < \lambda < \infty$. When $\lambda \approx 1$, the result of equation (6) is:

$$
P(C_j/S_i) \approx \frac{P(C_j)}{P(C_j) + P(\bar{C}_j)} = P(C_j) \tag{7}
$$

The closeness of λ to 1.0 can be used, therefore, as a measure of the significance of the evidence to the result. If λ is close enough to 1, the corresponding evidence can be ignored by the computations without changing them considerably. In this program the value $\max\{\lambda; \frac{1}{\lambda}\}$ is used as a heuristic function for cutting down the amount of computations required, and some of the results of both the affected accuracy and the affected computation time are shown in Section 8.

It can be seen from this section that the reduction in computation time is limited. That is because of the additional time required for checking the conditions for implementing the approximation. This reduction, though, will be amplified if the propagated properties are probability distributions rather than simple probabilities. In this case the heuristic function can be computed from the means of the distributions for cutoff of the whole distribution if applicable.

7. COMPARISON BETWEEN THE CF AND THE BM METHODS

As discussed above, the CF model is an approximation of the BM model. What this experiment is about is measuring the goodness of the approximation

and the gains from using the approximation rather than the exact model.
The following theoretical comparison will give some insight to the computed
results.

1. Both methods use the cause frequency $P(C_j)$ as a prior frequency. The
    evidence-propagating parameter, though is different. While BM uses
    $P(S_i/C_j)$ for its propagation model, CF uses $P(C_j/S_i)$. Taking into con-
    sideration the data collection system, it seems more conceivable that
    data will be collected by causes (events) rather than by symptoms. The
    values of $P(C_j/S_i)$ will have to be calculated from the more basic
    values $P(S_i/C_j)$.
2. CF seems to be more complicated than BM. It requires intermediate com-
    putations (MD and MB) and additional equations.
3. Computationally CF is more efficient, since summations are used rather
    than multiplications as in BM. Another cause for its higher efficiency
    is the cut-off in computation when BM and CF reach a value of 1.0. The
    gains of this cut-off depend on the conditional probabilities $P(C_j/S_i)$.
    This efficiency disappears, though, if $P(C_j/S_i)$ has to be computed as
    noted in #1 above.

8. RESULTS

Figures 1 and 2 below represent a sample of the large amount of output
results that have been collected in order to support the conclusions of the
following section. Figure 1 shows, for non-reliable symptoms (i.e., high
probabilities for occurrence of spurious symptoms), how increased number of
potential symptoms affects the accuracy of both methods for a constant
number of causes. Figure 2 describes computation time as a function of its
main parameter, which is the product NS x NC. In both figures the effect
of using various heuristic functions for BM is shown.

9. CONCLUSIONS

The main conclusions of the experiment are as follows:

a. From the point of view of accuracy, as seen in Figure 1, the CF model
    is inappropriate for calculations in engineering or in complex systems.
    One of its drawbacks is that the results are less accurate than those
    of BM model in all cases. But even more important is that while addi-
    tional evidence (e.g., larger numbers of symptoms or more symptoms per
    cause) always improves the results created by BM, the effect is very
    often the opposite on CF. The CF method can be viewed as a method
    which introduces noise with each additional input, which accumulates
    and eventually confuses the results carried by the input signals.
b. CF is especially sensitive to spurious symptoms, since it does not have
    a mechanism for reducing either MD or MB. This feature is inherent to
    the assumption that belief and disbelief are two independent parameters
    that can be carried out separtely in the calculations.
c. Another source of confusion, which affects CF more than BM, is the
    existence of causes with identical, or partially identical symptoms.
d. Both CF and BM have the same asymptotic time complexity, which is
    O(NC x NS). CF's computation time is smaller by a large factor of
    between 1.8 and 12, depending on the probabilities $P(C_j/S_i)$. This
    advantage disappears if we consider the computation time of $P(C_j/S_i)$,
    which will most probably not be treated as basic data, but rather as
    computed results from $P(S_i/C_j)$ and $P(C_j)$.
e. An alternative way of reducing computation time is proposed by virtue
    of a heuristic function λ. This alternative has a limited effect in

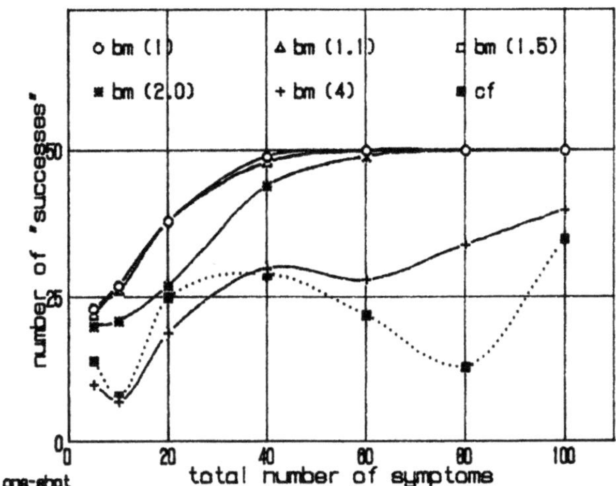

Figure 1.   Number of Successes as a Function of NS

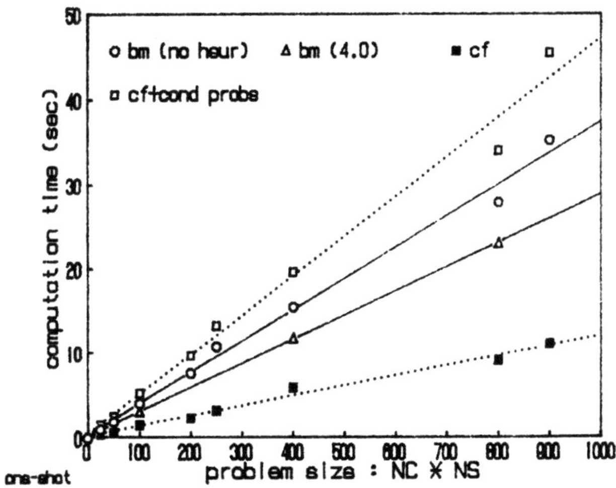

Figure 2.   Time Complexity as a Function of Problem Size

the current application, but may be more useful if probability distributions are propagated, rather than simple probabilities.

f. Additional factors that may affect the resutls are the distribution of
the input data and the ratio NC/NS. Both have a negligible or no
effect on time complexity. Their effect on accuracy is considerable
only if a low value is used for the heuristic function. The results
will always be accurate when the data has a very sharp distribution
(i.e., some symptoms will almost always be detected following a given
cause, while others will be rarely detected), and ratio NC/NS close to
1, as opposed to bad accuracy for rectangularly distributed symptoms
with NC/NS << 1.

10. ACKNOWLEDGMENT

The author is thankful to Professors Richard Kork and David Okrent for
thier encouragement and for their helpful discussion and comments.

The work in this study is supported in part by DOE contract No.
DE-FGO3-86-sf-16610.

11. REFERENCES

1. Duda, R.O., P.E. Hart, N.J. Nilsson, "Subjective Bayesian Methods for
   Rule Based Inference System," AFIPS Conference Proceedings, Vol. 45,
   (1976 National Computer Conference), pp. 1075-1082.
2. Shortliffe, E.H., "A Model of Inexact Reasoning in Medicine," Chapter
   11 in _Rule-Based_ _Expert_ _Systems_, B.G. Buchanan and E.H. Shortliffe,
   Eds., Addison-Wesley, 1984.
3. Zadeh, L.A., "Fuzzy Sets as a Basis for a Theory of Possibility," Fuzzy
   Sets and Systems, 1:3-28, 1978.
4. Shafer, G., "A Mathematical Theory of Evidence," Princeton University
   Press, 1976.
5. Bayes, T., "Essay Towards Solving a Problem in the Doctrine of
   Chances," Biometrica, 45:293-315 (1958) (Reproduction of 1763 paper).
6. Winkler, R.L. and W.L. Hays, "Statistics: Probability, Inference, and
   Decision," Holt, Rinehart and Winston, 1970.
7. Pearl, J., "Distributed Bayesian Processing for Belief Maintenance in
   Hierarchical Inference Systems," UCLA-ENGL-CSL-82-11 (Jan 1982).
8. Pearl, J., "A Constraint-Propagation Approach to Probabilistic
   Reasoning," Proceedings of Workshop on Uncertainty and Proability in
   AI, UCLA, Aug. 14-16, 1985, pp. 31-42.

ATTENDEES

Robert G. Abboud
Commonwealth Edison

Russell H. Adams
Virginia Power

Thomas E. Albert
SAIC

Ken J. Allen
Argonne Nat. Lab.

A. D. Alley
General Electric

V. Andersen
RISO Nat. Lab.

David G. R. Anderson
Ontario Hydro

John Anderson
AECL CANDU

Hakan Andersson
Studsvik Energiteknik AB

Ronald Antionja
Bechtel National, Inc.

George Apostolakis
UCLA

Shimshon Arueti
UCLA

Nicholas Avouris
JRC-ISPRA

Bridget Baird
Connecticut College

Claudio Balducelli
ENEA

Frederic Bastenaire
Tractebel

Herner Bastl
GRS

S. Basu
Ontario Hydro

Russ Beckmeyer
Dupont-SRL

Leo Beltracchi
US NRC

John A. Bernard
MIT

Wayne M. Berry
TVA

Thorbjorn Bjorlo
Halden Project

Paul Blanch
Northeast Utilities

Dennis C. Bley
Pickard Lowe & Garrick

William V. Botts
EI International

D. B. Braithwaite
Argonne Nat. Lab.

Michael A. Bray
EG&G Idaho

Laural L. Briggs
ANL - RAS

John S. Brtis
Sargent & Lundy

David H. Bullington
Arkansas Power & Light

Bernd Burger
University of Stuttgart

Art Butt
Mgmt. Analysis Co.

Joe Byrd
DuPont (SRL)

Carl Cooper
EG&G Idaho

C. Cacciabue
CEC JRC Ispra

David Cain
EPRI

Greg Candy
Combustion Engineering

Collins P. Cannon
Cannon Tech.

N. Scott Cannon
Westinghouse Hanford

Reed B. Carlson
ANL

Robert Carlton
Odetics

Roger G. Carter
Morton Thiokol

V. Charyulu
Idaho State Univ.

Patrick Y. C. Chen
Pickard Lowe & Garrick

G. H. Chisholm
Argonne Nat. Lab.

Alan Christie
Westinghouse

L. J. Christensen
Argonne Nat. Lab.

Maurin Au Congres
EDF

Gail Cordes
Intermountain Tech.

Dan Corsberg
EG&G Idaho

Kevan Crawford
University of Utah

Diane Croson
WINCO

Dirk Dahlgren
Sandia Lab.

Kerstin Dahlgren
Swedish Nuc. Pwr
  Inspectorate

Curt Dahm
IBM

John E. Dale
GE-Nuclear Energy

Linda Daniels
Impell Corp.

Russ Dedrickson
Georgia Power, Plant
  EI Hatch

Didier Delaigue
Framatome

Brent Dixon
EG&G Idaho

J. Michael Doster
NC State University

Patrick Duggan
Expert Ease

Jim Dukelow
Westinghouse Hanford

Robert M. Edwards
Penn State Univ.

Mokhtar Elatrash
Idaho State Univ.

Dr. Robert Engelmore
Stanford University

Steven H. Engle
Expert-EASE Systems

Steven Epstein
Mgmt. Analysis Co.

Joseph J. Feeley
University of Idaho

Lothar Felkel
GRS Forschungsgelande

Larry Fennern
General Electric

Evan Filby
WINCO

Eugene Filshtein
Combustion Engineering

Donald Flemming
Idaho State Univ.

Stan Focht
Amer. Nuclear Insurers

M. S. Foster
Advanced Nuclear Fuels

George Freund
SAIC

Bjorn Frogner
Expert-EASE

Minoru Fujii
JAERI

Phillippe Gaussot
Electricite de France

M. Gaussens
Framentec

Floyd Gelhaus
EPRI

David Gertman
EG&G Idaho

Daniel S. Giroux
Argonne Nat. Lab.

Gerry Golden
Argonne Nat. Lab

Sallie Gordon
University of Idaho

Jim Gross
WINCO

Rod Grow
URA

Sergio Guarro
Lawrence Livermore Lab.

Dale Gunter
Tech. Applications

Michael A. S. Guth
Oak Ridge Nat. Lab

J. C. Haire
EG&G Idaho

Dr. Brian Hajek
Ohio State

John W. Hallam
NSSS

Siavash Hashemi
Ohio State

Hideo Hatshui
JAERI

Steve Hetzel
TVA

Gary Hicks
Impell Corp.

Yuichi Higashikama
Hitachi, Ltd.

Mike Hitz
Argonne Nat. Lab

Li-Wei Ho
Inst. of Energy Research

Erik Hollnagel
CRI A/S

Steve Hollwarth
Tech. Applications

Hank Honeck
Computer Applications Tech.

Yasuhiro Horibe
Mitsubishi Research Inst.

G. K. Hurst
Quadrex

Max Hymas
Idaho State University

K. Imoto
General Electric Co.

Dr. A. Jaeschke
Kfk-IDT

Terry Jamieson
ECS Power Systems

Jim Jenkins
US NRC

C. E. Johnson
WINCO

Harry W. Johnson
Toledo Edison

Larry Johnson
EG&G Idaho

Harry Julian
Volian Enterprises

Haarla Jyrki
Research Center of Finland

John A. Kamel
Bechtel Western Power

Ari Kautto
Imatran Voima Ov.

John Keller
RPI Student

Tom Kerlin
Univ. of Tenn.

Takashi Kiguchi
Hitachi

Kil Koo Kim
UCLA

Ronald King
Argonne Nat. Lab

M. Kitamuru
Tokoku Univ.

Ellie Kurrasch
Odetics, Inc.

Robert B. Lang
MSU Sys. Services, Inc.

J. S. Larson
Halden Project

T. K. Larson
EG&G Idaho

H. J. Leder
Interatom

John C. Lee
Univ. of Michigan

Wayne Lehto
ANL-West

Monte Leng
Quadrex

Tai Seng Leong
Bechtel

David Levinskas
Idaho State Univ.

John Lewellen
US DOE

L. M. Lidsky
MIT

Richard Lindsay
Argonne Nat. Lab

Dan Line
EG&G Idaho

Rick Loeffler
Cleveland Illuminating

Amit Majumdar
Idaho State Univ.

Debu Majumdar
DOE-ID

Henry Makowitz
EG&G Idaho

Pierre Malvache
CEA

Lucas Mampaey
Tractebel

Robert Martin
Texas A&M

John McCarthy
Impell Corp.

Gordon Medford
SAIC

Orville Meyer
EG&G Idaho

Dr. Don Miller
Ohio State

James E. Mims
SAIC

Hajime Minowa
Kobe Steel

Mohammad Modarres
Univ. of Maryland

Patrick Moeaert
Tractebel

Reed Monson
Argonne Nat. Lab.

K. Monta
Nippon Atom Ind. Group

Michael Montgomery
Nuclear Fuel Ind.

Glen Mortensen
EG&G Idaho

Jack Mott
EI International

Michael Moulin
EDF-REAM

Ken Mulligan
Gould, Inc.

S. Murin
Electricite de France

Joseph A. Naser
EPRI

Dr. B. Nassersharif
Texas A&M Univ.

Bill Nelson
EG&G Idaho

Stephen Nicolosi
Battelle-Columbus

Masahide Nishio
Toshiba

Yoshiaki Oka
Univ. of Tokyo

Michael Oren
PUNGS

Bob Orendi
Westinghouse

Robert Osborne
Westinghouse

Pedro Otaduy
ORNL

Tom Ott
Arkansas Power

F. Owre
Halden Project

Randall H. Pack
General Physics

Donald H. Page
NASA

Mr. Paliwal
KHU

Jacques Panossian
Framatome

Alex Parlos
Texas A&M

Bryan Patterson
New Brunswick Power

Orpet Peixoto
Stanford

Bill Petrick
Nuclear Software Services

L. E. Pierre
EDF

Francois G. Pin
ORNL

Pete Planchon
ANL

Dr. B. Puetter
GRS

Magdi Ragheb
Univ. of Illinois

James R. Reiff
NY Power Authority

Bernard Reygrobellet
EDF

Roger S. Reynolds
Miss. State Univ.

B. Ricard
EDF

Richard Rigg
ANL

Douglas Roberts
LANL

Allen H. Robinson
Oregon State Univ.

Jim Robinson
ORNL

Howard Rohm
DOE

Peter Rzasa
Combustion Engineering

J. I. Sackett
Argonne Nat. Lab.

Hiroshi Sakamoto
Toshiba

Adela Salame-Alfie
RPI

Tetsu Sato
Toshiba

Susanne Schell
ANF

Wayne Schettle
BCSR

Bill Schlegelmilch
PSE&G

Fritz Schmidt
IKE Univ. of Stuttgart

Hyn Schuh
Bechtel

Morris Schwarzblat
IJE Mexico

Henry P. Seager
Penn. Power & Light

Mark C. Serres
Impell Corp.

Eric Shaffer
General Electric

Yoshikurra Shinohara
JAERI

Urs U. Siegfried
Leibstadt Nuclear Plant

Harve E. Sills
AECL

Lee L. Simmons
GTS

Gary A. Sly
EI Services

Devin Smith
SUNY

Robert L. Smith
EG&G Idaho

Joseph Somsel
Pacific Gas & Electric

Yves Souchet
CEA

Kevin St. John
Yankee Atomic

Jeffery Staffon
Argonne Nat. Lab.

John Stasenko
Ohio State

J. P. Steelman
Baltimore Gas & Electric

Rex C. Stratton
Battelle

Al Sudduth
Duke Power Co.

Tak Sung
Ebasco Services

Shigeru Takamiya
Toshiba America

Jacques Thibault
EDF

Baptiste Thomas
CEA

Bob Touchton
Tech. Applications

Lloyd Tripp
Odetics, Inc.

Robert E. Uhrig
Univ. Tenn/ORNL

Teretti Visuri
Finnish Press Corps

David Wait
Rockwell International

Vicki Walther
Impell Corp.

George Westrom
Odetics, Inc.

John White

G. H. Williams
CEGB

Allan R. Wilson
Gilbert Commonwealth

Hike Winter
Iowa State

David Wiser
SIAC

Naoyuki Yamada
Ohio State

Ching-Dhuan Yao
Taiwan Power Co.

Dixon Yee
Univ. of California

Lone-Yee Yee
UCLA

Shimon Yifta
SOREQ Nuclear Center

Kazuo Yoshida
JAERI

Bruce Zimmerman
Westinghouse Hanford

Dr. Gilles Zwingelstein
EDF

Tony Brown
Idaho State University

Dr. David Laning
Intellicorp

Betty Haire
Registrations

Madge Lindsay
Authors Desk

Catherine Majumdar
WINCO

Karen Sackett
Guest Program Chairman

Bechtel Power Corp.

Definicon Systems, Inc.

Expert-EASE Systems, Inc.

Gold Hill Computers, Inc.

Honeywell Inc., TID

Strategic Software Planning

MIX
Papier aus verantwortungsvollen Quellen
Paper from responsible sources
FSC® C105338

If you have any concerns about our products,
you can contact us on
ProductSafety@springernature.com

In case Publisher is established outside the EU,
the EU authorized representative is:
**Springer Nature Customer Service Center GmbH
Europaplatz 3, 69115 Heidelberg, Germany**

Printed by Libri Plureos GmbH
in Hamburg, Germany